A TEXTBOOK ON

MODELLING AND FINITE ELEMENT ANALYSIS

A TEXTBOOK

ON

MODELLING AND FINITE ELEMENT ANALYSIS

By

Ganesh S. Hegde

Department of Mechanical Engineering
R.V. College of Engineering,
Mysore Road, Bangalore
Karnataka

UNIVERSITY SCIENCE PRESS

(An Imprint of Laxmi Publications Pvt. Ltd.)

An ISO 9001:2008 Company

BENGALURU • CHENNAI • COCHIN • GUWAHATI • HYDERABAD
JALANDHAR • KOLKATA • LUCKNOW • MUMBAI • RANCHI • NEW DELHI
BOSTON (USA) • ACCRA (GHANA) • NAIROBI (KENYA)

A TEXTBOOK ON MODELLING AND FINITE ELEMENT ANALYSIS

Printed and bound in India
Typeset at Sukuvisa Enterprises, Delhi
First Edition : 2016
ISBN 978-93-83828-48-7

PUBLISHED IN INDIA BY

UNIVERSITY SCIENCE PRESS
(An Imprint of Laxmi Publications Pvt. Ltd.)

An ISO 9001:2008 Company
113, GOLDEN HOUSE, DARYAGANJ,
NEW DELHI - 110002, INDIA
Telephone : 91-11-4353 2500, 4353 2501
Fax : 91-11-2325 2572, 4353 2528
www.laxmipublications.com info@laxmipublications.com

Branches

✆	Bengaluru	080-26 75 69 30	
✆	Chennai	044-24 34 47 26,	24 35 95 07
✆	Cochin	0484-237 70 04,	405 13 03
✆	Guwahati	0361-254 36 69,	251 38 81
✆	Hyderabad	040-27 55 53 83,	27 55 53 93
✆	Jalandhar	0181-222 12 72	
✆	Kolkata	033-22 27 43 84	
✆	Lucknow	0522-220 99 16	
✆	Mumbai	022-24 91 54 15,	24 92 78 69
✆	Ranchi	0651-220 44 64	

C—10003/015/03

Printed at : Repro Knowledgecast Limited, Thane

This Book is Dedicated

to

Our Dear Parents

CONTENTS

PREFACE

This book has introductory materials on FEM. The book covers the basic principles and formulations related to finite element method. The book is written with lucid and simple way keeping in mind all categories of readers. The readers interested to collect basic, starter informations on FEM would be most benefited. The full coverage of theory and problems would make the reading interesting to the readers of mechanical engineering. The attractive pedagogy with highlights of the chapters and the question banks and the short viva-voce questions would make reading interesting and preparative. The self explanatory diagrams and solution to typical problems make this book different.

The Chapter-1 deals with the general introduction to FEM and attempts to give an identification to the topic. Chapter-2 covers the requirements of prerequisites in the form of numerical techniques used in FEM. Chapter-3 offers material on variational approaches that laid foundation for FEM development. The basics and introduction to theory of elasticity needed in FEM are dealt in Chapter-4. ID bar element analysis with examples is the material of Chapter-5. The FEM formulation of beam element is offered as study material in Chapter-6. The formulation related to 2D elements is the reading stuff in Chapter-7. The Chapter-8 provides detailed informations on higher order element used in FEM and convergence criteria with material on patch test. The utility of FEM in heat transfer analysis is briefed in Chapter-8.

The question bank collected from the exam papers of universities is provided at the end of the book. The short viva-voce question bank that guides the preparation for FEM laboratory is also listed at the end of the book. The points that should be remembered to acquire basic knowledge in FEM are listed at the end of each Chapter.

The authors are thankful to those who directly or indirectly helped and supported to bring out this book. The constructive criticism from the readers on the improvization of the book would be well received. The authors would be delighted if the book serves the interest of the students and readers effectively.

—Authors

Chapter 1

Introduction to Finite Element Method

मलयाचल संसर्मात्।
न वेणु: चन्दनायते।।
अंत: सार विहीनानाम्।
रस: केनोपजायते।।

[One should have the potential in himself because it is not possible to lend the potential by others.]

This is a introductory chapter on Finite Element Method which gives basic familiarity of understanding the concepts. The general questions of what and why of FEM are dealt in detail to give an overview of form and structure of the method. Comparison with the other methods originated prior to FEM, highlights the supremacy of FEM. Before knowing the methodology of workouts of finite element procedures it is imperative to know the definition and objectives that lead to the development of FEM. The term Finite Element was first coined by civil engineering scientists who started the procedure for structural engineering applications. In fifty years of development FEM has stretched its arms in different interdisciplinary areas of engineering like design, heat transfer, fluid flow, dynamics and electrical engineering etc., the adaptability is enhanced by flexibility of approach, accuracy and reliability of results and speed of generation. The development of fast digital computers has supplemented the growth of FEM to a commendable extent.

This introductory deliberation on FEM spans the following issues in simple terms for the beginners in FEM.

- Definition and description of continuum.
- Subdivision of the domain into discrete system.

- Features of numerical methods—FEM and FDM.
- Modules in FEM—input, solution and output modules.
- Steps in linear static analysis.
- Basic element types in 1D, 2D, and 3D.
- Engineering design and FEM.
- Computer Aided Design (CAD) and FEM.
- Comparison and contrasts in continuum, FDM and FEM.
- Why there is need for FEM—objectives and goals.
- Advantages and disadvantages of FEM.

FEM being filled with more of advantages, has become general purpose computer aided procedure, that is widely adopted and accepted in mechanical engineering analysis, design and optimization.

1.1 INTRODUCTION

The solid mechanics, strength of materials and the structural mechanics have contributed to the exact solution to the simple problems with easy loadings, geometry and ideal boundary conditions with the aid of natural and physical laws. The practical situations in civil and mechanical engineering designs have structures, systems and parts that are complex in shapes, indeterminant in boundary and combined by many types of loads. Such cases do not exact analytical solutions or rather difficult to arrive at the exact solutions. Mechanics have the answer to the simple geometries like cantilever beams, simply supported beams, and fixed beams with simple loads like point loads, uniformly distributed loads (UDL), uniformly varying loads, couples and torques. A real world physical problem are all three dimensional with the combination of above mentioned boundary conditions and loads. Further the mechanics methods do not tell about the state of stress in the material within the close vicinity of the loads and supports in a continuous structure or the component. A structure or a system wholly considered for analysis as a complete body with the governing equations derived from the physical laws of nature can be called as '**continuum**'. The geometric, the material and the environmental non-linearity are common in all continuum examples of practical importance, which have to be solved by simplifications and assumptions to arrive at the analytical solutions with the idealization error in the analysis.

The natural behaviours of the parts, systems and structures are governed by physical laws in terms of spatial variables, environmental effects and the material properties, that are formulated by algebraic, differential and integral equations. For a practical three dimensional problem such equations pose greater difficulty or impossibility to solve in a continuous domain. But by dividing the continuum into large number of simple elements, known as discretization makes task easy. The method of analysing element by element of the discretized structure and consolidating the result for the total structure, is known as **Finite Element Method (FEM)**.

In FEM a finite number of well-defined components render the continuous model into a discrete system by the process of discretization. The true solution to the continuum is closely approached by dividing the same into large number of **discretes** connected at the '**nodes**' forming the grid. The informations such as forces and displacement at the connected points help in establishing stress variation in the continuous system.

With the advent in fast digital computers it is possible to solve a continuum problem subdivided into large number of discrete finite elements. Larger the number of elements the solution obtained closely matches the exact solution with the minimization of error and convergence of the result. In FEM the subdivision is finite and definite so that the problem defined with mathematical relations applied to discrete elements makes the solution procedure simple. The gross effect is the summation of the effects on the discretes that closely approach the analysis of the continuum as a whole system. The tools and techniques available in mathematics and the number of subdivisions of the continuum are limited by the capabilities of the digital computers to handle large amout of data. The digital computer has to have higher processing capacity and memory to solve large and geometrically orthotropic problems.

The governing differential equations for a discrete system are approximated by the mathematicians and applied scientists into various numerical methods such as 1. finite difference approximations, 2. weighted residual procedures, 3. functional approximations with rules of stationary conditions etc. The contributions of Argyris and Turner in 1940s, gave initial start to the development of FEM in structural problems of civil engineering. It started with the analysis of continuum as a discrete system consisting of finite number of elements. It was Clough who coined the term "Finite Element" in 1960s to give a new dimension to FEM. Hence FEM is a general discretization process of a continuum system with application of expressions of differential and integral calculus formulated in the mathematical logic.

The following sections of this chapter highlight the definition and features, explained briefly, of a continuum, discrete system and the numerical methods, mainly the finite difference and finite element methods. Never-the-less the FEM procedure of analysis of linear static elastic problem is also covered.

1.2 CONTINUUM

A rigid body or a physical system is made of materials fitted to a required shape that is continuous with infinite number of small discrete elements, is known as **continuum**. Hence, for a continuum, the degrees of freedom defining the displacement behaviour are infinite. A continuum system can be described by the mathematical statements containing differentials or/and integrals for an infinitesimal implying infinite in number in a continuum. Hence assemblage of inifinite number of small elements form a continuous system considered to be continuum. The exact solution to continuum problem, is obtained by the following considerations.

- *Simplification of geometry*: The tapered and curved geometries are simplified into linear geometry, equivalent but different in dimension.
- *Simplified Boundary Conditions*: Either or more of possible six degrees of freedom (three translation and three rotations) are constrained to define the boundary.
- *Linear variation in material property*: The technological properties of the materials made to behave linearly.
- *Simplification of environmental effects*: The complex loads are transformed into one or more of the types—viz. pointed load, UDL, UVL, torsion load, bending load, thrust load and a couple.

The solutions in solid mechanics of continua are possible with the preceding assumptions and further the following laws are applied to formulate the governing equations.

1. Newton's Laws of motion
2. Continuity equation

3. Equilibrium equations
4. Stress-strain relation
5. Castigliano theorems etc.

1.3 DISCRETE SYSTEM

A system is divided into similar or dissimilar smaller elements to form a discrete system. More the number of elements in the discrete system, the solution converges to exact solution. The excessive refinement of discretization can lead to higher cost of analysis. Hence an optimum descretization has to be worked out within the specified accuracy of result. When a continuous system is discretized there is no need to simplified geometry, as there is provision to accommodate the inclined and curved geometry by triangular and curved elements respectively in FEM. Three dimensional problems can also be easily handles using 3D elements. Complex loading situations can also be handled with ease. Accommodation of combination of materials, orthotropy in material composition with the systems discretized into smaller elements.

A structural civil engineer adapts the following procedure in analysing the problems of discrete nature.

- First establishes the force-displacement relationship for each element in the structure.
- Establishes local equilibrium at each connecting point (node) of the structure.
- With well defined procedure assembles all the elements in the discrete system.
- The equations are formulated with system matrix and displacement/load vectors.
- The matrix equation is solved for unknown displacement vector using various numerical methods suiting the size of the problem.

Before the emergence of FEM, the civil structures were analyzed as the assemblage of bars and beams using matrix analysis of structures. In such analysis the constitutive equations were solved in the matrix form.

1.4 THE NUMERICAL METHODS

The commonly used and adapted techniques in the numerical analysis of the equations governing the discrete structures are mainly of two types.

- Finite Difference Method (FDM)
- Finite Element Method (FEM)

The finite difference method is not dealt in detail as it is not in the scope of this book. The following paragraphs briefly highlight the features of the two methods, in general.

1.4.1 Finite Difference Method (FDM)

The FDM is briefed in the following few steps:

- The conservation that governs the physical behaviour, is expressed in the differential form.
- A discritized grid forms the solution domain. The grid can be structured or unstructured.
- The partial derivatives (time dependent or/and space dependent) of the governing PDE, are approximated at the nodal point of the grid.

- This leads to one algebraic equation at each node of the grid. The unknowns are the values at the node considered and the neighbouring certain nodes.
- The approximation to partial derivative is based on the Taylor series expansion of the function. The forward difference, backward difference or central difference schemes are generally being considered in the approximation.

There are certain disadvantages in using FDM approximation of numerical technique.

1. The solutions are obtained at the nodal points in the grid but not between the grid points as FDM is the pointwise approximation of the governing PDE.
2. The inclined and the curved geometries are meshed with stepped grid formation.
3. To enforce conservation special care has to be exercised in terms of convergence, stability and accuracy of result. Smaller the grid size the convergence is ensured. There has to be specific relation between time step and space step to ensure the stability of the solution with decreasing or constant error.

1.4.2 Finite Element Method (FEM)

The solution domain is divided into finite set of discrete areas or volumes called finite elements. In two dimensions the finite elements are triangles or quadrilaterals or axisymmetric. In 3D, they are tetrahedrons or hexahedrons. The degrees of freedom are assumed at the nodal points. In FEM, the solution in terms of displacements, is approximated by linear shape function within each element. The trial function for displacement field, assumed should ensure convergence across the element boundaries. The solution to the governing equation is obtained by multiplying by the weighting function before being integrated over the entire domain. The displacement obtained at the nodal points are interpolated to get displacement and stress distribution throughout the element. Hence FEM is a piecewise approximation. The dominant advantage of FEM is the ability to deal with non-linear geometries and orthotropic materials. The curved geometries are modelled with curved elements in FEM. The arbitrary geometries are discretized with unstructured grids. The elements can be further subdivided to refine the mesh that leads to convergent and accurate solutions.

Modules in FEM

In finite element analysis, the physical problem has to be subjected to the following modular steps.

- **Input Stage**

  To define the mathematical model of the physical problem different aspects to be described at the input stage are:

  1. The geometric model
  2. The material specification
  3. The load application
  4. Boundary condition incorporation
  5. Kinematic description

  The geometric modelling establishes the solution domain. The type of material and material properties are to be specified. Types of loads are to be specified on the nodes. The degrees of freedom are to be constrained at the boundary. The type of analysis describing the mechanism are to be given as input.

- **Solution Stage**

  The algorithms in the solver accepts the input and generates output. The procedural steps are:

  1. Generation of shape functions.
  2. Obtaining elemental stiffness and mass matrix using shape functions.
  3. Assemblage of elemental matrices.
  4. Formulation of algebraic equations.
  5. Solving for displacement and forces at the nodes with the aid of boundary conditions.
  6. Interpolation of the results at nodes to get solution in between the connecting points of the elements.
  7. Assessment of accuracy and convergence.

- **Output Analysis Stage**

  The processed information after solver is given out as output which is to be analysed, which is

  1. Graphical display of distribution of deformation and stress.
  2. Interpretation of the result obtained.
  3. Refinement of the model grid.
  4. Improvement on design and optimization.

1.5 STATIC ANALYSIS IN FEM

The problems in elastic continua require the generation of solution in terms of stress and strain. The two dimensional real world physical problems include plane stress or plane strain distribution, axisymmetrical solids, plate bending and shells. The three dimensional problems are solved by usage of tetrahedron and hexahedron finite elements in the mesh generation. The difficulty arising out of interconnections between finite elements, and the isolated and imaginary boundaries are overcome by approximations in the following way. The steps in the FEM static elastic analysis procedure, are outlined in Fig. 1.1.

1.5.1 Finite Element Analysis Process

The finite element analysis process is briefed in the following steps that highlight general procedure.

- Discretize the object and select the element type.
- Select a displacement function that closely approximate the deformation pattern.
- Define strain/displacement and stress/strain relation from the physics governing the problem.
- Derive element stiffness matrix and equation for the element type chosen.
- Assemble to form global stiffness matrix and equation that represent the discrete system. Introduce boundary condition representing the constraints.
- Solve the global equations for unknown degrees of freedom, and the nodal forces.
- Solve for element stresses and strains for all the elements in the mesh (discretized system).
- Interprete the results.

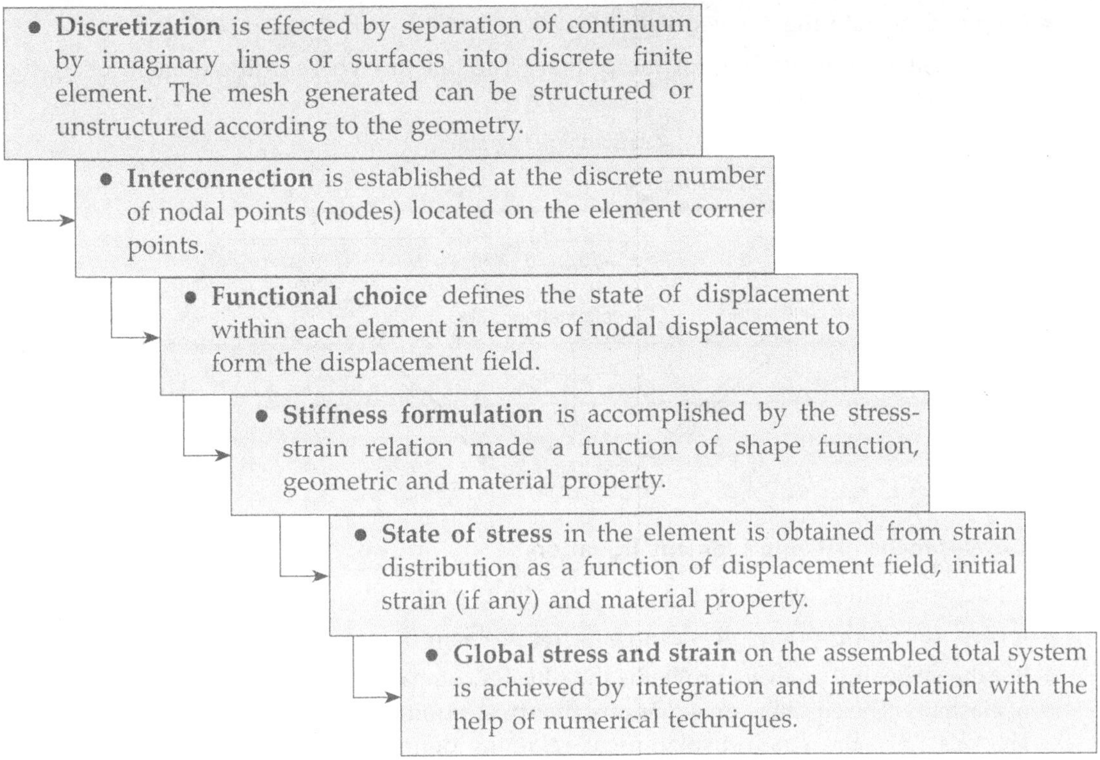

Fig. 1.1. Steps in static analysis.

1.5.2 Steps in FEM

The previously mentioned procedure of finite element analysis gets moulded into six basic important steps that should be followed by the readers to under-stand FEM in brief.

- **Step 1: Discretization**

  The problem domain is geometrically modelled and discretized into collection of simple shapes called elements. The elements are connected at points called Nodes.

- **Step 2: Development of element equations**

  The element equations are formulated using the physics of the problem. Typically used method are Galerkin's method or variational principle.

- **Step 3: Assembly**

  The element equations for each element in FEM mesh are assembled into set of global equations that model the properties of entire system.

- **Step 4: Application of boundary conditions**

  The solution cannot be obtained unless the boundary conditions are imposed. They reflect the known values in the domain of primary variables. Applying boundary conditions modifies global equations.

- **Step 5: Solve for primary unknowns**

  The modified global equations are solved for primary unknowns at the nodes.

- **Step 6: Calculate the derived variables**

  It is possible to interpolate for the primary variable anywhere in the domain using the calculated nodal values.

Relationship Diagram

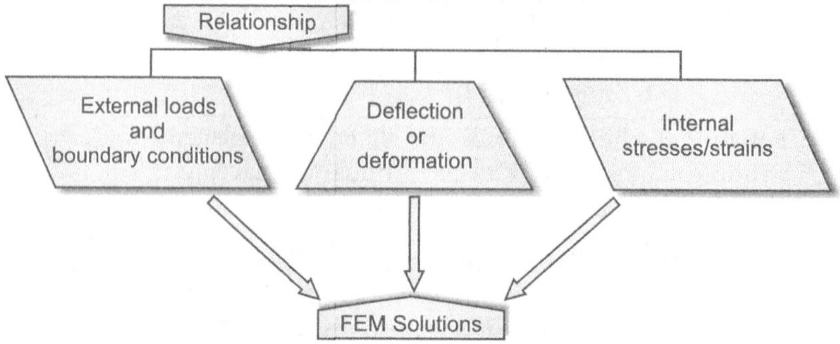

1.5.3 Development of Finite Element Equation

The few basic generalities required to develop the finite element equations can be stated as:

- Incorporate appropriate physics to suit the problem domain under consideration.
- For the structural analysis problems the physics comes from strength of materials, theory of elasticity. The equilibrium equations, the stress-strain relations governed by generalized Hooke's Law, the strain-displacement relations define the physics of the problem in structural mechanics. In heat transfer analysis the Fourier Law of heat conduction defines the basic physics to be followed.
- FEM equations are developed commonly by
  1. Direct method
  2. Method of variational virtual work
  3. Weighted residual method.

     Direct Method: This is based on direct reasoning and elementary principles leading to understanding of simple problems. This is limited to simple determinate cases.

     Method of variational Virtual work: This is based on the concept of virtual displacement leading to relations between internal and external virtual work and their balances and minimization of system potential energy for equilibrium.

     Weighted Residual Method: This starts with governing differential equation. With special mathematical manipulations 'weak form' solutions are developed and FEM equations are obtained. When no variational formulations are possible such methods are best suited. For stress analysis problems Rayleigh-Ritz, Galerkin's and WRM give identical solution as that of variational methods.

1.5.4 Top-Down Steps

1. Setup a global geometric model in world co-ordinate system of frame.
2. Discretize the object into nodal mesh.
3. The differential and integral equations describing the physics is to be discretized using finite difference and summation techniques.

4. Using (2) and (3) write the global equations with stiffness matrix times the vector of nodal values that are unknown. The element equations are assembled to form global equations as:

$$[K]\{U\} = \{F\}$$

   $[K]$ is the stiffness or the property matrix.

   $\{U\}$ is the nodal displacement vector.

   $\{F\}$ is the known nodal force vector.

   In this equation the unknown displacement vector is to be determined by numerical techniques.

5. Solve for the nodal values of the deformation and unknown nodal forces by using the constraints/boundary conditions. In case of
   - Static Analysis – nodal values at equilibrium
   - Dynamic Analysis – nodal values at next time step.
6. Interpolate values between the nodal co-ordinates

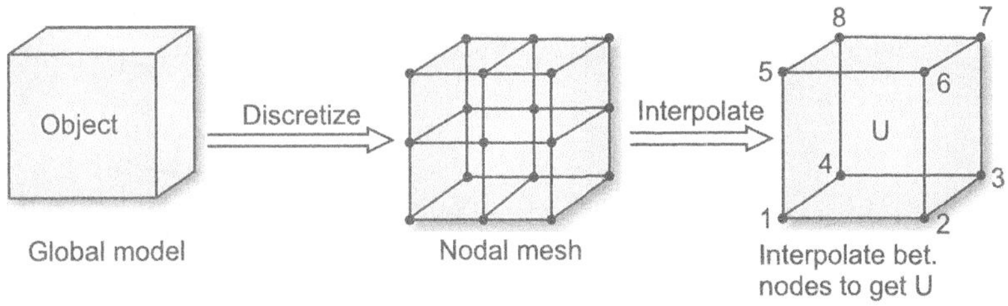

1.5.5 Bottom-Up Steps in FEM

- Nodes are the point masses connected with springs.
- A continuum equation is solved for nodes and interpolated for the intermediate points.
- A collection of nodes forms the element.
- A collection of elements forms the object.

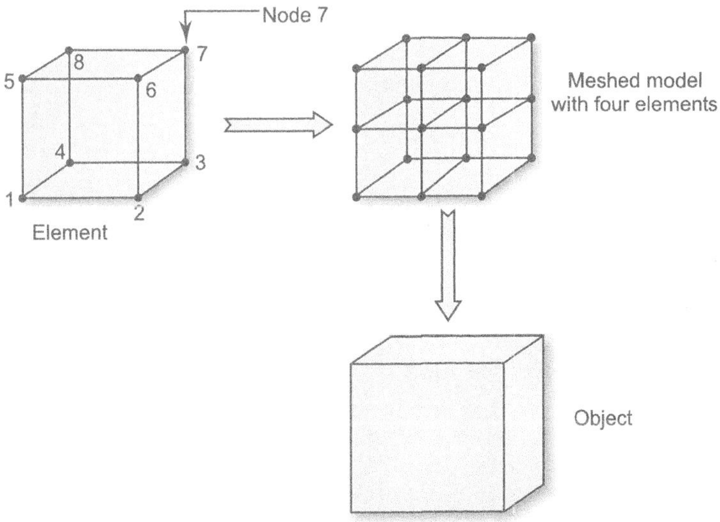

TABLE 1.1. Finite Element Types

One Dimensional Elements	Bar Element	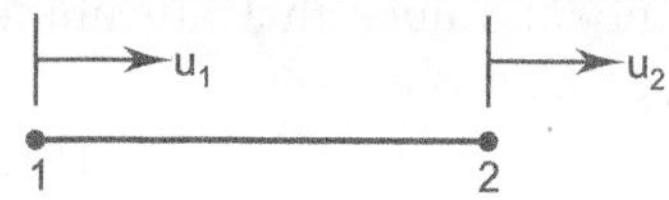
	Beam Element	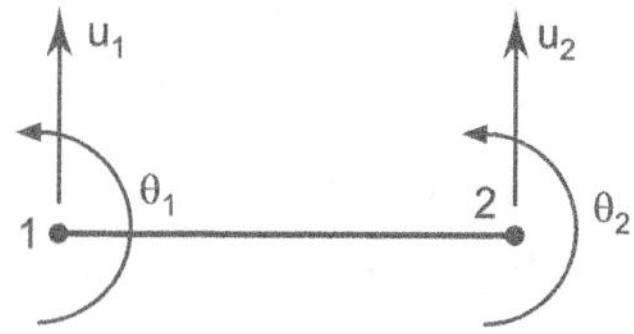
	Truss Element	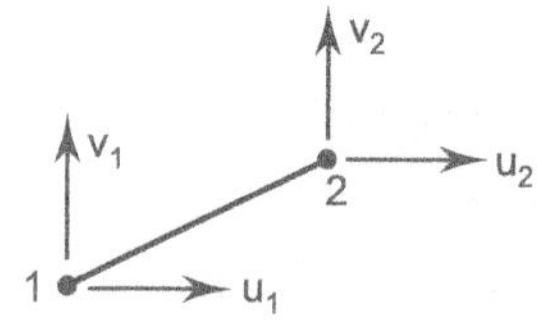
Two Dimensional Elements	Triangular Element	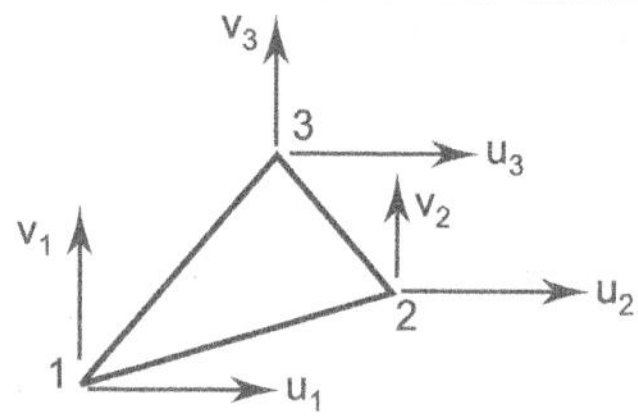
	Quadrilateral Elements	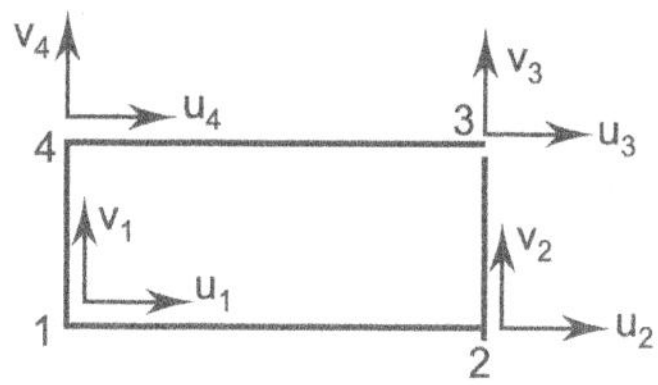
	Axisymmetric Element	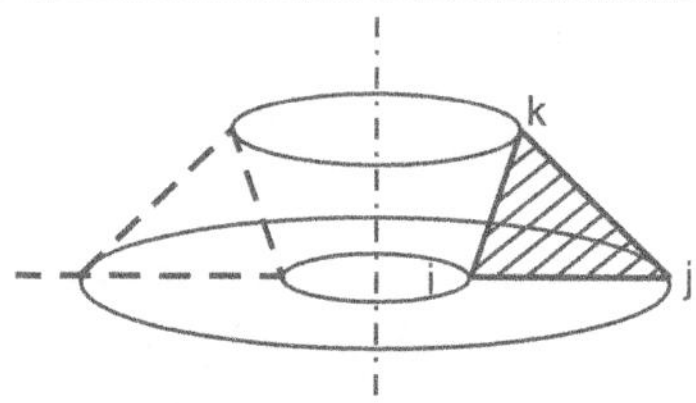
Three Dimensional Elements	Tetrahedron Element	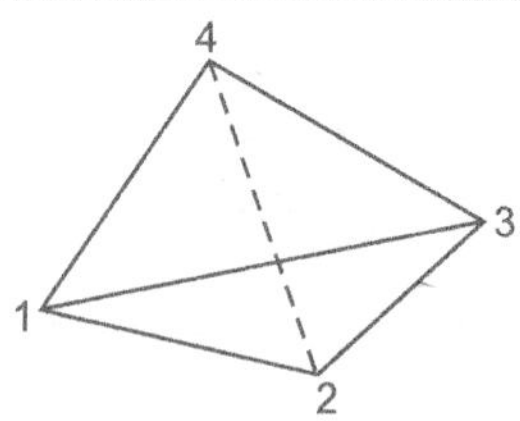
	Hexahedron (Brick) Element	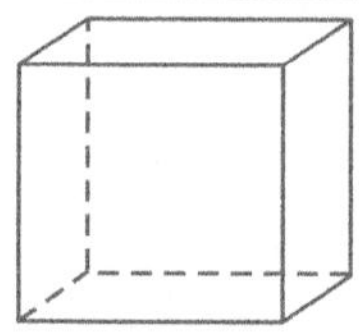

1.6 FEM AND ENGINEERING DESIGN

Engineering design, mainly is concerned with synthesis of geometrical dimension of the system given the input and output specifications. If the material properties and permissible deformations form the inputs, the loads, moments and torques form the outputs. With this it is possible to arrive at the initial design. Whether the initial design obtained by the application of design rules is satisfactory or not, is checked using FEM analytical procedures. The finite element analysis (FEA) aids in determining the point of maximum stress in the system, which helps in further improvization to be effected in the design. With informations of stress and strain obtained from FEA, it is possible to carryout modifications in design to streamline the stress flow at the critical regions of stress concentration. Optimization of structures and parts can also be done in the region where the stress is minimum. Decreasing the dimension in the less stress distributed regions and stiffening the higher stress regions, can be accomplished to produce a better design without affecting the function and performance of the system. This can lead to uniform stress-strain distribution in the system designed. After design modifications the design is resubmitted for FEA. The continuous refinement in the design is possible with successive application of FEA and assessment of results. By knowing the maximum stress in the part and comparing it with the permissible (design) stress, it is possible to calculate the factor of safety. Factor of safety is one parameter tool available to the design engineer that gives an idea of design solidarity. With the conventional design rules it is not possible to exactly estimate the induced stress in the system or the part for the given complex loading and boundary conditions. The location of the weakest point in the structure becomes difficult or rather impossible with the conventional design methodology. With proper modelling and by providing correct material property inputs, FEA can give the exact (near to) solution to stress and strain in the design of a engineering system. Needless to say that FEA is made imperative on all designs before they are manufactured and proved.

1.7 FINITE ELEMENT ANALYSIS AND CAD

Finite element analysis is a computer-based analytical procedure that estimates stress and strain for the given geometry and loading. The initial starter design gives geometry and boundary conditions, and environmental effects to be considered for FEA. Within the accuracy of results needed by the user the FE packages generate the mesh and the solutions re-assessed by automatic refinement of the meshes. With the results of linear static elastic, dynamic and non-linear analysis the basic shapes get modified within the design constraints, to get a CAD model. The FEA gives informations useful to CAD engineers as

- The maximum stress and the point of its occurrence.
- The maximum deformation and its location in the system.
- The eigen values and eigen vectors.

These informations talk about the criticality of the parameters relevant the CAD engineer. The results provide informations about the localization of stress leading to the stress concentration in the system. FEA provides insight to the CAD engineer about the design resulting in following modifications.

- Stream-lining the stress flow lines by changing the geometry where the stress concentration is indicated by FEA.
- Usage of higher grade material that has material property better than the maximum principal stress in the system.
- Avoidance of resonance by lowering or highering the system natural frequency.
- Stiffening the structure to optimize the stress distribution within the acceptable limit.
- Identify the weakest point, locally stressed load and support points and exercise the needed modifications.
- Estimating the factor of safety for the system when the material strength is known and tune the system in that direction.

For adapting the FEA results in CAD models there is the need for following considerations:

- Accurate geometric modelling of the system.
- Selection of suitable finite element.
- Anticipation of correct boundary conditions.
- Realistic load estimation and application.
- Providing proper technological material property.
- Interpretation of results obtained from right assessment.

However the accuracy of result and convergence are dependent on mathematical model, technique and size of grid selected.

1.7.1 FEM Modelling and Simulation

(*a*) Physical FEM

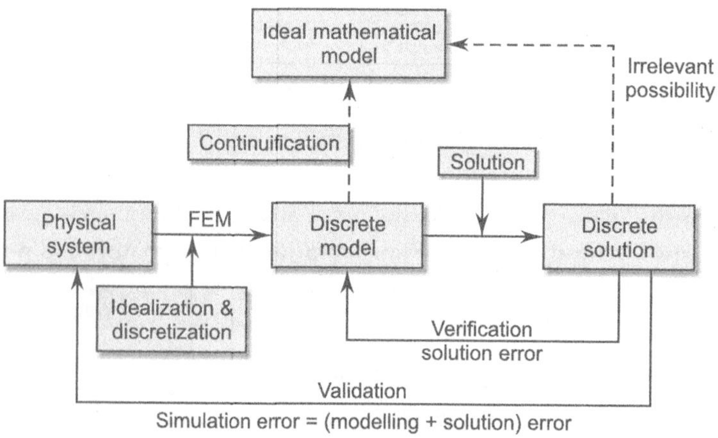

(*b*) Mathematical FEM

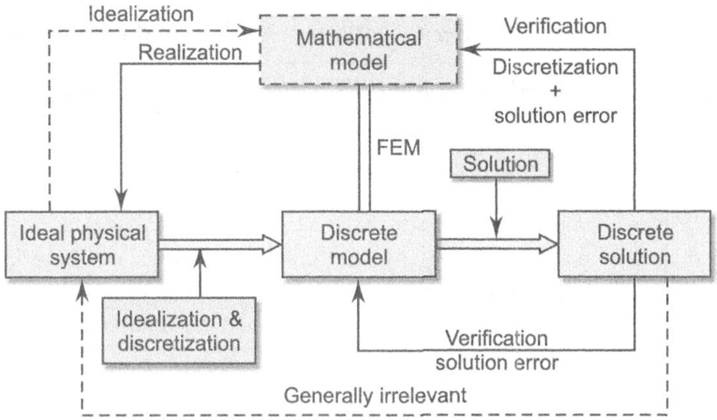

(*c*) Where FEM fits?

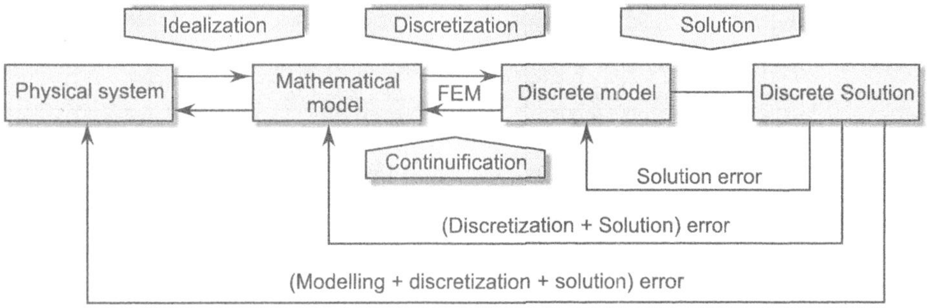

1.7.2 FEM Analysis — Process Flow

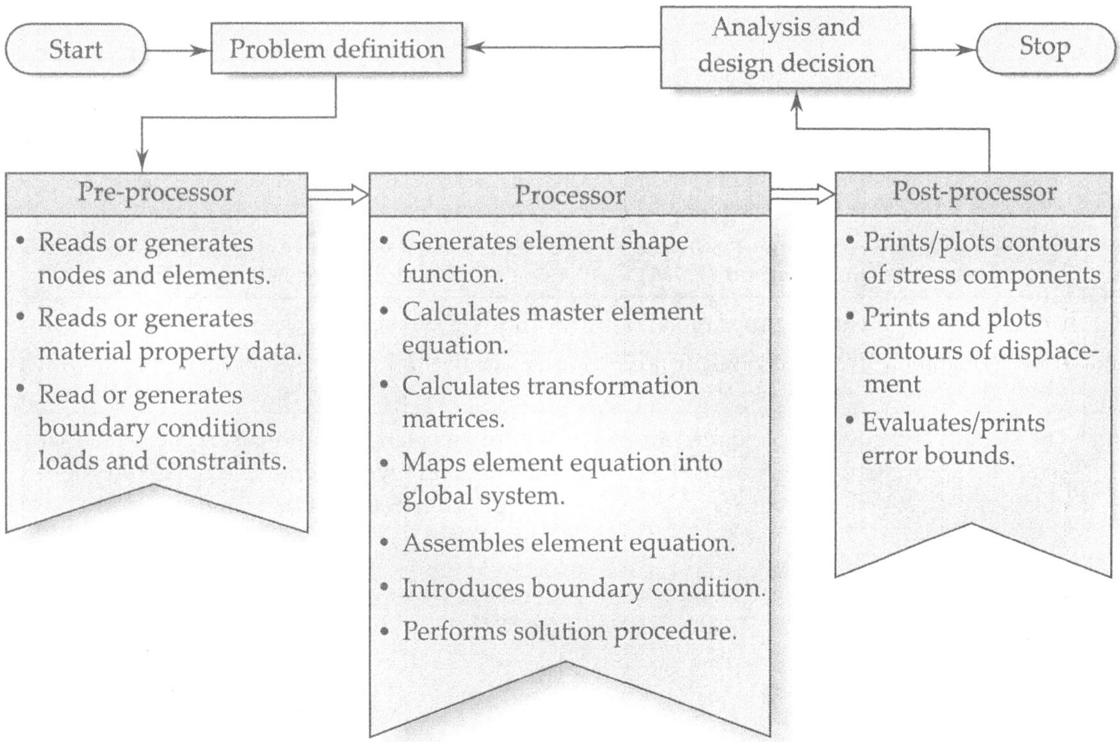

1.8 CONTINUUM–FEM–FDM

TABLE 1.2. Comparison between Continuum Method and FEM approach

Basis of Comparison	Continuum Method	FEM Approach
• Mathematical model	Formulation of the equations governing the physical behaviour is exact.	The governing equations of mechanics applied to finite elements are exact.
• Solution generation	For the class of problem, the formed equations can be solved accurately.	Because of discretization and round off efforts the solution obtained is approximate.
• Class of problem	Standard case of problems like simply supported, cantilever beams are easily solved.	Any type of problem with all dimensions are solved with ease.
• Cases of higher complexity	Drastic assumptions on shape, boundary and loading are needed to formulate the problem.	Various complexities have answer with reasonable accuracy without any assumptions.
• Anisotropic material property	In classical methods it is difficult to analyze when material is anisotropic.	Handling the considerations of material anisotropy is easy.
• More than one material	Systems with parts having different materials types are highly difficult to deal with.	Heterogenity in component materials has an easy answer in this approach.
• Orthotropic materials	Assumptions in classical mechanics speak about the analysis with homogeneous materials.	Material orthotropy is not a great problem to handle in FEM.
• Geometric non-linearity	The non-linear geometries are transformed to equivalent linear shapes before analysis.	Non-linear geometries are modelled with curved elements and unstructured grids.

TABLE 1.3. FEM versus FDM

	Finite Difference Method (FDM)	Finite Element Method (FEM)
1.	Dominated by pointwise approximation to the governing equation describing the physical phenomena.	This is prominent by piecewise approximation to the physical behaviour of the model.
2.	The sloping and curved boundaries are given stepped approximation as shown below.	There are triangular and curved elements that aid in mesh generation describing the non-linear geometries as shown under

	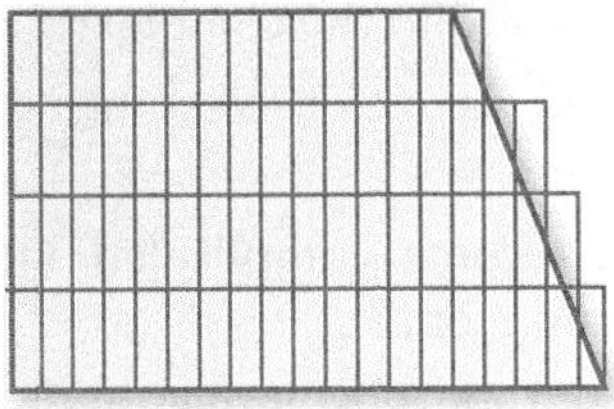 Sloping geometry 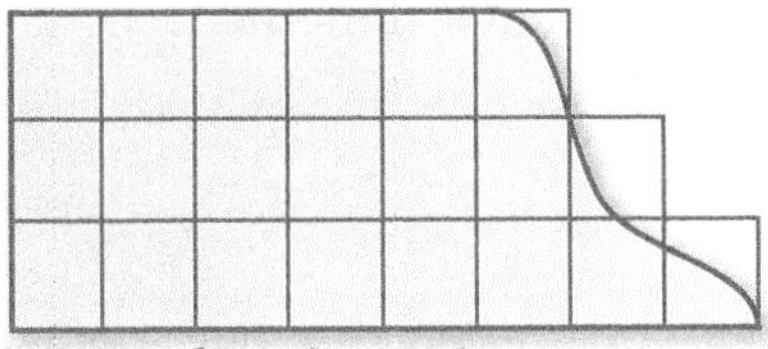 Curved geometry	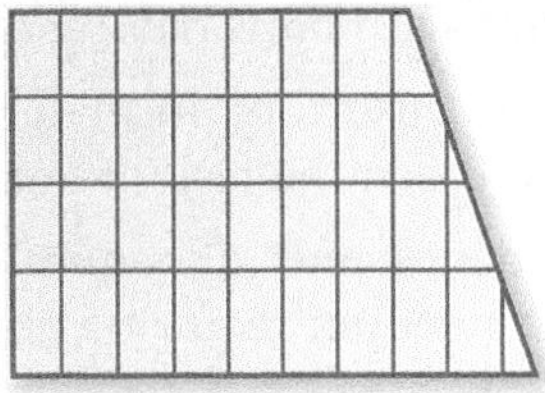 Sloping geometry 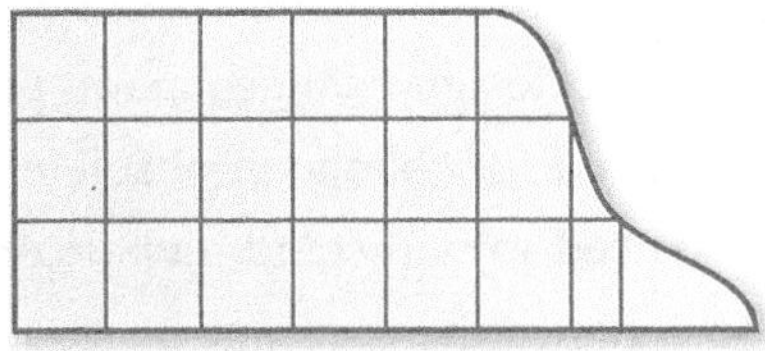 Curved geometry
3.	Ensures continuity at only the nodal points in the grid.	Ensures continuity of solution at nodes as well as imaginary boundaries in the grid.
4.	The deflection, stresses and forces can be computed at only the nodal points.	FEM gives values at node directly and at other points by interpolation.
5.	The FDM uses Taylor series expansion to evaluate partial derivatives.	FEM uses the polynomial function to describe the displacement function that ensures convergence of solution.
6.	For fair approximation the number of nodes in the grid have to be more.	With relatively less number of grid points good approximation to solution is achieved.
7.	Highly complex problems are not comfortably handled. Only simple and medium complex problems can be solved.	All and any type of complex problem are acceptable for analysis through FEM.

1.9 THE NEED FOR FEM

The foregone sections have dealt in brief the disadvantages of analyzing the continuous system and advantages of discrete systems giving fair approximation to the solutions. The difficulties in handling the complex real world physical applications as a continuum has led to the development of a computer aided simplified procedure in the form of Finite Element Method (FEM). The need for FEM is the result of the following considerations of the problem faced in finding solution for a continuous domain.

- Three dimensional complex shapes would have to be simplified with equivalent analyzable shapes.
- Irregular and combined load effects would have to be idealized with solvable pattern of load distribution.
- The indeterminant boundary conditions would have to be transformed into determinant boundary constraints.
- The mathematical relations pose difficulty in solution for a continuous domain, with its complex form.

- The systems with parts having multiple material types add to the complexity of analysis.
- Addition of anisotropy in material composition would have to be assumed as isotropic material to easen the task of analysis in the continuum.
- The kinematics and dynamics of the mechanism become complicated with infinite degrees of freedom requiring simplification and elimination.

Further some of the subjective aspects that render the need for FEM imperative, are

- The difficult and complicated experimental procedures.
- The fast design changes made possible by using fast digital computers.
- The need for change in obsolete designs because of low product life cycle.
- The need to withstand global competition.
- To enhance the accuracy of result and quality of product.
- The need for optimizing design to reduce the material cost of the components and systems.
- To incorporate higher safety and reliability in the design.
- To minimize and eliminate the inevitability and to easen the processing and manufacture of the components used in the products developed.

1.10 FEM—ITS MERITS AND DEMERITS

Wide spectrum of advantages derived from analysis through FEM and limited disadvantages in its usage and application, have made FEM very popular in civil and mechanical engineering applications. With the advent in fast digital computers and with the decedance in design categories it has become inevitable to use FEM packages. Higher cost component of usage of FEM in design is overridden by the flexibility and quality of design favoured by the realistic insight into the complexity of the problem. Accessibility to complex shapes, difficult environmental effects (loads) and indeterminant fixation (boundary) makes the modelling of the real world physical problem easy in FEM. Specification of material property on the elemental basis favours the analysis of the system with material non-linearity. For better accuracy, the FEM has the provision of discretizing with higher order elements (to be discussed later in the book). It is easy to understand the localization effects of the load on the physical domain with concentrated meshes. Without basic knowledge of FEM the interpretation of results and effecting modifications can go waste. However discretization, approximation in governing equations and round-off error in numerical techniques contribute errors in the solution obtained.

1.10.1 Advantages of FEM

Here are the list of advantages derived out of analysis by FEM.

- FEM can handle irregular geometries in a convenient manner with help of it elements library, when closed form solutions are non-existant.
- The spacing of nodes need not follow a pattern or rule. When the nodes are not equally spaced the mesh is said to be the graded mesh. Under the loads and near the supports the grid can be finer so that the localization effect can be easily analyzed. But in mechanics of materials the stress distribution near the load is assumed to be uniform.

- To handle non-homogenous and anisotropic material provision is made in the material property inputs.
- Various types of boundary conditions can be imposed automatically, in the formulation of the model.
- The accuracy of result is improved by selecting the higher order elements provided by special facilities in FEM packages.
- Analysis by FEM does not require explicit governing equations representing the behaviours of physical phenomena.
- FEM method can be used to solve problems in design, structural analysis, heat transfer, fluid flow, electrical circuit analysis, and dynamic analysis.

1. Can easily handle complex geometries
    - This is the heart and power of FEM.
2. Can handle complex analysis types
    - Vibration
    - Transients
    - Non-linear
    - Heat transfer
    - Fluid flow
3. Can handle complex loading
    - Node based loading (point loads)
    - Element based loading (pressure, thermal, inertial forces)
    - Time or Frequency based loading.
4. Can handle complex restraints
    - Indeterminate structures can be analysed.
5. Can handle bodies with non-homogeneous materials.
    - Each element in the model can be assigned with a different set of material properties.
6. Can handle bodies with non-isotropic materials.
    - orthotropic
    - anisotropic
7. Special material effects can be handled.
    - Temperature dependent properties
    - Plasticity
    - Creep
    - Swelling
8. Special geometric effects can be handled
    - Large displacements
    - Large rotations
    - Contact (gap) conditions

1.10.2 Disadvantages of FEM

- Requires a sophisticated FEA packages which are very costly.
- The large size problems would have large number of elements. To handle large size of data high speed digital computers are needed. High speed computers are costly because of which small players tend to adopt conventional procedures in design and analysis.
- The justification and interpretation of results need expertise and proper judgement with knowledge of FEM.
- Without proper understanding of various procedures in FEM the exercises and workouts would follow GIGO (Garbage In, Garbage Out) philosophy. The confidence in the input can only produce justified outputs.
- Obtained results are case specific. Closed form solutions are not possible.
- FEM is the approximation to the mathematical model defined for a system.
- Experience and judgement are required to construct a good FEM model.
- A powerful computer and reliable FEM software package are essential.
- Input and output data may be large and quite tedious to handle and interpret.
- Numerical problems
  1. Computer gives finite number of significant digits.
  2. Round off and error accumulation can happen.
  3. Stiff and flexible element choices are situation dependent.
- Susceptible to user introduced modelling errors.
  1. Poor choice of element types.
  2. Distorted elements.
  3. Geometry not adequately modelled.
- Certain effects are not automatically taken care.
  1. Buckling effect.
  2. Large deflection and rotations.
  3. Material non-linearity.
  4. Other non-linearities.

1.11 SOURCES OF ERRORS IN FEM

Three main sources of error in a typical FEM solution to an engineering problem are

1. Discretization error
2. Formulation error
3. Numerical error

1. The Discretization Error

This error results in transforming the continuum (physical) domain into finite element model. This can be related to boundary shapes and boundary conditions. This error decreases with the refinement of the grid.

2. Formulation Error

- This error results from the choice of the elements that do not precisely describe the physical system behaviours.
- Such elements are called ill-conditioned or mathematically unsuitable elements.
- The elements used to represent linear displacement variation induce formulation error when used to represent quadratic or cubic variation in displacement.

3. Numerical Error

- The numerical techniques used to represent the FEM equations (differential/integral) would have truncation error and round off errors.
- Numerical error is a problem for vendor and developer.
- Incorrect specification of physical quantity like Young's Modulus, Poisson's ratio results in numerical error contributed by the user.

1.11.1 Measure of Accuracy

The error in FEM solutions can be defined as the difference between exact solution and FEM solutions.

$$\text{Error} = |(\text{Exact solution}) - (\text{FEM solution})|.$$

Convergence is the limit of error approaching zero. By increasing the number of elements and the order of approximation it is possible to reduce the error. But this consumes higher computational time and efforts. The convergence based on number of elements is called 'h-convergence'. The convergence based on order of approximation is called 'p-convergence'. By exercising h- and p-convergences the accurate solutions can be generated.

1.12 APPLICABILITY OF FEM

1. Structural/Stress Analysis	⇒	Mechanical FEA
2. Fluid flow, Heat transfer	⇒	Heat transfer FEA
3. Electromagnetic Fields	⇒	Electro-mechanical FEA
	⇒	Electro-magnetical FEA
4. Accoustics	⇒	Accoustical FEA
5. Impact/crash Analysis, Fluid structure interaction	⇒	Multiphysics

1.12.1 How Can FEM Help the Design Engineers?

The FEM offers important advantages to design engineers that are highlighted in the following ways.

- FEM is easily applied to complex, irregular-shaped objects composed of several different materials and problems with difficult boundary conditions.
- Applicable to steady state, time dependent and eigen value problems.
- Applicable to linear and non-linear problems.

- Wide applicability to solid mechanics, fluid mechanics chemical reactions, electro-magnetics, bio-mechanics heat transfer and accoustics problems.
- General purpose FEM packages available, can be easily executed on microcomputers like PC's, and work stations.
- Inputs from CAD programs can be used in modelling and mesh generation useful to FEM.
- Pre- and post-processors are user friendly with features like GUI interfaces, automeshing and sophisticated graphics.

1.12.2 How Can FEM Help the Design Organization?

The important business advantages to the design organizations using FEM packages are:

- Reduced testing and redesign costs, thereby shortening product development time.
- Identifies issues in design before tooling is committed.
- Refinement of component design establishing compatibility with other component design.
- Optimize performance before prototyping.
- Discover design problems before failure complaints.
- Favour engineering judgements without resorting to 'turning the crank' with trial and error efforts.

1.13 FACILITIES IN GENERAL PURPOSE PROGRAMS

In the sophisticated general purpose programs for finite element analysis the facilities generally observable are:

- Structural linear
- Structural non-linear
- Structural contact/common boundary
- Structural dynamics
- Structural Buckling
- Thermal analysis
- CFD analysis
- Electro-magnetic-low frequency
- Electro-magnetic-high frequency
- Field and coupled field analysis
- Solver
  - Iterative solver
  - Sparse
  - Frontal
  - Explicit
  - Implicit
- Pre-processing
- Post-processing
- General features

Advantages of General Purpose Programs

- Easy input—user friendly pre-processor
- Solves many types of problems with multiphysics
- Modular design-fluid, dynamics, heat etc.
- Runs easily on PC's
- Relatively low cost

Disadvantages

- High development cost
- Less efficient than smaller custom programs
- Offen proprietary. Use of access to codes is limited.

POINTS TO BE REMEMBER

- **Continuum:** A continuous domain with boundary considered for the analysis of environmental effects.
- **Discretes:** A domain divided into simple sub domains.
- **Finite elements:** Collection of simple sub divisions representing the continuum.
- **Mesh:** Group of finite elements separated by imaginary lines or surfaces.
- **Nodes:** The points of connection between the elements.
- **Structured Mesh:** Mesh containing one type of finite elements.
- **Unstructured Mesh:** Mesh containing more than one type of element.
- **Shape Function:** An expression relating the displacement field with displacements at nodes.
- **Errors in FEM**
  1. Discretization Error – error due to approximation of the domain by sub division.
  2. Solution Error – error due to mathematical model chosen.
  3. Round-off error – error due to numerical computation and limitation of computer.
- **Accuracy:** Difference between exact solution and solution by finite element method.
- **Convergence:** Increase in accuracy with increase in number of elements in the mesh.
- **Element types**
  1. 1D elements — Bar element, Beam element truss element.
  2. 2D elements — Triangular and quadrilateral elements. Axisymmetric elements.
  3. 3D elements — Tetrahedron and hexahedron (brick) elements.

EXERCISE

1. Define the following

   (a) Continuum (b) Discrete system (c) Mesh
2. What are the considerations to be made in solving the problem by continuum method?

3. Explain the procedure adapted by structural engineer in analysing the problem of discrete nature.
4. What are the various steps to be followed in Finite Difference Method (FDM)?
5. Define and explain Finite Element Method.
6. What are different modules in FEM and what are their involvements?
7. Explain the elastic static analysis using FEM.
8. Give few examples of various types of elements used in FEM.
9. Discuss the application of finite element method in Engineering Design.
10. How Computer Aided Design (CAD) has a dependent bearing with Finite Element Analysis (FEA)?
11. Compare and contrast FEM with the classical continuum method.
12. Differentiate FDM from that of FEM.
13. Explain the need for FEM in Engineering.
14. List and explain the advantages and disadvantages of finite element method.
15. What are different types of errors possible in FEM? Explain.
16. How sloping and curved geometries are handled in FDM and FEM?

Chapter 2

NUMERICAL TECHNIQUES IN FINITE ELEMENT METHOD

बहूनामप्यसाराणाम्।
समवायो दुरासदः॥
तृणैरारभते रज्जुः।
तेन नागोपि बध्यते॥ (सुभाषितावलि)॥

[*There is lot of strength in unity. It is possible to tie the elephant from the rope made of grass elements.*]

Since finite element method is a totally computer oriented approach, the handling of equations is completely based on solving the simultaneous equations in the matrix form. Hence it is the pre-requisite to learn for the FEM engineers to learn the terminology, the algebra, solution methods and the numerical methods related to matrices. This chapter attempts to throw light on the following important issues.

- Terminology of different matrices definitions.
- The algebraic operations pertaining matrices.
- The determinants, inversion and calculus of matrices.
- The solution methods such as
  - Gauss elimination method
  - Cholesky's method
  - LU decomposition method
  - Tri-Diagonal Matrix Algorithm (TDMA)
- The numerical integration by Gauss Quadrature.

The formulations of the governing equations representing the dynamic and static behaviours of stress and strain in the mechanical systems and civil structure in finite element analysis are included with matrix equations that are solved by various numerical methods like Gauss elimination, Cholesky's method, etc., that are direct or interactive, as they get supported by implementation in computer analytical softwares. Needless to mention that this is an imperative prerequisite for understanding the theory and concepts by engineers successfully implementing and carrying out research in FEM.

2.1 MATRICES

- **Matrix:** A rectangular array of numbers arranged in m rows and n columns is called a matrix. For example

$$A = \begin{bmatrix} a_{11} & a_{12} & a_{13} & a_{14} \\ a_{21} & a_{22} & a_{23} & a_{24} \\ a_{31} & a_{32} & a_{33} & a_{34} \end{bmatrix}$$

  is a 3 × 4 matrix. There are three rows (m) of elements and four columns (n) of elements in $[A]$.

- **Square Matrix:** A square matrix is one in which the number of rows is equal to number of columns. It is referred to as an $n \times n$ matrix or a matrix of order n. For example

$$A = \begin{bmatrix} a_{11} & a_{12} & a_{13} \\ a_{21} & a_{22} & a_{23} \\ a_{31} & a_{32} & a_{33} \end{bmatrix}$$

  is a 3 × 3 matrix or $[A]$ is a matrix of order 3.

- **Symmetric Matrix:** A square matrix is said to be symmetric if the elements of the upper right half by mirroring the matrix about the diagonal. In a symmetric matrix $a_{ij} = a_{ji}$ for $i \neq j$. For example

$$A = \begin{bmatrix} 1 & -2 & 3 \\ -2 & 1 & 1 \\ 3 & 1 & 1 \end{bmatrix} = \text{symmetric matrix.}$$

- **Trace:** The sum of the diagonal elements of a square matric is called the trace. Hence

$$\text{Trace } [A] = \sum a_{ij} \text{ for } i = j$$

  For the matrix above

$$\text{Trace } [A] = 1 + 1 + 1 = 3$$

- **Singular Matrix:** If the determinant (to be discussed later in section 2.2) is zero the matrix is said to be singular.

- **Row Matrix:** A row matrix $m = 1$. For example

$$[B] = [b_1\ b_2\ b_3]$$

- **Column Matrix:** For a column matrix $n = 1$. For example

$$[C] = \begin{bmatrix} c_1 \\ c_2 \\ c_3 \end{bmatrix}$$

- **Zero Matrix**: The zero matrix is one in which all the elements in the matrix are zero

$$[0] = \begin{bmatrix} 0 & 0 & 0 \\ 0 & 0 & 0 \end{bmatrix}$$

- **Unit Matrix:** The unit matrix is a square matrix in which the diagonal elements from the top left to the bottom right are unity with all other elements equal to zero.

  i.e. $a_{ij} = 1$ for $i = j$ and $a_{ij} = 0$ for $i \neq j$. For example

$$[I] = \begin{bmatrix} 1 & 0 & 0 \\ 0 & 1 & 0 \\ 0 & 0 & 1 \end{bmatrix}$$

- **Diagonal Matrix:** A square matrix having elements a_{ij} for $i = j$ along the diagonal with all other element equal to zero (*i.e.*, $a_{ij} = 0$ for $i \neq j$) is a diagonal matrix. For example

$$[a_{ij}] = \begin{bmatrix} a_{11} & 0 & 0 \\ 0 & a_{22} & 0 \\ 0 & 0 & a_{33} \end{bmatrix}$$

  is a diagonal matrix of order 3.

- **Transpose of a Matrix:** The transpose A' of a matrix A is one in which rows and columns of A are interchanged. For example

$$A = \begin{bmatrix} a_{11} & a_{12} & a_{13} \\ a_{21} & a_{22} & a_{23} \end{bmatrix} \qquad [A'] = \begin{bmatrix} a_{11} & a_{21} \\ a_{12} & a_{22} \\ a_{13} & a_{23} \end{bmatrix}$$

  Hence transpose of a_{ij} is a_{ji}.

- **Minor:** A minor of a matrix A, M_{ij} is formed by deleting i^{th} row and j^{th} column of the determinant of given matrix.

  Let

$$A = \begin{bmatrix} a_{11} & a_{12} & a_{13} \\ a_{21} & a_{22} & a_{23} \\ a_{31} & a_{32} & a_{33} \end{bmatrix} \text{ be the given original matrix}$$

$$M_{23} \text{ of } [A] = \begin{bmatrix} a_{11} & a_{12} \\ a_{31} & a_{32} \end{bmatrix} = (a_{11} \times a_{32} - a_{12}\, a_{31})$$

- **Co-factor:** The co-factor C_{ij} is equal to the signed minor given by $(-1)^{i+j} M_{ij}$. For the previous example

$$C_{23} = (-1)^{2+3} M_{23} = [-M_{23}]$$

- **Adjoint Matrix:** The adjoint matrix of a square matrix A is the transpose of co-factor matrix of A.

  Let the cofactor matrix of A be

$$[C_{ij}] = \begin{bmatrix} C_{11} & C_{12} & C_{13} \\ C_{21} & C_{22} & C_{23} \\ C_{31} & C_{32} & C_{22} \end{bmatrix}$$

$$adj\,[A] = [C'_{ij}] = \begin{bmatrix} C_{11} & C_{21} & C_{31} \\ C_{12} & C_{22} & C_{32} \\ C_{13} & C_{23} & C_{33} \end{bmatrix}$$

- **Inverse of a Matrix:** The inverse A^{-1} of a matrix A satisfies the relationship

$$[A^{-1}]\,[A] = [I]$$

The inversion of a matrix is dealt in detail in section (2.4).

- **Orthogonal Matrix:** A orthogonal matrix is one which satisfies the relationship

$$[A']\,[A] = [I]$$

From the definition of inverse matrix it is clear that for an orthogonal matrix $A^{-1} = A'$ (transpose of A).

2.2 DETERMINANT OF A MATRIX

A determinant of a matrix of second order can be numerically evaluated as

$$D = \begin{bmatrix} a & b \\ c & d \end{bmatrix} = ad - bc.$$

To determine the determinant of the third order matrix is expressed in terms of the co-factors of the matrix. For example

$$D = \begin{vmatrix} 2 & 1 & 3 \\ 4 & 2 & 1 \\ 3 & 0 & 5 \end{vmatrix} = 2(-1)^{1+1}\begin{vmatrix} 2 & 1 \\ 0 & 5 \end{vmatrix} + 1(-1)^{2+1}\begin{vmatrix} 4 & 1 \\ 3 & 5 \end{vmatrix} + 3(-1)^{3+1}\begin{vmatrix} 4 & 2 \\ 3 & 0 \end{vmatrix}$$

$$= 20 - 17 + (-18) = -15$$

Properties of Determinants:

1. Interchange of any two columns and rows changes the sign of the determinant.
2. If two rows or two columns are identical the determinant is zero.
3. Any row or column may be multiplied by a constant and added to another row or column without changing the value of the determinant.

2.3 MATRIX ALGEBRA

- **Addition:** Two matrices with same number of rows and columns may be added by summing the corresponding elements. For example

$$\begin{bmatrix} 1 & 2 & 3 \\ 3 & 2 & 1 \end{bmatrix} + \begin{bmatrix} 2 & 3 & 4 \\ 4 & 3 & 2 \end{bmatrix} = \begin{bmatrix} 3 & 5 & 7 \\ 7 & 5 & 3 \end{bmatrix}$$

i.e.,
$$[A_{ij}] + [B_{ij}] = [C_{ij}]$$

- **Multiplication:** The product two matrices A and B is another matrix C. For matrix multiplication to happen columns of A must be equal to rows in B. Hence

$$[C_{ij}] = [A_{ik}]\,[B_{kj}]$$

The elements of $[C_{ij}]$ are given by

$$C_{ij} = \sum a_{ik} \cdot b_{kj}$$

For example

Let
$$A = \begin{bmatrix} 1 & 2 & 3 \\ 2 & 3 & 4 \\ 3 & 4 & 5 \end{bmatrix} \text{ and } B = \begin{bmatrix} 1 & 3 \\ 2 & 4 \\ 3 & 5 \end{bmatrix}$$

$$AB = \begin{bmatrix} 14 & 26 \\ 20 & 38 \\ 26 & 50 \end{bmatrix}$$

It may be observed that $C_{32} = 3 \times 3 + 4 \times 4 + 5 \times 5 = 50$.

The post multiplication of a matrix by a column matrix results in a column matrix.

For example

$$\begin{bmatrix} 1 & 2 & 1 \\ 2 & 3 & 2 \\ 3 & 4 & 3 \end{bmatrix} \begin{bmatrix} 1 \\ 2 \\ 3 \end{bmatrix} = \begin{bmatrix} 8 \\ 14 \\ 20 \end{bmatrix}$$

Premultiplication of a matrix by a row matrix (or transpose of a column matrix results in row matrix).

$$[1\ 2\ 3] \begin{bmatrix} 1 & 2 & 1 \\ 2 & 3 & 2 \\ 3 & 4 & 3 \end{bmatrix} = [10\ 20\ 10]$$

It may be observed that $AB \neq BA$

To show $C = AB$ and $C' = B'A'$

Consider the example

$$A = \begin{bmatrix} 1 & 2 \\ 3 & 4 \end{bmatrix} \qquad B = \begin{bmatrix} 3 & 4 \\ 5 & 6 \end{bmatrix}$$

$$AB = \begin{bmatrix} 13 & 16 \\ 29 & 36 \end{bmatrix} = [C]$$

$$B'A' = \begin{bmatrix} 3 & 5 \\ 4 & 6 \end{bmatrix} \begin{bmatrix} 1 & 3 \\ 2 & 4 \end{bmatrix} = \begin{bmatrix} 13 & 29 \\ 16 & 36 \end{bmatrix} = C'$$

It may be observed that C = transpose of $[C']$.

2.4 INVERSION OF A MATRIX

Consider a square matrix

$$A = \begin{bmatrix} a_{11} & a_{12} & a_{13} \\ a_{21} & a_{22} & a_{23} \\ a_{31} & a_{32} & a_{33} \end{bmatrix}$$

The inverse of A is defined by

$$A^{-1} = \frac{1}{|A|} \, adj\, A$$

Procedure of Finding Inverse

1. Find the determinant of the given square matrix.
2. Find the minors of A.
3. Supply the sign $(-1)^{i+j}$ to the minors to find the co-factors.
4. Find the adjoint matrix as the transpose of co-factor matrix.
5. Obtain the inverse matrix by dividing the adjoint matrix from the determinant of the matrix.
6. The check is $[A^{-1}]\,[A] = [I]$.

Example 2.1. *Find the inverse of A given by*

$$A = \begin{bmatrix} 1 & 1 & 1 \\ 1 & 2 & 2 \\ 1 & 2 & 3 \end{bmatrix}$$

Solution.

1. Determinant of $A = 1\,(6-4) - 1\,(3-2) + 1\,(2-2) = 1$
2. Minors of A

$$M_{11} = \begin{vmatrix} 2 & 2 \\ 2 & 3 \end{vmatrix} = (6-4) = 2 \qquad M_{12} = \begin{vmatrix} 1 & 2 \\ 1 & 3 \end{vmatrix} = (3-2) = 1$$

$$M_{13} = \begin{vmatrix} 1 & 2 \\ 1 & 2 \end{vmatrix} = (2-2) = 0 \qquad M_{21} = \begin{vmatrix} 1 & 1 \\ 2 & 3 \end{vmatrix} = (3-2) = 1$$

$$M_{22} = \begin{vmatrix} 1 & 1 \\ 1 & 3 \end{vmatrix} = (3-1) = 2 \qquad M_{23} = \begin{vmatrix} 1 & 1 \\ 1 & 2 \end{vmatrix} = (2-1) = 1$$

$$M_{31} = \begin{vmatrix} 1 & 1 \\ 2 & 2 \end{vmatrix} = (2-2) = 0 \qquad M_{32} = \begin{vmatrix} 1 & 1 \\ 1 & 2 \end{vmatrix} = (2-1) = 1$$

$$M_{33} = \begin{vmatrix} 1 & 1 \\ 1 & 2 \end{vmatrix} = 2 - 1 = 1$$

3. Co-factor matrix

$$C_{ij} = \begin{bmatrix} 2 & -1 & 0 \\ -1 & 2 & -1 \\ 0 & -1 & 1 \end{bmatrix}$$

4. Adjoint matrix $Adj\,[A] = C'_{ij}$

$$Adj\,[A] = \begin{bmatrix} 2 & -1 & 0 \\ -1 & 2 & -1 \\ 0 & -1 & 1 \end{bmatrix}$$

5. Inverse of A

$$A^{-1} = \frac{1}{|A|} Adj\,[A]$$

$$= \frac{1}{1}\begin{bmatrix} 2 & -1 & 0 \\ -1 & 2 & -1 \\ 0 & -1 & 1 \end{bmatrix}$$

6. Check $[A^{-1}]\,[A] = [I]$

$$\begin{bmatrix} 2 & -1 & 0 \\ -1 & 2 & -1 \\ 0 & -1 & 1 \end{bmatrix}\begin{bmatrix} 1 & 1 & 1 \\ 1 & 2 & 2 \\ 1 & 2 & 3 \end{bmatrix} = \begin{bmatrix} 1 & 0 & 0 \\ 0 & 1 & 0 \\ 0 & 0 & 1 \end{bmatrix}$$

It may be observed that

$$[A^{-1}]\,[A] = [I]$$

$$\frac{Adj\,[A]}{|A|} \cdot [A] = [I]$$

$$|A|\,[I] = [A]\,Adj\,[A].$$

Differentiation of a Matrix

To find the derivative of a matrix, we differentiate each element of the matrix in the conventional manner. For example, if

$$[A] = \begin{bmatrix} x & x^2 & 3 \\ 3x & 2x^2 & x^3 \\ x^2 & 4 & 2x \end{bmatrix}$$

then,
$$\frac{\partial}{\partial x}[A] = \begin{bmatrix} 1 & 2x & 0 \\ 3 & 4x & 3x^2 \\ 2x & 0 & 2 \end{bmatrix}$$

Integration of a Matrix

Matrix integration is in line, but reverse of differentiation where-in each element is integrated. For example, if

$$[a] = \begin{bmatrix} 1 & 2x & 0 \\ 3x & 3x^2 & 4x \\ 2x & 0 & 2 \end{bmatrix}$$

then,
$$\int [a]\,dx = \begin{bmatrix} x & x^2 & 1 \\ 3 & x^3 & 2x^2 \\ x^2 & 1 & 2x \end{bmatrix} + [c]$$

where $[c]$ is the constant of integration. For definite integrals, each term is evaluated for the limits of integration given.

2.5 EIGEN VALUE AND EIGEN VECTOR

Eigen value is the measure of natural frequency of vibration *i.e.*, the eigen value symbolized by λ is the square of the natural frequency. Hence $\lambda_i = \omega^2_i$. Eigen vector is the displacement vector X_i corresponding to the respective eigen value λ_i. Eigen vectors represent the mode shapes of vibration. For a given matrix $[A]$ the eigen values are obtained by

$$|[A] - \lambda\,[I]| = 0 \qquad ...(2.1)$$

For an n ordered matrix, there are n eigen values and n eigen vectors. The equation (2.1) is derived from the equation of motion for free vibration. The free vibration is characterized by

$$[K]\,X + [M]\ddot{X} = 0$$

or

$$\{[K] - \omega^2\,[M]\}\,X = 0 \qquad ...(2.2)$$

pre-multiplying equation (2.2) by M^{-1}

$$\{[M^{-1}\,K] - 1\,M^{-1}\,M]\,X = 0 \qquad ...(2.3)$$

or $|A - \lambda\,[I]\,| = 0$. This is same as equation (2.1) where $A = M^{-1}\,K$ and X is the eigen value X is the displacement (eigen) vector. $[I]$ is the unity matrix.

Example 2.2. *Find the eigen values and the eigen vector for the matrix*

$$A = \begin{bmatrix} 1 & 2 \\ 3 & 4 \end{bmatrix}$$

$$A - \lambda\,[I] = \begin{bmatrix} 1-\lambda & 2 \\ 3 & 4-\lambda \end{bmatrix}$$

$$\begin{vmatrix} 1-\lambda & 2 \\ 3 & 4-\lambda \end{vmatrix} = 0$$

or

$$(1 - \lambda)\,(4 - \lambda) - 6 = 0$$

$$\lambda^2 - 5\,\lambda + 4 - 6 = 0 \qquad ...(2.4)$$

Solution. Equation (2.4) is the quadratic equation in λ. The solutions to λ are

$$\lambda_1 = \frac{5 + \sqrt{25+8}}{2} = 5.3723$$

$$\lambda_2 = \frac{5 - \sqrt{25+8}}{2} = -\,0.3723$$

Eigen vectors are obtained by substituting λ in equation (2.3) and solving for X.

$$\begin{bmatrix} -4.37 & 2 \\ 3 & -1.37 \end{bmatrix} \begin{Bmatrix} X_1^1 \\ X_2^1 \end{Bmatrix} = 0$$

Hence

$$\begin{Bmatrix} X_1^1 \\ X_1^1 \end{Bmatrix} = \begin{Bmatrix} 0.457 \\ 0.457 \end{Bmatrix} \{X_2^1\}$$

and

$$\begin{bmatrix} 5.37 & 2 \\ 3 & 4.37 \end{bmatrix} \begin{bmatrix} X_1^2 \\ X_2^2 \end{bmatrix} = 0$$

$$X_1^2 = 0.3724\,[X_2^2]$$

By assuming X_2^1 and X_2^2 to be 1.

$$\begin{Bmatrix} X_1^1 \\ X_2^1 \end{Bmatrix} = \begin{Bmatrix} 0.457 \\ 1.0 \end{Bmatrix} \quad \text{and} \quad \begin{Bmatrix} X_1^2 \\ X_2^2 \end{Bmatrix} = \begin{Bmatrix} 0.372 \\ 1 \end{Bmatrix}$$

2.6 METHODS OF SOLUTION OF LINEAR EQUATIONS

The simultaneous linear system of equations can be converted into matrix equation in short. The co-efficients of the terms in the simultaneous equations form a square matrix. For example, if the simultaneous equations are

$$\begin{aligned} a_{11}x_1 + a_{12}x_2 + \ldots\ldots a_{1n}x_n &= b_1 \\ a_{21}x_1 + a_{22}x_2 + \ldots\ldots a_{2n}x_n &= b_2 \\ \vdots \quad\quad \vdots \quad\quad\quad \vdots \quad &\quad \vdots \\ a_{n1}x_1 + a_{n2}x_2 + \ldots\ldots a_{n}x_n &= b_n \end{aligned} \qquad \ldots(2.5)$$

then the matrix equation takes the form

$$[A]\,[X] = [B] \qquad \ldots(2.6)$$

where

$$[A]=\begin{bmatrix} a_{11} & a_{12} & \cdots & a_{1n} \\ a_{21} & a_{22} & \cdots & a_{2n} \\ \vdots & & & \vdots \\ a_{n1} & a_{n2} & \cdots & a_{nn} \end{bmatrix} \quad \text{and} \quad X = \begin{bmatrix} x_1 \\ x_2 \\ \vdots \\ x_n \end{bmatrix}, \quad B = \begin{bmatrix} b_1 \\ b_2 \\ \vdots \\ b_n \end{bmatrix}$$

The matrix, $[A]$ can be a full matrix, sparse matrix, a banded matrix or a symmetric matric. Several solution algorithms are proposed to solve the matrix equation of the type (2.5), where the vector $[X]$ is to be determined given the matrix $[A]$ and the vector $[B]$. One simple way to solve is by inversion of $[A]$ *i.e.*, $[X] = [A^{-1}]\,[B]$. This is handy when the order of [A] is small. Hence the other direct methods proposed are

- Gauss-Elimination
- LU Decomposition
- Cholesky's Method
- Tri-diagonal Method

Each of these methods is presented in detail with examples.

2.6.1 Gauss-Elimination Method

Gauss-Elimination method is one of the basic methods of solving linear system of algebraic equations. In this procedure the elements of the matrix are modified to reduce the size of large system of equations to smaller ones. It is more convenient to describe the algorithm in the matrix form. If, for example

$$[A] = \begin{bmatrix} a_{11} & a_{12} & a_{13} & \cdots & a_{1n} \\ a_{21} & a_{22} & a_{23} & \cdots & a_{2n} \\ \vdots & \vdots & \vdots & \cdots & \vdots \\ a_{n1} & a_{n2} & a_{n3} & \cdots & a_{nn} \end{bmatrix}$$

is the matrix formed by the co-efficients of linear system of equations, then the reduced form of the matrix is

$$[U] = \begin{bmatrix} A_{11} & A_{12} & A_{13} & \cdots & A_{1n} \\ 0 & A_{22} & A_{23} & \cdots & A_{2n} \\ \vdots & \vdots & \vdots & \cdots & \vdots \\ 0 & 0 & 0 & \cdots & A_{nn} \end{bmatrix}$$

- Algorithm:
  1. The first row of $[U]$ remains same as $[A]$.
  2. The element a_{21} is replaced with zero. This is accomplished by multiplying first row of [A] by a_{21}/a_{11} and subtracting with second row elements in the process all the elements of second row and the elements of the forcing vector on the right hand side get modified.
  3. The other elements of the first column $a_{31}, a_{41} \ldots a_{n1}$ are treated similarly. For example, to eliminate a_{i1} the first row of the matrix is multiplied with a_{i1}/a_{11} and substracted from the ith row. Hence all the elements below a_{11} are eliminated. After the completion of this process none of the equations 2, 3, 4,...n contain the variable x_1.
  4. The procedure is repeated for all the columns. This results in the matrix modified to upper triangular matrix of the form $[U]$.
  5. All the elements of $[U]$ except the first row differ from those of the original matrix $[A]$.
  6. The last equation contains only one variable x_n and is easily solved as

$$x_n = \frac{B_n}{A_{nn}} \qquad \text{...(2.7)}$$

  7. The last but one, equation contains x_{n-1} and x_n, and once x_n is known, x_{n-1} can be easily solved for.
  8. The ith equation yields x_i, as

$$x_i = \frac{B_i - \sum_{i+1}^{n} A_{ik} x_k}{A_{ii}} \qquad \text{...(2.8)}$$

All x_k appearing in the summation are known, hence equation (2.8) is easily solvable for x_k.

Example 2.3. *Solve the following linear system of equations by Gauss Elimination method.*

$$\begin{aligned} 2x + 3y + 4z &= 5 \\ 3x + 4y + 5z &= 6 \\ 4x + 5y + 6z &= 7 \end{aligned}$$

Solution. The matrix form of equation is

$$\begin{bmatrix} 2 & 3 & 4 \\ 3 & 4 & 5 \\ 4 & 5 & 6 \end{bmatrix} \begin{bmatrix} x \\ y \\ z \end{bmatrix} = \begin{bmatrix} 5 \\ 6 \\ 7 \end{bmatrix}$$

i.e. $[A]\,[X] = [B].$

- Elimination of a_{21} (= 3) is done by multiplying first row with 3/2, and subtracting from 2nd row. This results in

$$\begin{bmatrix} 2 & 3 & 4 \\ 0 & -0.5 & -1 \\ 4 & 5 & 6 \end{bmatrix} \begin{bmatrix} x \\ y \\ z \end{bmatrix} = \begin{bmatrix} 5 \\ -1.5 \\ 7 \end{bmatrix}$$

- The Elimination of $a_{31} = 4$ is done by multiplying first row with 4/2 and subtracting from third row, resulting in

$$\begin{bmatrix} 2 & 3 & 4 \\ 0 & -0.5 & -1 \\ 0 & -1 & -2 \end{bmatrix} \begin{bmatrix} x \\ y \\ z \end{bmatrix} = \begin{bmatrix} 5 \\ -1.5 \\ -3 \end{bmatrix}$$

- To eliminate $a_{32} = -1$ multiply the second row by –1/–0.5 and subtract from third row.

$$\begin{bmatrix} 2 & 3 & 4 \\ 0 & -0.5 & -1 \\ 0 & 0 & 0 \end{bmatrix} \begin{bmatrix} x \\ y \\ z \end{bmatrix} = \begin{bmatrix} 5 \\ -1.5 \\ 0 \end{bmatrix}$$

- Hence $$z = 0.$$
- $$y = \frac{-1.5-(-1)0}{-0.5} = 3$$

and $$x = \frac{5-(3)3-4(0)}{2} = -2$$

Hence the roots are $(x, y, z) = (-2, 3, 0)$.

2.6.2 Cholesky's Method of Solution

The matrix equation representation of simultaneous equations to be solved, is,

$$[A]\{X\} = \{C\} \qquad \text{...(2.9)}$$

This may be solved by pre-multiplying the equation (2.9) by the inverse of $[A]$, as

$$\{X\} = [A^{-1}]\{C\}$$

Cholesky's method avoids the necessity of inverting the matrix $[A]$. The solution for $\{X\}$ is made available by successive algebraic steps.

Cholesky's method depends on converting the original equation (2.9) to the form,

$$[T]\{X\} = \{K\} \qquad \text{...(2.10}a\text{)}$$

where, for an third order of matrix $[A]$,

$$[T] = \begin{bmatrix} 1 & t_{12} & t_{13} \\ 0 & 1 & t_{23} \\ 0 & 0 & 1 \end{bmatrix}$$

is an upper matrix with unit diagonal elements and the equation (2.10*a*) is in the form

$$\begin{bmatrix} 1 & t_{12} & t_{13} \\ 0 & 1 & t_{23} \\ 0 & 0 & 1 \end{bmatrix} \begin{Bmatrix} x_1 \\ x_2 \\ x_3 \end{Bmatrix} = \begin{Bmatrix} k_1 \\ k_2 \\ k_3 \end{Bmatrix} \qquad \text{...(2.10}b\text{)}$$

The elements of $\{X\}$ from equation (2.10*b*) are simply found by backward substitution as follows.

$$\begin{aligned} x_3 &= k_3 \\ x_2 + t_{23}x_3 &= k_2 \quad \Rightarrow \quad x_2 = k_2 - t_{23}x_3 \\ x_1 + t_{12}x_2 + t_{13}x_3 &= k_1 \quad \Rightarrow \quad x_1 = k_1 - t_{12}x_2 - t_{13}x_3 \end{aligned} \qquad \text{...(2.11)}$$

Hence with the values of the elements of $[T]$ and $\{k\}$ the solution $\{X\}$ is straight-forward.

- To know $[T]$ and $\{k\}$

Multiply the equation (2.10*a*) by lower triangular matrix

$$[L] = \begin{bmatrix} l_{11} & 0 & 0 \\ l_{21} & l_{22} & 0 \\ l_{31} & l_{32} & l_{33} \end{bmatrix}$$

as in the following manner,

$$[L]\,[T]\,\{X\} = [L]\,\{K\} \qquad ...(2.12)$$

For this equation to equal the original equation, the following relations must exist.

$$[A] = [L]\,[T]$$
$$[C] = [L]\,\{K\} \qquad ...(2.13)$$

Hence we write expanded form of equations (2.13) as

$$\begin{bmatrix} a_{11} & a_{12} & a_{13} \\ a_{21} & a_{22} & a_{23} \\ a_{31} & a_{32} & a_{33} \end{bmatrix} = \begin{bmatrix} l_{11} & l_{11}t_{12} & l_{11}t_{13} \\ l_{21} & (l_{21}t_{12}+l_{22}) & (l_{21}t_{13}+l_{22}t_{23}) \\ l_{31} & (l_{31}t_{12}+l_{32}) & (l_{31}t_{13}+l_{32}t_{23}+l_{33}) \end{bmatrix}$$

$$\begin{Bmatrix} c_1 \\ c_2 \\ c_3 \end{Bmatrix} = \begin{Bmatrix} l_{11}k_1 \\ l_{21}k_1 + l_{22}k_2 \\ l_{31}k_1 + l_{32}k_2 + l_{33}k_3 \end{Bmatrix}$$

Equating the elements in these equations for $[A]$ and $\{C\}$ we have

$$l_{11} = a_{11}$$
$$l_{21} = a_{21}$$
$$l_{31} = a_{31}$$
$$l_{11}\,t_{12} = a_{12} \qquad \therefore \qquad t_{12} = \frac{a_{12}}{a_{11}}$$
$$l_{11}\,t_{13} = a_{13} \qquad \therefore \qquad t_{13} = \frac{a_{13}}{a_{11}}$$
$$l_{21}\,t_{12} + l_{22} = a_{22} \qquad \therefore \qquad l_{22} = \frac{a_{22} - a_{21} \cdot a_{12}}{a_{11}}$$
$$l_{21}\,t_{13} + l_{22}\,t_{23} = a_{23} \qquad \therefore \qquad t_{23} = \frac{a_{23} - (a_{21} \cdot a_{13})/a_{11}}{l_{22}}$$
$$l_{31}\,t_{12} + l_{32} = a_{32} \qquad \therefore \qquad l_{32} = a_{32} - a_{31} \cdot a_{12}/a_{11}$$
$$l_{31}\,t_{13} + l_{32}\,t_{23} + l_{33} = a_{33} \qquad \therefore \qquad l_{33} = a_{33} - \frac{a_{31} \cdot a_{13}}{a_{11}} - l_{32}t_{23}$$
$$l_{11}\,k_1 = C_1 \qquad \therefore \qquad k_1 = \frac{C_1}{a_{11}}$$
$$l_{21}\,k_1 + l_{22}\,k_2 = C_2 \qquad \therefore \qquad k_2 = \left[-\left(\frac{a_{21}}{a_{11}}\right)c_1 + c_2\right]\frac{1}{l_{22}}$$
$$l_{31}\,k_1 + l_{32}\,k_2 + l_{33}\,k_3 = C_3 \qquad \therefore \qquad k_3 = [c_3 - a_{31}k_1 - l_{32}k_2] \cdot \frac{1}{l_{33}}$$

Thus, the elements of the matrices $[L]$, $[T]$ and $\{K\}$ are available in terms of known elements of $[A]$ and $[C]$. Substituting these values in equations (2.11) would provide the elements of $\{X\}$. It is evident Cholesky's method has set of algebraic operation involved in it without resorting to inversion.

Example 2.4. *Solve the system of equations given by Cholesky's method.*

$$2x + 3y + 4z = 5$$
$$3x + 4y + 5z = 6$$
$$4x + 5y + 6z = 7$$

Solution.

$$[A] = \begin{bmatrix} a_{11} & a_{12} & a_{13} \\ a_{21} & a_{22} & a_{23} \\ a_{31} & a_{32} & a_{33} \end{bmatrix} = \begin{bmatrix} 2 & 3 & 4 \\ 3 & 4 & 5 \\ 4 & 5 & 6 \end{bmatrix}$$

And $$\{C\} = \begin{Bmatrix} 5 \\ 6 \\ 7 \end{Bmatrix} = \begin{Bmatrix} c_1 \\ c_2 \\ c_3 \end{Bmatrix}$$

From the method of Cholesky

$$l_{11} = 2, \quad l_{21} = 3, \quad l_{31} = 4$$

$$t_{12} = \frac{a_{12}}{a_{11}} = 3/2, \quad t_{13} = \frac{a_{13}}{a_{11}} = 4/2 = 2$$

$$l_{22} = \frac{a_{22} - a_{21} \cdot a_{12}}{a_{11}} = 4 - \frac{3 \times 3}{2} = -0.5$$

$$t_{23} = (a_{23} - a_{12} \cdot t_{13}) \frac{1}{l_{22}} = (5 - 3 \times 2) \frac{1}{(-0.5)} = 2$$

$$k_1 = \frac{c_1}{a_{11}} = \frac{5}{2} = 2.5, \quad k_2 = [c_2 - a_{21} \cdot k_1] \frac{1}{l_{22}} = 3$$

$$k_3 = \frac{1}{l_{33}} [c_3 - a_{31}k_1 - l_{32}k_2] = \frac{1}{l_{33}} [7 - 4 \times 2.5 - (-1)\,3] = 0$$

$$l_{32} = a_{32} - \frac{(a_{31} \cdot a_{12})}{a_{11}} = 5 - \frac{4 \times 3}{2} = -1$$

$$l_{33} = a_{33} - l_{31}\, t_{13} - l_{32}\, t_{23} = 6 - 4 \times 2 - (-1)\,2 = 0$$

Hence, the roots are obtained as

$$z = k_3 = 0$$
$$y = k_2 - t_{23}\, z = 3 - 2\,(0) = 3$$
$$x = k_1 - t_{12}\, y - t_{13}\, z = 2.5 - (1.5)3 - 2\,(0) = -2$$

Hence, the solution vector $\{X\} = \{-2, 3, 0\}$.

2.6.3 LU Decomposition Method

The matrix $[A]$ formed from the co-efficients of terms of the simultaneous equations should be non-singular to decompose into lower, $[L]$ and $[U]$, upper triangular form of matrices, as given by

$$[L] = \begin{bmatrix} l_{11} & 0 & 0 & \dots & 0 \\ l_{21} & l_{22} & 0 & \dots & 0 \\ \vdots & & & & \vdots \\ l_{n1} & l_{n2} & l_{n3} & \dots & l_{nn} \end{bmatrix} \quad [U] = \begin{bmatrix} 1 & u_{12} & u_{13} & \dots & u_{1n} \\ 0 & 1 & u_{23} & \dots & u_{2n} \\ \vdots & & & & \vdots \\ 0 & 0 & 0 & & 1 \end{bmatrix}$$

The matrix $[L]$ and $[U]$ are the factors of $[A]$

So that
$$[A] = [L]\,[U] \qquad \text{...(2.14)}$$

Hence,
$$[A]\{X\} = [L]\,[U]\{X\}$$
$$[A]\{X\} = [L]\{Y\} = \{C\}$$

Comparing the product of $[L]$ and $[U]$ with the elements of $[A]$

$$l_{11} = a_{11}, \qquad l_{21} = a_{21}, \ldots\ l_{n1} = a_{n1}$$

$$u_{12} = \frac{a_{12}}{a_{11}}, \qquad u_{13} = \frac{a_{13}}{a_{11}}, \ldots\ u_{1n} = \frac{a_{1n}}{l_{11}}$$

The procedure can be continued to obtain all the elements of $[L]$ and $[U]$ in terms of elements of $[A]$.

Example 2.5. *The same example considered previously in Gauss-Elimination and Cholesky's method is used to solve by LU decomposition method.*

Solution. Hence
$$[A] = \begin{bmatrix} 2 & 3 & 4 \\ 3 & 4 & 5 \\ 4 & 5 & 6 \end{bmatrix} \qquad C = \begin{Bmatrix} 5 \\ 6 \\ 7 \end{Bmatrix}$$

$$L = \begin{bmatrix} 2 & 0 & 0 \\ 3 & -0.5 & 0 \\ 4 & -2 & 6 \end{bmatrix} \qquad U = \begin{bmatrix} 1 & 1.5 & 2 \\ 0 & 1 & 2 \\ 0 & 0 & 1 \end{bmatrix}$$

$$[L]\{Y\} = \{C\}$$

i.e.,
$$\begin{bmatrix} 2 & 0 & 0 \\ 3 & -0.5 & 0 \\ 4 & -2 & 6 \end{bmatrix} \begin{Bmatrix} y_1 \\ y_2 \\ y_3 \end{Bmatrix} = \begin{Bmatrix} 5 \\ 6 \\ 7 \end{Bmatrix}$$

Hence
$$\{y_1, y_2, y_3\} = \{5/2, 3, 1/2\}$$

Also
$$[U]\{X\} = \{Y\}$$

i.e.,
$$\begin{bmatrix} 1 & 1.5 & 2 \\ 0 & 1 & 2 \\ 0 & 0 & 1 \end{bmatrix} \begin{Bmatrix} x \\ y \\ z \end{Bmatrix} = \begin{Bmatrix} 5/2 \\ 3 \\ 1/2 \end{Bmatrix}$$

Hence
$$z = \frac{1}{2}, \qquad y + 2\left(\frac{1}{2}\right) = 3$$

$\therefore$
$$y = 2$$

$$x + 1.5\,(2) + 2\,(1/2) = 5/2 \qquad \therefore \quad x = -\,1.5$$

Hence it is led to a different valid solution for the variables of the simultaneous equation

$$\{X, Y, Z\} = \{-\,1.5, 2, 0.5\}.$$

2.6.4 Tri-Diagonal Matrix Algorithm (TDMA) Method

Consider a case of simultaneous equations that form in $[A]$ matrix to have three diagonals of elements and other elements are zero. Let the matrix be of the form

$$[A] = \begin{bmatrix} d_1 & -a_1 & 0 & \ldots & 0 \\ -b_2 & d_2 & -a_2 & \ldots & 0 \\ 0 & -b_3 & d_3 - a_3 & \ldots & 0 \\ \vdots & & & & -a_{n-1} \\ 0 & 0 & 0 & \ldots & b_n d_n \end{bmatrix}, \quad \{X\} = \begin{Bmatrix} x_1 \\ x_2 \\ \vdots \\ x_n \end{Bmatrix}, \quad \{C\} = \begin{Bmatrix} c_1 \\ c_2 \\ \vdots \\ c_n \end{Bmatrix}$$

The solutions to variables {X} by the TDMA procedure is given by the recurrence relationship,

$$x_j = A_j\, x_{j+1} + C'_j \qquad ...(2.15)$$

where
$$A_j = \frac{a_j}{d_j - b_j\, A_{j-1}} \qquad ...(2.16)$$

and
$$C'_j = \frac{b_j C'_{j-1} + C_j}{d_j - b_j A_{j-1}} \qquad ...(2.17)$$

The values to x_j are obtained in the reverse order (x_n, x_{n-1}... x_1) by means of recurrence formula (2.15). The success of TDMA depends on the provision of boundary values x_{n+1} and x_0. This method has more useful application in computational fluid dynamics (*CFD*).

Example 2.6. *A system of simultaneous equations assumed to demonstrate the TDMA procedure are*

$$2w + 3x = 4$$
$$4w + 5x + 6y = 7$$
$$7x + 8y + 9z = 10$$
$$10y + 11z = 12$$

Solution. The given simultaneous equations have the diagonal coefficients written as

$$a_j = \{-3, -6, -9, 0\}$$
$$b_j = \{0, -4, -7, -10\}$$
$$d_j = \{2, 5, 8, 11\}$$
$$c_j = \{4, 7, 10, 12\}$$

The solution {X} is obtained using recurrence relationships to obtain the values of

$$A_j = \{-3/2, 6, -9/50, 0\}$$
$$C'_j = \{2, 1, 3/50, 1.239\}$$

Order of generation of roots are reverse

$$z = 1.239$$
$$y = (-9/50)\,(1.239) + 3/50 = -0.163$$
$$x = 6\,(-0.163) + 1 = 0.022$$
$$w = -3/2\,(0.022) + 2 = 1.967$$

It may be verified whether {X} satisfy system of equations.

2.7 NUMERICAL INTEGRATION

Mathematically integration is the inverse operation of differentiation.

If
$$\frac{d}{dx}[I(x)] = g(x)$$

then
$$I(x) = \int g(x)\,dx$$

In reality integral of a function $g(x)$ called the integrand between the limits x_1 and x_2 is equivalent to the net area between integrand and the x-axis as shown in Fig 2.1.

$$I(x) = \int_a^b g(x)\,dx$$

Many a time the function $g(x)$ cannot be integrated to give closed form solution to $I(x)$. This provides motivation to find the approximate solution by numerical methods. *e.g.*, solution to = $(1 + xy)$ cannot be obtained exactly by regular integration methods. In such cases the integral is obtained by numerical integration methods.

In the given definite limit the integral is equivalent to a summation operation.

$$\int_{x_1}^{x_2} g(x)\,dx = \lim_{\Delta x \to 0} \sum_{x_1}^{x_2} g(x)\,\Delta x$$

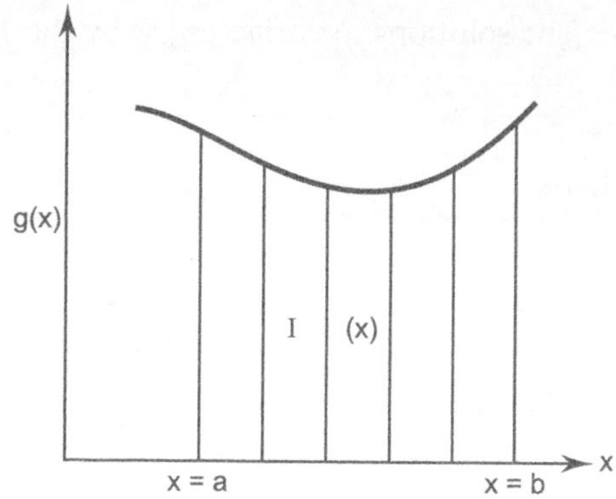

Fig. 2.1. Integration.

Approximate methods for determining the integrals are mostly based on the idea of obtaining the area between integrand and the axis representing the independent variable.

Numerical Integration Methods:

The basic numerical integration methods are derived in the form of

- Trapezoidal Rule-Linear
- Simpson's Rule
  - –1/3 Rule-Quadratic
  - –3/8 Rule-Cubic
- Gaussian Quadrature

2.7.1 Trapezoidal Rule

It is a method of straight line approximation. It uses two end points to form the straight line. Hence it is called linear rule with $n = 1$. The integral is estimated to be the area under the straight line.

$$I(x) = \int_a^b f(x)\,dx \approx \sum_0^1 C_i f(x_1)$$

$$= C_0 f(x_0) + C_1 f(x_1)$$

$$= \frac{n}{2}\,[f(x_0) + f(x_1)]$$

Where $n = \dfrac{b-a}{2}$

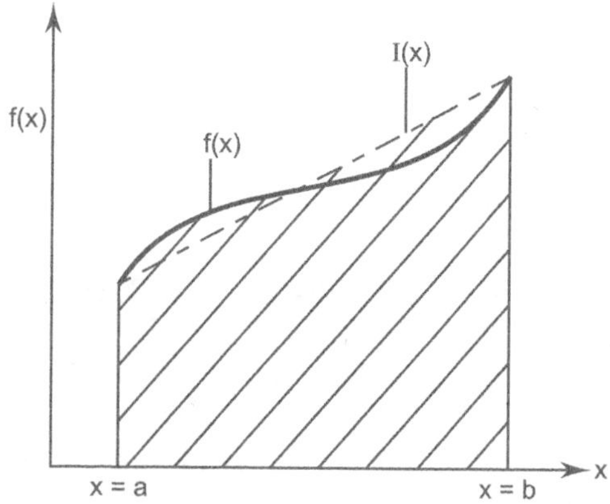

Fig. 2.2. Trapezoidal rule.

Example

Evaluate the integral $\int_0^4 xe^{2x}\,dx$ by Trapezoidal rule and compare with the exact solution.

Exact solution

$$\int_0^4 xe^{2x}dx = \left[\frac{x}{2}e^{2x} - \frac{1}{4}e^{2x}\right]_0^4$$

$$= \frac{1}{4}e^{2x}(2x-1)\Big|_0^4 = 5216.926$$

Solution by Trapezoidal Rule

$$I = \int_0^4 xe^{2x}\,dx \approx \frac{4-0}{2}[f(0)+f(4)]$$

$$= 2\,(0 + 4e^8)$$

$$= 23847.66$$

Error,
$$\in = \frac{5216.926 - 23847.66}{5216.926} = -357.12\%.$$

2.7.2 Simpson's 1/3 Rule

This method approximates the integral function by a parabolic estimation. It requires three points on the curve representing the integrand. The integral is estimated to be the area under the Para-bola passing through these point. Hence it is called a quadratic rule with $n = 2$. Refer Fig. (2.3) for 1/3 rule.

$$\int_a^b f(x)\,dx \approx \sum_{i=0}^{2} C_i f(x_i)$$

$$= C_0 f(x_0) + C_1 f(x_1) + C_2 f(x_2)$$

$$= \frac{n}{3}[f(x_0) + 4f(x_1) + f(x_2)]$$

where
$$n = \frac{b-a}{2}$$

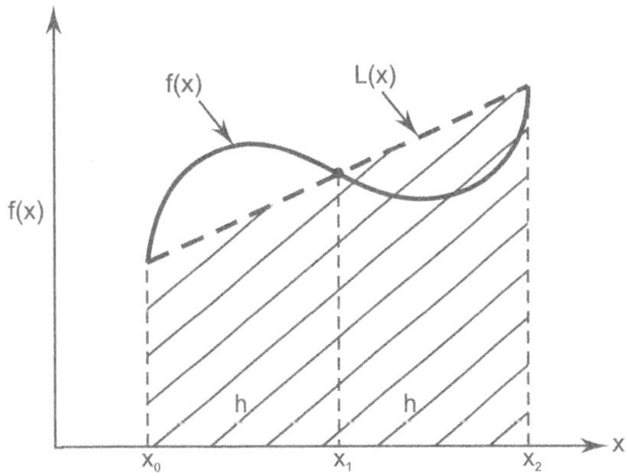

Fig. 2.3. Simpons's 1/3 rule.

The interval is exactly divided into two equal segments as shown in Fig 2.3.

Example 2.7. *Evaluate the integral* $\int_0^4 xe^{2x}dx$ *by Simpson's 1/3 rule and compare with exact solution.*

Solution.
$$I = \int_0^4 xe^{2x}dx \approx \frac{n}{3}[f(0) + 4f(2) + f(4)]$$

$$= \frac{(4-0)/2}{3}[0 + 4(2e^4) + 4e^8] = 8240.411$$

Error,
$$\in = \frac{5216.926 - 8240.411}{5216.926} = -57.96\%$$

2.7.3 Simpson's 3/8 Rule

This approximates the integral function by using a cubic polynomial. It is called the cubic rule with $n = 3$. Here the number of points required are four. The points are so chosen to divide the interval equally into three segments. The integral is the area under the cubic curve passing through these four points. Refer Fig. (2.4) for 3/8 rule

$$\int_a^b f(x)\,dx = \sum_{i=0}^{3} C_i\, f(x_i)$$

$$= C_0 f(x_0) + C_1 f(x_1) + C_2 f(x_2) + C_3 f(x_3)$$

$$= \frac{3h}{8}\,[f(x_0) + 3f(x_1) + 3f(x_2) + f(x_3)]$$

where $n = \dfrac{b-a}{3}$.

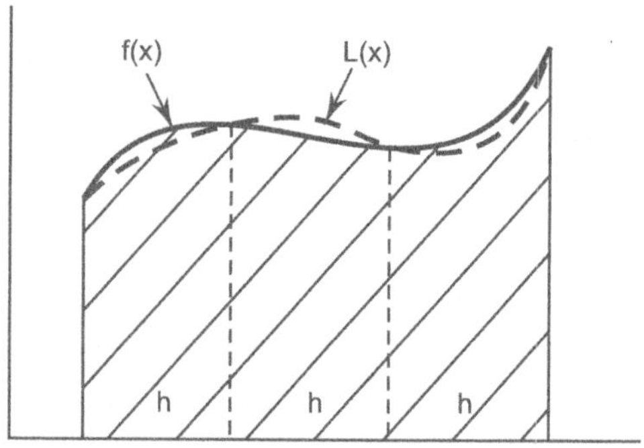

Fig 2.4. Simpson's 3/8 rule.

Example 2.8. *Solve for the integral* $\int_0^4 xe^{2x}\,dx$ *by Simpson's 3/8 rule and compare with the exact solution.*

$$I = \int_0^4 xe^{2x}\,dx \approx \frac{3h}{8}\,[f(0) + 3f(4/3) + 3f(8/3) + f(4)]$$

$$= \frac{3(4/3)}{8}\,[0 + 3(19.189) + 3(552.34) + 11923.83]$$

$$= 6819.209$$

Solution. Comparison with exact solution

$$\in = \frac{5216.926 - 6819.209}{5216.926} = -30.71\%$$

2.8 GAUSS QUADRATURE

Consider a function $f(x)$, the integral of which is to be evaluated over the limits from –1 to 1. If the exact integral of $f(x)$ is not possible to obtain but the finite value of $\int f(x)$ exists, In such cases the Numerical Integration gives the approximate value for the integral. Numerical integration refers to the determination of integral using the values of $f(x)$ at distinct points specified in the interval (–1, 1). Numerical techniques that exist to evaluate the integral

$$I = \int_{-1}^{1} f(x)\,dx \qquad \text{...(2.18)}$$

are Trapezoidal rule, Simpson's 1/3rd and 3/8th rule and Gauss Quadrature. In Gauss Quadrature, the main scope and focus, proper choice of the weighting sample as well as sample points to evaluate the function provides considerable accuracy with very few points of integration. Hence in this method of numerical integration using N points exact answer to a polynomial of $(2N - 1)$ degree can be achieved. Hence for a more specific case of integration.

$$I = \int_a^b f(x)\,dx = \frac{b-a}{2} \int_{-1}^{1} f(y)\,dy \qquad \text{...(2.19)}$$

Where y is related to x by $\frac{2(x-a)}{b-a} - 1$ the integral is evaluated by the Gauss Quadrature as

$$I = \sum_{i=1}^{N} W_i\, f(x_i) \qquad \text{...(2.20)}$$

The equation (2.20) is the one-dimensional N point formula of Gauss Quadrature.

2.8.1 One Dimensional 1 Point Analysis

The function to be evaluated has only one variable and the sample point to be chosen is one as well. The polynomial for approximation, for $N = 1$ is $(2N - 1) = 1$, is of the degree one.

Let the function $f(x)$ be described by the curve as shown in Fig. 2.5. The area under the curve between the assumed limits gives the integral value of the function.

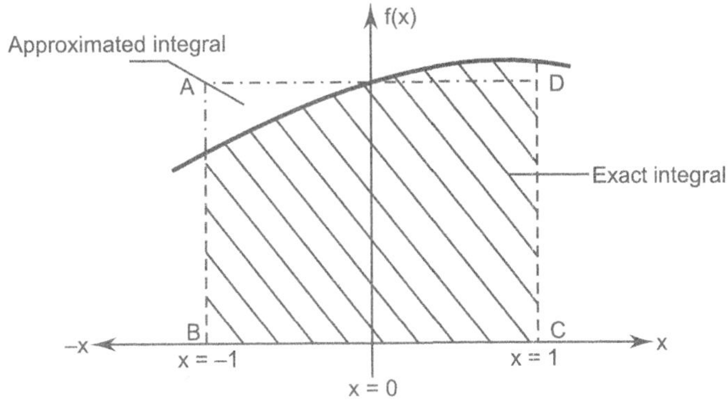

Fig 2.5. One dimensional integration.

The approximate area $ABCD$ is given by the value of the function $f(x)$ at $x = 0$ [*i.e,* $f(0)$] and the length of the rectangle (*i.e.,* = 2).

Hence the integral

$$\int_{-1}^{+1} f(x)\,dx = f(0)\,.\,x\Big|_{-1}^{+1} = 2f(0) \qquad \text{...(2.21)}$$

By comparing equation (2.21) with that Gauss Quadrature in equation (2.20) it is evident that

$$W_1 = 2 \quad \text{and} \quad x_1 = 0$$

Proof. The proof of the equation (2.21) can be presented by assuming the polynomial of degree $(2N - 1) = 1$ for $N = 1$

As $$f(x) = a_0 + a_1 x$$

$$I = \int_{-1}^{+1} f(x)\,dx = \int_{-1}^{+1} (a_0 + a_1 x)\,dx \qquad \text{...(2.22)}$$

$$= a_0 x + \frac{a_1 x^2}{2}\Bigg|_{-1}^{+1}$$

$$= 2a_0 + \frac{a_1}{2}[(+1)^2 - (-1)^2]$$

$$I = 2a_0 \qquad \text{...(2.23)}$$

But $\hat{I} = w_1 f(x_1)$

$$= w_1 [a_0 + a_1 x_1] \qquad \text{...(2.24)}$$

But the error, e between I and $\hat{I}$ is given by the difference, as

$$e = I - \hat{I}$$

$$= 2a_0 - w_1 a_0 - w_1 a_1 x_1 \qquad \text{...(2.25)}$$

For the error to be minimum

$$\frac{\partial e}{\partial a_0} = \frac{\partial e}{\partial a_1} \qquad \text{...(2.26)}$$

Hence $2 - w_1 = 0$ and $-w_1 x_1 = 0$

or $w_1 = 2$ and $x_1 = 0$

it is established that

$$I \approx w_1 f(x_1) = 2 f(0)$$

where $w_1 = 2$ is the sampling weight and $x_1 = 0$ is the sampling weight in one dimensional one point formula.

Example 2.9. *Given $f(x) = x^5 + x^4 + x + 8$, evaluate the integral between the limits –1 to +1 using one Gauss point. And compare with the exact value.*

$$I = \int_{-1}^{+1} f(x)\,dx = \int_{-1}^{+1} (x^5 + x^4 + x + 8)\,dx$$

Solution. Using one Gauss point formula

$$I = w_1 f(x_1) \text{ where } w_1 = 2 \qquad x_1 = 0$$

Hence $I = 2 \times (8) = 16$

By direct integration

$$I = \frac{x^6}{6} + \frac{x^5}{5} + \frac{x^2}{2} + 8x\Bigg|_{-1}^{+1}$$

$$= 0 + \frac{2}{5} + 0 + 16 = 16.4$$

Answer 1. By Gauss Quadrature, $I = 16$

2. By direct integration, $I = 16.4$.

2.8.2 One Dimensional 2 Point Quadrature

In this analysis of numerical integration for evaluating integral, the function has one variable and Gauss quadrature is evaluated with two weighting and two sampling points. To be specific the integral takes the form.

$$I = \int_{-1}^{+1} f(x)\,dx = \sum w_i f(x_i) = w_1 f(x_1) + w_2 f(x_2) \qquad \text{...(2.27)}$$

where w_1 and w_2 are the sample weights and x_1, x_2 are the sample points.

To evaluate w_1, w_2 and x_1, x_2

Since $N = 2$ and $(2N - 1) = 3$ the polynomial function to be assumed is $f(x) = a_0 + a_1 x + a_2 x^2 + a_3 x^3$ a third degree polynomial.

The integral is of the form, for direct integration

$$I = \int_{-1}^{+1} f(x)\,dx = \int_{-1}^{+1} (a_0 + a_1 x + a_2 x^2 + a_3 x^3)\,dx$$

$$= a_0 x + \frac{a_1 x^2}{2} + \frac{a_2 x^3}{3} + \frac{a_3 x^4}{4} \Big|_{-1}^{+1}$$

$$= 2a_0 + \frac{a_1}{2}(0) + \frac{a^2}{3}(2) + \frac{a_3}{4}(0)$$

$$I = 2a_0 + \frac{2a_2}{3}$$

The integral in the 2 point Gauss quadrature

$$\hat{I} = W_1(a_0 + a_1 x_1 + a_2 x_2^2 + a_3 x_3^3) + W_2(a_0 + a_1 x_2 + a_2 x_2^2 + a_3 x_2^3)$$

The error function is

$$e = I - \hat{I}$$

For minimization of error

$$\frac{\partial e}{\partial a_0} = \frac{\partial e}{\partial a_1} = \frac{\partial e}{\partial a_2} = \frac{\partial e}{\partial a_3} = 0$$

This leads to the formation of equations

$$2 - w_1 - w_2 = 0$$

$$-w_1 x_1 - w_2 x_2 = 0$$

$$2/3 - w_1 x_2^2 - w_2 x_2^2 = 0$$

$$-w_1 x_2^3 - w_2 x_2^3 = 0$$

These set of non-linear equations are solved to get

$$w_1 = 1 \quad \text{and} \quad w_2 = 1$$

$$x_1 = \frac{1}{\sqrt{3}} \quad \text{and} \quad x_2 = \frac{1}{\sqrt{3}}$$

Example 2.10. $\int_{-1}^{+1} \left(3e^x + x^2 + \frac{1}{x+2}\right) dx$ *is given. Integrate using 2 point Gauss quadrature.*

Solution. $I = w_1 f(x_1) + w_2 f(x_2)$

where $$w_1 = w_2 = 1 \qquad x_1 = -x_2 = -\frac{1}{\sqrt{3}}$$

$$I = 1\left(3e^{x_1} + x_1^2 + \frac{1}{x_1+2}\right) + 1\left(3e^{x_2} + x_2^2 + \frac{1}{x_2+2}\right)$$

$$= 2.72 + 5.945$$

$$= 8.665.$$

Example 2.11. *Evaluate* $\int_{-1}^{+1} (x^5 + x^4 + x + 8)\,dx$ *by 2 point quadrature.*

Solution.

$$I = w_1 f(x_1) + w_2 f(x_2)$$

$$= 1(x_1^5 + x_1^4 + x_1 + 8) + 1(x_2^5 + x_2^4 + x_2 + 8)$$

$$= 1\left[\left(-\frac{1}{\sqrt{3}}\right)^5 + \left(-\frac{1}{\sqrt{3}}\right)^4 + \left(\frac{1}{-\sqrt{3}}\right) + 8\right] + \left[\left(\frac{1}{\sqrt{3}}\right)^5 + \left(\frac{1}{\sqrt{3}}\right)^4 + \left(\frac{1}{\sqrt{3}}\right) + 8\right]$$

$$= 7.581 + 8.753$$

$$= 16.334.$$

2.8.3 One Dimensional 3 Points Quadrature

In a three point Gauss Quadrature the integral is approximated over three points for a function that has only one variable. Naturally there are three sample weights and three sampling points with the degree of the polynomial amounting to $(2N - 1) = 5$. Hence the trial polynomial is of the form

$$f(x) = a_0 + a_1 x + a_2 x^2 + a_3 x^3 + a_4 x^4 + a_5 x^5 \qquad ...(2.28)$$

The Gauss Quadrature takes the form

$$I = w_1 f(x_1) + w_2 f(x_2) + w_3 f(x_3)$$

where $$x_1 = 0.77459, \qquad x_2 = 0.0 \qquad x_3 = 0.0$$

$$w_1 = 0.5555, \qquad w_2 = 0.8888 \qquad w_3 = 0$$

The values of w_i and x_i are obtained by minimization of error as done in the previous two cases.

Example 2.12. *Evaluate the integral* $I = \int_0^4 x(1-x)^3\,dx$ *by 1 point, 2 points and 3 points Gauss Quadrature.*

Solution. 1. One point formula: Since the limits $a = 0$ and $b = 4$ are different from (–1, 1) as useful,

$$I = \int_a^b f(x)\,dx = \frac{b-a}{2} \int_{-1}^{1} f(y)\,dy$$

where $$y = \frac{2(x-a)}{(b-a)} - 1$$

Hence
$$y_1 = \frac{2(x_1-4)}{4} - 1 = \frac{2(0-4)}{4} = -2-1=-3$$
$$w_1 = 2$$
$$I = \frac{b-a}{2}\int_{-1}^{1} y(1-y)^3 dy$$
$$= w_1 f(y_1) = 2 \times 2[(-3)(1+3)^3]$$
$$= -768$$

2. Two point formula for the given example has $w_1 = w_2 = 1$ and $x_1 = -x_2 = -\frac{1}{\sqrt{3}}$

Transformation of variable to accommodate the limits of integration
$$y_1 = -y_2 = \frac{2\left(-\frac{1}{\sqrt{3}}-4\right)}{4} - 1 = -3.2886$$

The integral is
$$I = \frac{b-a}{2}[w_1 f(y_1) + w_2 f(y_2)]$$
$$= \frac{4-0}{2}[1\{-3.2886(1+3.2886)^3\} + 1\{3.2886(1-3.2886)^3\}]$$
$$= -597.6747$$

3. Three point formula application to integral evaluation has,
$$w_1 = 0.5555, \quad w_2 = 0.8888, \quad w_3 = 0.0$$
and
$$x_1 = -0.77459, \quad x_2 = x_3 = 0$$
The transformed variables
$$y_1 = \frac{2(-0.77459-4)}{4} - 1 = -3.3873$$
$$y_2 = y_3 = 0$$
Gauss Quadrature formula for the integral is
$$I = \frac{b-a}{2}[w_1 f(y_1) + w_2 f(y_2) + w_3 f(y_3)]$$
$$= \frac{4-0}{2}[0.5555\{-3.3873(1+3.3873)^3\} + 0 + 0]$$
$$= -572.102.$$

2.8.4 Two Dimensional Analysis

Infinite element we come across the problems with integrals having two variables (x and y). The 2D elements are used to model the surfaces. The displacement function in such case are having two variables. For example quadrilateral element shown in Fig. 2.6 has two dimension along x and y axes.

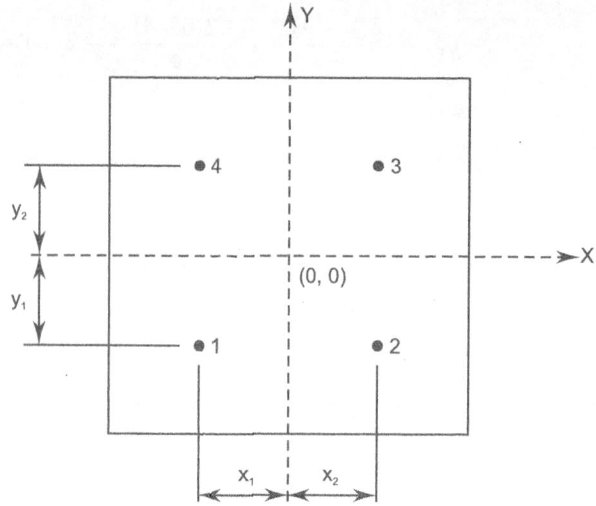

Fig 2.6. Quadrilateral element.

The integral to be evaluated for the two dimensional problem has the form

$$I = \int_{-1}^{+1} f(x, y)\, dx\, dy \qquad ...(2.29)$$

The formula for two point Gauss Quadrature is given in the form

$$I = \sum_{i=1}^{2} \sum_{j=1}^{2} w_i w_j\, f(x_i, x_j) \qquad ...(2.30)$$

In the expanded form of equation (2.30)

$$I = w_1[w_1\, f(x_1, y_1) + w_2\, f(x_1, y_2)] + w_2[w_1\, f(x_2, y_1) + w_2\, f(x_2, y_2)] \qquad ...(2.31)$$

Further it can be shown by error optimization that

$$w_1 = w_2 = 1$$

$$x_1 = -\frac{1}{\sqrt{3}} \text{ and } x_2 = \frac{1}{\sqrt{3}}$$

$$y_1 = -\frac{1}{\sqrt{3}} \text{ and } y_2 = \frac{1}{\sqrt{3}}$$

- For two dimensional three point Gauss Quadrature the sample weights are

$$w_1 = \frac{1}{3} = w_2 = w_3$$

the sample points are

$$x_1 = 0.0, \quad x_2 = 0.5 \quad \text{and} \quad x_3 = 0.5$$
$$y_1 = 0.5, \quad y_2 = 0.5 \quad \text{and} \quad y_3 = 0.0$$

Example 2.13. *Evaluate the following integral using 2 points Gauss quadrature formula, and 3 point Gauss Quadrature.*

$$\int_{-1}^{+1}\int_{-1}^{+1} (x^2 + 2xy + x^3)\, dx\, dy$$

Solution. The weights are $w_1 = w_2 = 1$

The sample points are $x_1 = -x_2 = -\frac{1}{\sqrt{3}}$ $(= -0.57735)$

$$y_1 = -y_2 - \frac{1}{\sqrt{3}}\ (-0.57735)$$

The functionals are

$$f(x_1, y_1) = \left[\left(-\frac{1}{\sqrt{3}}\right)^2 + 2\left(-\frac{1}{\sqrt{3}}\right)\left(-\frac{1}{\sqrt{3}}\right) + \left(-\frac{1}{\sqrt{3}}\right)^3\right]$$

$$= 0.8075$$

$$f(x_1, y_2) = \left[\left(-\frac{1}{\sqrt{3}}\right)^2 + 2\left(-\frac{1}{\sqrt{3}}\right)\left(\frac{1}{\sqrt{3}}\right) + \left(\frac{1}{\sqrt{3}}\right)^3\right]$$

$$= -0.1409$$

$$f(x_2, y_1) = \left[\left(\frac{1}{\sqrt{3}}\right)^2 + 2\left(\frac{1}{\sqrt{3}}\right)\left(-\frac{1}{\sqrt{3}}\right) + \left(-\frac{1}{\sqrt{3}}\right)^3\right]$$

$$= -0.5258$$

$$f(x_2, y_2) = \left[\left(+\frac{1}{\sqrt{3}}\right)^2 + 2\left(\frac{1}{\sqrt{3}}\right)\left(\frac{1}{\sqrt{3}}\right) + \left(\frac{1}{\sqrt{3}}\right)^3\right]$$

$$= 1.1924$$

Using equation (2.31)

$$I = 1[1(0.8075) + 1(-0.1409)] + [1(-0.5258) + 1(1.1924)]$$
$$= 1.3332.$$

3 points Gauss Quadrature

The sample weights are $w_1 = w_2 = w_3 = 1/3$

The sample points are $x_1 = 0.0 \quad x_2 = 0.5 \quad x_3 = 0.5$

$$y_1 - 0.5 \quad y_2 - 0.5 \quad y_3 = 0.0$$

Integral using Gauss Quadrature is

$$I = w_1[f(x_1y_1)w_1 + w_2 f(x_1, y_2) + w_3 f(x_1y_3)] + w_2[f(x_2y_1)$$
$$w_1 + f(x_2, y_2)\, w_2 + w_3 f(x_2, y_3)] + w_3[w_1 f(x_3y_1)$$
$$+ w_2 f(x_3, y_2) + w_3 f(x_3y_3)] \qquad ...(2.32)$$

$$f(x_1, y_1) = 0$$
$$f(x_1, y_2) = 0$$
$$f(x_1, y_3) = 0$$

$$f(x_2, y_1) = 0.875$$
$$f(x_2, y_2) = 0.875$$
$$f(x_2, y_3) = 0.375$$
$$f(x_3, y_1) = 0.875$$
$$f(x_3, y_2) = 0.875$$
$$f(x_3, y_3) = 0.375$$

Hence $I = 0.7083 + 0.7083 = 1.4167.$

Examples on Gauss Elimination

Example 2.14. *Solve the following system of simultaneous equation by Gauss Elimination method.*

$$4x_1 + 6x_2 = 12$$
$$5x_1 + 5x_2 = 0$$

Solution. The problem in the resultant matrix form is

$$\begin{bmatrix} 4 & 6 & 12 \\ 5 & 5 & 0 \end{bmatrix}$$

Replace row_2 by (row_2 – 5/4 row_1) yields

$$\begin{bmatrix} 4 & 6 & 12 \\ 0 & -2.5 & -15 \end{bmatrix}$$

Solve for

$$-2.5x_2 = -15 \rightarrow x_2 = 6$$

By back substitution in

$$4x_1 + 6x_2 = 12$$

$\therefore$ $$4x_1 + 6\,(6) = 12$$

$$x_1 = (12 - 36)/4 = -6$$

$$\begin{bmatrix} x_1 \\ x_2 \end{bmatrix} = \begin{bmatrix} -6 \\ 6 \end{bmatrix}.$$

Example 2.15. *Consider the following system of equation for solution by Gauss elimination*

$$\begin{array}{r} 2x_1 - 4x_2 + x_3 = 6 \\ 3x_1 - x_2 + x_3 = 11 \\ x_1 + x_2 - x_3 = -3 \end{array} \Rightarrow \begin{bmatrix} 2 & -4 & 1 & 6 \\ 3 & -1 & 1 & 11 \\ 1 & 1 & -1 & -3 \end{bmatrix}$$

Solution. Forward elimination steps

$$row_2 \leftarrow row_2 - (3/2)\,row_1$$
$$row_3 \leftarrow row_3 - (1/2)\,row_1$$

These two operations lead to the result

$$\begin{bmatrix} 2 & -4 & 1 & 6 \\ 0 & 5 & -0.5 & 2 \\ 0 & 3 & -1.5 & -6.0 \end{bmatrix}$$

$row_3 = row_3 - (3/5)\,row_2$ operation yields

$$\begin{bmatrix} 2 & -4 & 1 & 6 \\ 0 & 5 & -0.5 & 2 \\ 0 & 0 & -1.2 & -7.2 \end{bmatrix}$$

By backward substitution

$$x_3 = -7.2/-1.2 = 6$$
$$x_2 = [2 + 0.5\,(6)]\,/\,5 = 1$$
$$x_1 = [6 - 1\,(6) + 4\,(1)]\,/2 = 2$$

Hence the solution vector is

$$\begin{bmatrix} x_1 \\ x_2 \\ x_3 \end{bmatrix} = \begin{bmatrix} 2 \\ 1 \\ 6 \end{bmatrix}.$$

Example 2.16. *The upward velocity of a rocket is given by table-A for different time instances.*

TABLE-A

Time, t sec	*5*	*8*	*12*
Velocity, v m/s	*106.8*	*177.2*	*279.2*

The velocity data is approximated by a polynomial function

$$v\,(t) = a_1\,t^2 + a_2\,t + a_3 \qquad 5 \le t \le 12$$

Find the velocity of the rocket at t = 6, 7.5, 9, 11 sec.

Solution. The polynomial function relating velocity with time is given as

$$v\,(t) = a_1\,t^2 + a_2 t + a_3 \quad 5 \le t \le 12$$

The matrix template with three different time slots results with

$$\begin{bmatrix} t_1^2 & t_1 & 1 \\ t_2^2 & t_2 & 1 \\ t_3^2 & t_3 & 1 \end{bmatrix} \begin{bmatrix} a_1 \\ a_2 \\ a_3 \end{bmatrix} = \begin{bmatrix} v_1 \\ v_2 \\ v_3 \end{bmatrix}$$

Using t and v values from table-A

$$\begin{bmatrix} 25 & 5 & 1 \\ 64 & 8 & 1 \\ 144 & 12 & 1 \end{bmatrix} \begin{bmatrix} a_1 \\ a_2 \\ a_3 \end{bmatrix} = \begin{bmatrix} 106.8 \\ 177.2 \\ 279.2 \end{bmatrix}$$

In the matrix form of system of equations

$$\begin{bmatrix} 25 & 5 & 1 & 106.8 \\ 64 & 8 & 1 & 177.2 \\ 144 & 12 & 1 & 279.2 \end{bmatrix}$$

Step 1. Forward elimination process

$\text{Row}_2 - \text{Row}_1 \left(\frac{64}{25}\right)$ yields

$$\begin{bmatrix} 25 & 5 & 1 & 106.8 \\ 0 & -4.8 & -1.56 & -96.21 \\ 144 & 12 & 1 & 279.2 \end{bmatrix}$$

Step 2. The operation of forward elimination yields,

$\text{Row}_3 - \text{Row}_1 \left(\frac{144}{25}\right)$ results in

$$\begin{bmatrix} 25 & 5 & 1 & 106.8 \\ 0 & -4.8 & -1.56 & -96.21 \\ 0 & -16.8 & -4.76 & -336.0 \end{bmatrix}$$

Step 3. Forward elimination operation for row_3

$\text{Row}_3 - \text{Row}_2 \left(\frac{-16.8}{-4.8}\right)$ yields

$$\begin{bmatrix} 25 & 5 & 1 & 106.8 \\ 0 & -4.8 & -1.56 & -96.21 \\ 0 & 0 & 0.7 & 0.735 \end{bmatrix}$$

By back substitution of variables

$$0.7a_3 = 0.735 \Rightarrow a_3 = 1.05$$

$$-4.8a_2 - 1.56\, a_3 = -96.21$$

$$a_2 = [-96.21 + 1.56(1.05)]/(-4.8) \Rightarrow a_2 = 19.7$$

$$25\, a_1 + 5a_2 + a_3 = 106.8$$

$$a_1 = [106.8 - 1.05 - 5\,(19.7)]/25 = a_1 = 0.29$$

Hence the generalized coefficient vector is

$$[a_1, a_2, a_3]^T = [0.29. 19.7, 1.05)^T$$

The polynomial that describes the velocity of flight is

$$V(t) = 0.29\, t^2 + 19.7t + 1.05$$

$$V(6) = 0.29\,(6)^2 + 19.7\,(6) + 1.05 = 129.69 \text{ m/s}$$

$$V(7.5) = 0.29\,(7.5)^2 + 19.7\,(7.5) + 1.05 = 165.1 \text{ m/s}$$

$$V(9) = 0.29\,(9)^2 + 19.7\,(9) + 1.05 = 201.8 \text{ m/s}$$

$$V(11) = 0.29\,(11)^2 + 19.7\,(11)\, 1.05 = 252.8 \text{ m/s.}$$

POINTS TO BE REMEMBER

- **Matrix:** A rectangular array of numbers arranged in rows and columns is a matrix.
- **Symmetric Matrix:** If $a_{ij} = a_{ji}$ then 'a' is a symmetric matric.
- **Trace:** Sum of the diagonal elements of a square matrix.

$$\text{Trace } [A] = \sum a_{ij} \text{ for } i = j$$

- **Determinant:**

$$det \begin{bmatrix} a & b \\ c & d \end{bmatrix} = \begin{vmatrix} a & b \\ c & d \end{vmatrix} = ad - bc$$

- **Singular Matrix:**

  If $det\ [A] = 0$. A is a singular matrix.
- **Unity Matrix:** A square matrix with

$$[A] = a_{ij} = [I]$$

- **Inverse Matrix:** Inverse of a matrix A^{-1} satisfies the relation

$$[A^{-1}]\,[A] = [I]$$

- **Orthogonal Matrix:** A matrix is said to be orthogonal if

$$[A']\,[A] = [I]$$

- **Transpose of a Matrix:** Interchange of the rows into column results in transpose.

$$[A'] = [a_{ji}]$$

- **Addition of Matrices:**

$$[C] = [A] + [B]$$

  where $C_{ij} = a_{ij} + b_{ij}$
- **Multiplication of Matrices:**

$$[C_{ij}] = [A_{ik}]\,[B_{kj}]$$

  where $C_{ij} = \sum a_{ik} \cdot b_{kj}$
- **Minor:**

$$[A] = \begin{bmatrix} a & b \\ c & d \end{bmatrix} \quad \text{Minor of } A = [M] = \begin{bmatrix} d & c \\ b & a \end{bmatrix}$$

- **Co-factor:** The co-factor matrix elements are given by

$$[C] = (-1)^{i+j}\,[M_{ij}]$$

- **Adjoint of a Matrix** is the transpose of the co-factor matrix.
- The inverse of the matrix is given by dividing the adjoint matrix by the determinant of the matrix.
- **Eigen Value:** The values of the natural frequency of the system is termed the eigen value.
- **Eigen Vector:** The mode shapes for a particular eigen value is the eigen vector.

$$|[A] - \lambda\,[I]| = 0 \text{ gives the } \lambda \text{ (eigen values).}$$

  where $A = [M^{-1}\,K]$ where M is the mass matrix and K is stiffness matrix.
- **Methods of solution**
  1. Inversion
  2. Gauss elimination method
  3. Cholesky's method
  4. LU decomposition method
  5. Tri-diagonal Matrix (*TDMA*) method.
- Numerical Integration, sample weights and sample points

	One dimensional			Two dimensional	
	One point	Two points	Three points	Two points	Three points
		1	0.55555	1	1/3
W	2		0.88888		1/3
		1	0.0	1	1/3
X_i	0	$-\frac{1}{\sqrt{3}}$	0.77459	$-\frac{1}{\sqrt{3}}, \frac{1}{\sqrt{3}}$	0.0, 0.5, 0.5
			0.0		
Y_i		$+\frac{1}{\sqrt{3}}$	0.0	$-\frac{1}{\sqrt{3}}, \frac{1}{\sqrt{3}}$	0.5, 0.5, 0.0

EXERCISE

1. Evaluate $\int_0^1 \frac{1}{(1+x)} dx$ using Gauss Quadrature two point formula and compare the result with exact solution for accuracy.
2. Given $f(z) = z^5 + z^4 + z + 8$, evaluate the integral between the limits –1 to +1 using 1 and 2 Gaussian points. Also compare with exact value.
3. Integrate using Gauss Quadrature.

   (*a*) $\int_{-1}^{+1} \left(3e^x x^2 + \frac{1}{x+2}\right) dx$. Use two Gauss points

   (*b*) $\int_{-1}^{+1} \int_{-1}^{+1} x^2 y^2 dx\, dy$. Use 2 × 2 Gauss points
4. Derive an expression for one dimensional one point formula and two point formula by using integration by Gauss Quadrature approach and analyze the accuracy.
5. If $N = [\xi, 1-\xi^2]$, evaluate $\int_{-1}^{+1} N^T N d\xi$.
6. Solve the following set of equations by

   (*a*) Gaussian Elimination method (*b*) Cholesky's method

   (*c*) LU Decomposion method (*d*) Inversion method.

$$2x + y + 4z = 16$$
$$x + 2y + 3z = 14$$
$$3x + y + 2z = 11$$

7. Solve the following set of equations by *TDMA* method

$$2p + q = 16$$
$$p + 2q + 3r = 14$$
$$q + 2r + 3s = 11, \quad 4r + 2s = 7$$

8. Find the inversion of the matrix given

$$[A] = \begin{bmatrix} 3 & 2 & 1 \\ 2 & 1 & 0 \\ 4 & 5 & 6 \end{bmatrix}$$

9. Show that the given matrix is orthogonal matrix

$$\begin{bmatrix} 0.6 & -0.8 & 0 \\ -0.8 & -0.6 & 0 \\ 0 & 0 & 1 \end{bmatrix}$$

10. Find the eigen values and the eigen vectors for the equation

$$[K - \lambda M]\,\{X\} = \{B\}$$

where

$$K = \begin{bmatrix} 2 & 3 & 4 \\ 3 & 4 & 5 \\ 4 & 5 & 6 \end{bmatrix}, \qquad M = \begin{bmatrix} 1 & 2 & 3 \\ 2 & 3 & 4 \\ 3 & 4 & 5 \end{bmatrix}, \qquad B = \begin{Bmatrix} 1 \\ 2 \\ 3 \end{Bmatrix}$$

given that $x_3 = 0$.

11. Explain the Gauss Quadrature for one dimensional analysis using 2 point approximation.
12. Explain the error optimization in Gauss Quadrature.
13. Given the integral $I = \int_1^2 \int_0^4 (e^x + y)\,dx\,dy$

find the integral by two point Gauss Quadrature.

14. Using two point Gaussian Quadrature formula, evaluate the following integrals.

(i) $\int_{-1}^{+1} \int_{-1}^{+1} (r^2 + 2rs + s^2)\,dr\,ds$ (ii) $\int_0^1 x^2\,dx.$

Chapter 3

Variational Principles and Approximate Methods

वर्णयंति नराभासान्।
वाणीं लब्ध्वापि ये जनाः॥
लब्ध्वापि कामधेनुं ते।
लांभले विनियुंजते॥

[The knowledge and information in the hands of animal like people is like utilizing Kamadhenu (cow that gives whatever asked for) for the purpose of ploughing the field]

The variational calculus can be applied to continuous and discrete systems, especially to discretized systems with higher comfort. The exact solutions requires the satisfaction of compatibility and equilibrium conditions to be satisfied completely. When the exact solution does not exist, the approximate solution close to the exact is made possible by Rayleigh-Ritz method. The basis for this method is the minimum potential energy principle. More the number of Ritz constants in trial function more accurate is the solution. The Lagrange-Euler formulation of variational principle aids in establishing the governing differential equation of equilibrium describing the physical phenomenon. The physical problems like heat conduction and fluid flow cannot be addressed by Rayleigh-Ritz method as it is based on stationarity condition in potential energy functional. Such processes have the answer in weighted residual methods. Several weighted residual methods are – point collocation, subdomain collocation, least square and Galerkin's method. The Galerkin's method is popularly used in finite element method as it can be comfortably employed for elasticity and fluid/heat related problems. The virtual displacement approach gives suitable base for FE formulation of static analysis problems. The Newton's laws of motion, the work-energy principle are the basis for the principle of virtual work which helps to derive expressions for stress, strain and displacement in an elastic structure. This chapter deals with variational approach, minimum potential energy principle, Rayleigh-Ritz method, weighted residual methods, and principle of virtual displacement and virtual work with suitable illustrative examples from elasticity and heat transfer.

3.1 CONTINUOUS SYSTEMS—SOLUTION METHODS

The focus in the solution to continuous systems is in the consideration of a differential element rather than finite element in a discrete system. The objectives of the solution methods for continuous systems are highlighted by:

- Expressing element equilibrium needs in the form of the differential equations.
- Establishing constitutive relations.
- Satisfying interconnectivity requirements between the differential elements.
- Imposing boundary conditions and initial conditions (in dynamic analysis) before evaluating the solutions.

The governing differential equations are generated by two different approaches: (1) The differential formulation (direct method) (2) variational principle.

3.1.1 Differential Formulation

The system behaviours of continuous model with x and y as the displacement domain, and the unknown state variable u, can be assumed to be represented by second order partial differential equation as

$$a\frac{\partial^2 u}{\partial x^2} + b\frac{\partial^2 u}{\partial x \partial y} + c\frac{\partial^2 u}{\partial y^2} = \phi\left(x, y, \frac{\partial u}{\partial x}, \frac{\partial u}{\partial y}\right) \quad \text{...(3.1)}$$

This formulation establishes equilibrium and constitutive requirement of the differential elements in the state variable with respect to spatial domain. It is imperative that the equation (3.1) should satisfy the compatibility requirement (inter-connectivity between differential elements). The equation should satisfy continuity requirement to represent the assumed system. The boundary conditions and the initial conditions exercise needed constraint on the state variable to arrive at the solution. The co-efficient a, b and c of equation (3.1) decide the physical behaviour of the differential equation to be elliptic, hyperbolic or parabolic.

$$\Delta = b^2 - 4ac \begin{cases} < 0 \text{ .... elliptic} \\ > 0 \text{ .... hyperbolic} \\ = 0 \text{ .... parabolic} \end{cases}$$

Some of the examples for the pre-stated types of equation are

- **Laplace Equation**

$$\frac{\partial^2 \phi}{\partial x^2} + \frac{\partial^2 \phi}{\partial y^2} = 0 \quad \text{...(3.2)}$$

Here $a = 1, \quad b = 0, \quad c = 1$

$\Delta = 0 - 4\,(1)\,(1) = -4 < 0$, hence elliptic in behaviour. The seepage of water through the porous permeable soil is considered to obey the equation (3.2).

- **Wave Equation**

One dimensional wave equation can be written in the form

$$\frac{\partial^2 u}{\partial x^2} - \frac{1}{c^2}\frac{\partial^2 u}{\partial t^2} = 0 \quad \text{...(3.3)}$$

In the equation (3.3) the co-efficients are

$$a = 1, \quad b = 0 \quad \text{and} \quad c = -\frac{1}{c^2}$$

$\Delta = 0 - 4(1)\left(-\frac{1}{c^2}\right) = \frac{4}{c^2} > 0$, hence equation (3.3) representing the wave equation is hyperbolic.

- **Heat Conduction Equation**

One dimensional heat conduction equation as given by

$$\frac{\partial^2 T}{\partial x^2} - \frac{\rho c}{k}\frac{\partial T}{\partial t} = 0 \qquad \text{...(3.4)}$$

has the co-efficients $a = 1$, $b = 0$ and $c = 0$, which results in $\Delta = b^2 - 4\,ac = 0 - 4\,(1)\,(0) = 0$; hence parabolic in behaviour.

The elliptic equation with only space domain can be solved only with the boundary conditions. Because of this reason they are called boundary value problem. The hyperbolic and parabolic equations (3.3) and (3.4) include time variable defining the propogation problem. The solution to such an equation depends on the initial conditions. Hence they are called initial value problems.

In mechanics, the problems in static steady state equilibrium form the boundary value problems. The dynamic problems that have state variables depending on time also, are boundary–initial-value problems.

3.1.2 Variational Principle

The variational formulation in mechanics problems is based on the extremization of the state variables like deflection, which leads to the governing equilibrium differential equation. The variational principle dealing the extremum problem includes the locating of the functionals differentiated with respect to the state variables or the generalized co-ordinates a_i and equating the first derivatives to be zero. The stability is established by the sign and value of the second derivative. Let the functional

$$\pi = \pi(a_1, a_2, \ldots, a_n) \qquad \text{...(3.5)}$$

$$\delta\pi = \frac{\partial \pi}{\partial a_1}\,.\,\delta a_1 + \frac{\partial \pi}{\partial a_2}\,.\,\delta a_2 + \ldots + \frac{\partial \pi}{\partial a_n}\,.\,\delta a_n \qquad \text{...(3.6)}$$

In equation (3.6), the extremum of a_i (generalized co-ordinates are achieved when $\delta\pi = 0$. This can only be evidenced when

$$\frac{\delta \pi}{\partial a_i} = 0 \quad \text{for } i = 1, 2, \ldots\, n \qquad \text{...(3.7)}$$

The condition of maximization is established if the derivative $\delta^2\pi < 0$. This invokes a unstable stationarity in the functional π.

The condition of minimization is reached if the second derivative $\delta^2\pi > 0$ which leads to stable stationarity in the functional π.

The saddle point condition can be observed when the second derivative $\delta^2\pi = 0$ resulting in neutral equilibrium in the functional.

The variational formulation is the powerful mechanism for solving the continuous system and most effective aided by the generation of boundary conditions useful in variational approach. The steps involved in the variational approach to solve the continuous problems of elasticity are as follows.

- Establishing the total potential π which is also known as functional.
- If the order of the highest derivative is 'm' in the functional, such a problem is C^{m-1} variational problem.
- Identification boundary conditions of the problem. In structural mechanics there are two classes of boundary conditions–1. Essential Boundary conditions, 2. The natural boundary conditions. The geometric boundary conditions are prescribed by displacements and rotations. The natural boundary conditions correspond to the boundary forces and moments.
- Solving the equation of equilibrium with the exercise of stationarity and stability principle on the functional.

Advantages and Disadvantages of Variational Approach

The variational approach can comfortably applied to continuous and discrete systems. The examples, with simple physical geometry and boundary conditions, in continuous systems are given exact solution by variational approach. The more case with which variational approach can be applied to discrete systems with the same principle makes it highly effective in finite element method. More pronounced advantage of this approach is envisaged in the numerical solution to continuous systems. In discrete systems the equilibrium equations can be generated by adding the contribution of the elements of the functional, and then solving by stationarity condition. The elemental interconnectivity is automatically established by the exercise of stationarity condition which is the essence of this approach.

The main disadvantage of variational approach is evidenced in problem formulation providing less physical insight into the problem, unlike the solution by direct approach. Hence, after establishing the problem formulation by variational approach, system of equilibrium equations is to be physically interpreted critically. By this it is possible to identify the errors in the solution and gain better understanding of the physical meaning of the equilibrium equation.

Observations

Certain observations pertaining to variational approach are:

- Formulation of governing system equations is made easy in variational formulation by considering the functionals of scalar quantities like energy, potential, etc.
- The governing equations and boundary conditions are achieved more directly.
- Checking independently with additional insight is possible in variational approach.
- Large class of displacement function provides close approximate solution to real world problems.

3.2 PRINCIPLE OF MINIMUM POTENTIAL ENERGY

In the linear structural analysis with displacements as the state variables the functional π, is the total potential energy. That is

$$\pi = U + W \quad ...(3.8)$$

where, U is the strain energy or elastic potential energy due to the deformation of the structure and W is the potential energy due to the position of the loads.

For example.

A body of mass, m moves from a height h_1 to h_2 from a position A to B. The potential energy function.

$$\pi = U + W$$
$$= 0 - (+ \text{mg}) (+ \Delta H)$$

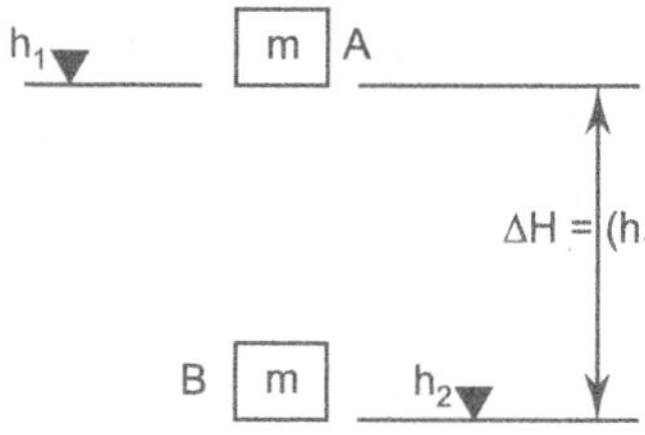

Fig. 3.1. Falling mass.

The strain energy at A of mass m as shown in Fig. 3.1 is zero.

The force due to mass m is $(+ mg)$ and the displacement is $(+ \Delta H)$.

The positive sign for the force and displacement is due the fact that displacement is taking place in the direction of the force. The loss of potential due to the displacement of the load dictates the negative sign for the potential W. Hence the total potential energy in structural mechanics problems can be written specifically as

$$\pi = U - W \quad ...(3.9)$$

Further in discrete systems, π is called the total potential energy 'function' and in continuous systems π is termed as 'functional', because of infinite degrees of freedom being attached to a continuous system.

In a continuous system, π the differentiated form of the function depending on the space variable and the second derivative, has to be integrated over the region to accomodate the infinite degrees of freedom.

The principle of minimum potential energy can be stated as,

"For all displacements of acceptable form, the value of potential energy is constant for the systems satisfying equilibrium condition".

That is

$$\pi = \text{constant}$$

and leads to $$\delta\pi = 0$$

For stable equilibrium the total potential energy function, π has to be minimum which is possible when $\delta^2\pi > 0$. It is already discussed that the unstable (maximum condition) and neutral equilibrium are possible when $\delta^2\pi < 0$ and $\delta^2\pi = 0$ respectively. The graphical representation can be given for stable, unstable and saddle point equilibrium as in Fig. 3.2.

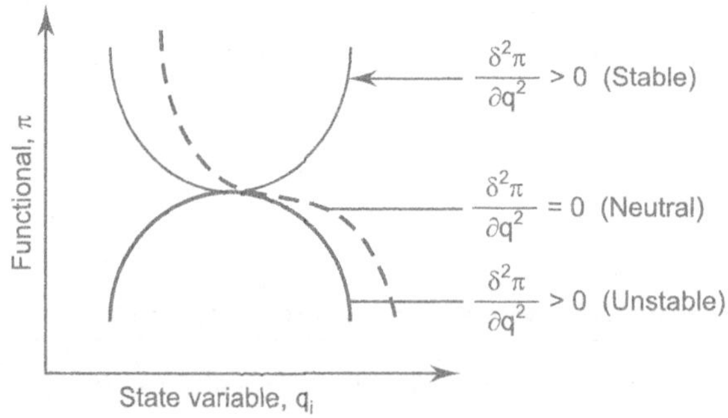

Fig. 3.2. Stable, unstable and neutral equilibrium.

3.2.1 Further Explanation on Principle of Minimum Potential Energy

The direct physical argument considered in FEM procedure is limited by the suitability to principle of virtual work. But FEM procedure is more general than this. The attempt is not only to solve determinate problems but also the indeterminate problems in engineering. General FEM procedure is concerned with the functional and statement of mathematical models in a weak (approximate) sense.

- **Strong form of a problem statement:** To describe a mathematical model we require.

1. The Governing Equations (G.E.)
2. The Boundary Conditions (B.C.)

If the GE and BC are satisfied at every point in the physical domain then the problem is said to be stated in the strong form.

For example, consider an axial bar element with a axial load P(x). The governing equation is obtained by equating the resistive force with the applied force. By Hooke's Law of stress and strain relations, the governing equation is

$$AE\frac{du}{dx} = p(x)$$

The boundary condition is

$$u(0) = a.$$

With this boundary condition it is possible to obtain the exact solution that satisfies the GE at every point.

- **Weak form of a problem statement:** A mathematical model to a problem is stated by an integral expression that implicitly contains the governing equations and the boundary conditions.

This integral expression is called the 'Functional' if the GE and BC are satisfied in an average (approximate) sense the problem is said to be stated in the weak form.

- **Potential Energy π:** The total potential energy of a system is defined as the difference in strain energy and the work potential.

$$\pi = \text{Strain energy} - \text{work potential}$$
$$= (U - WP)$$

- **Strain Energy (U):** Consider the stress strain diagram in the elastic region of deformation. The strain energy is given by the area shaded under the curve.

The strain energy density, u

$$u = \frac{\Delta U}{\Delta V} = \frac{1}{2}\sigma \in$$

Therefore

$$\Delta U = \frac{1}{2}\sigma \in \Delta V$$

Total strain energy,

$$U = \frac{1}{2}\int_V \sigma^T \in dV$$

- **Work Potential:** The work potential due to various forces acting on a body is given by

Work potential = work done by (body force + surface force + point loads)

For a conservative system,

$$WP = \int_V U^T f\, dV \text{.....Body force}$$

$$+ \int_V U^T T dV \text{.....Surface force}$$

$$+ \sum_i U_i^T P_i \text{.....Point load}$$

Now the total Potential Energy (Functional)

$$\pi = \frac{1}{2}\int \sigma^T \in dV - \int_V U^T f dV - \int_V U^T T dV - \sum U_i^T P_i$$

• **Strain Energy of a Linear Spring:** Consider a spring with an applied force F. If the Force displacement curve is a straight line the spring is said to be linear.

Let

$$F = \text{force on the spring}$$
$$x = \text{deflection of the spring}$$
$$k = \text{spring rate (stiffness)}$$

The differential strain energy of the spring for a small change in displacement, dx of the spring is

$$dU = Fdx$$

But for a linear spring $F = kx$.

Hence $$dU = kx\, dx.$$

Total strain energy of the spring

$$U = \int_0^x kx\, dx = \frac{1}{2}kx^2$$

and the work potential for a spring acted upon by force F and deflecting by x is

$$WP = Fx.$$

Hence the total potential energy or the functional for the linear spring is

$$\pi = U - WP = \frac{1}{2}kx^2 - Fx$$

• **Axial Bar Element:** For an axially loaded bar element

Axial strain, $$\in = \frac{du}{dx}$$

Axial stress, $$\sigma = E\in = \frac{Edu}{dx}$$

The strain energy density, $du = \frac{1}{2}\sigma \in = \frac{1}{2}E\left(\frac{du}{dx}\right)^2.$

Total strain energy $U = \int_V du = \int \frac{1}{2} E\left(\frac{du}{dx}\right)^2 dV.$

But the differential volume of bar of fixed cross-section A,

$$dV = Adx$$

Hence $$U = \frac{1}{2} AE \int \left(\frac{du}{dx}\right)^2 dx$$

The work potential,

$$WP = \int_0^L bu\, dx + Fu_{x=L}$$

Hence the total potential energy (Functional)

$$\pi = \frac{1}{2} AE \int_0^L \left(\frac{du}{dx}\right)^2 dx - \int_0^L bu\, dx - Fu_L$$

- **Principle of Minimum Potential Energy:** Further elaboration of minimum potential energy highlights the importance attached in FEM procedure.

"For a conservative system, of all kinematically admissible displacement fields, (those coresponding to equilibrium) extremize the total potential energy if the extremum condition is minimum, the equilibrium state is stable."

Among all admissible displacements that a body can have, one that minimizes the total potential energy of the body, satisfies strong formulation.

- **Admissible Displacements**: These are the displacements that satisfy certain minimum continuity requirements and also specified Dirichlet (displacement) boundary conditions.

Min/Max $\frac{\partial \pi}{\partial u_i} = 0$ for $i = 1, 2, \ldots$ all admissible displacements.

- **Axially Loaded Elastic Bar:** The differential equation governing the response of the axially loaded bar is, (by force balance)

$$\frac{d}{dx}\left(AE\frac{du}{dx}\right) + b = 0 \qquad 0 < x < L$$

where $A(x)$ = cross-section at section x.

$b(x)$ = body force distribution (force/unit length)

$E(x)$ = Young's modulus

$u(x)$ = deflection of the bar at distance x.

The equation is second order differential equation which requires two boundary conditions to solve.

$u = 0$ at $x = 0 \Rightarrow$ Dirichlet (displacement) boundary condition.

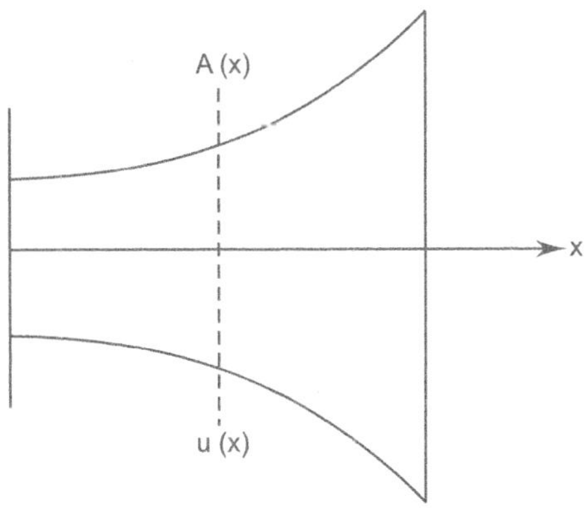

Fig. 3.3. Axially loaded elastic bar.

$AE\frac{du}{dx} = F$ at $x = L \Rightarrow$ Neumann (Force) boundary condition.

Differential Equation + BC = strong form solution to BVP.

Tonti Diagram of governing equations

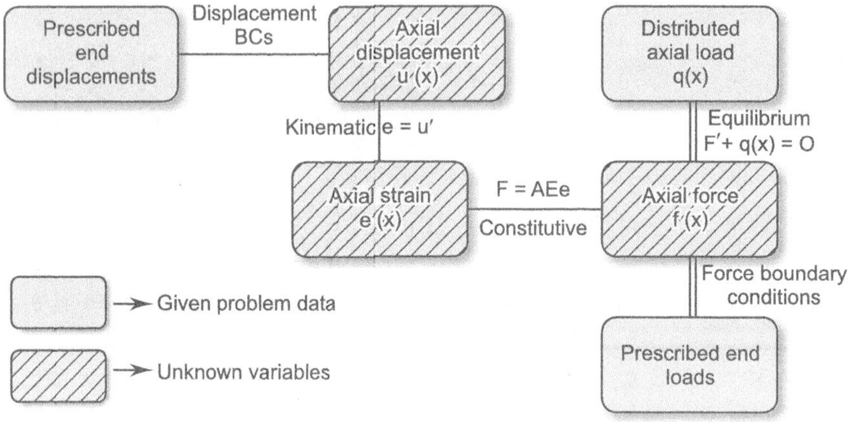

The governing equation of equilibrium is

$$AEe + qu(x) = 0 \Rightarrow AE\frac{du}{dx} + qu(x) = 0$$

The prescribed end displacement for a contilever beam is known as at $x = 0$, $u = 0$ and $u = u_L$ at $x = L$, which are known. The unknowns are the axial displacement, $u(x)$, the axial strain $\frac{du}{dx}$ and the axial force, $f(x)$. The axial force is obtained by distributed and end loads known as force boundary conditions.

Consider a generic problem is one dimension

$$\frac{d^2u}{dx^2} + x = 0 \qquad 0 < x < 1$$

The boundary conditions are

$$u = 0 \text{ at } x = 0$$
$$u = 1 \text{ at } x = 1$$

The exact analytical solution is

$$u(x) = -1/6x^3 + 7/6x$$

In this case the exact solution exist. But in most of the problems in FEM the exact solution cannot be obtained. The best way is to make a guess for $u(x)$ such that

1. It satisfies the boundary conditions
2. It satisfies the governing differential equation.

In this approximate solution strategy, the guess is made in the following way

$$u(x) = a_0\phi_0(x) + a_1\phi_1(x) + a_2\phi_2(x) + \ldots$$

where $\phi_i(x)$ are the known functions

a_i are the constants chosen so as to satisfy the boundary conditions and differential equation.

For an axial bar, the total potential energy with an exact solution, $u(x)$,

$$\pi(u) = \frac{1}{2}AE\int_0^L\left(\frac{du}{dx}\right)^2 dx - \int_0^L bu\,dx - Fu_L$$

The functional pertaining to admissible displacement $w(x)$ is

$$\pi(u) = \frac{1}{2}AE\int_0^L\left(\frac{dw}{dx}\right)^2 dx - \int_0^L bw\,dx - Fw_L$$

For example consider the DE of the form

$$AE\frac{d^2u}{dx^2} + b = 0 \qquad 0 < x < L$$

The boundary conditions are

$$u = 0 \text{ at } x = 0 \Rightarrow \text{Dirichlet}$$

$$AE\frac{du}{dx} = F \text{ at } x = L \Rightarrow \text{Neumann}$$

Assuming $AE = 1$, $b = 1$, $L = 1$, $F = 1$

The exact solution works out to be

$$u = 2x - \frac{x^2}{2}$$

The potential energy function corresponding to the exact solution is

$$\pi(u) = \frac{1}{2}\int_0^1\left(\frac{du}{dx}\right)^2 dx - \int_0^1 u\,dx - u(x=1) = -7/6$$

Assume an admissible function

$$W = x$$

The function arbitrarily chosen only needs to have first derivative and has to satisfy displacement boundary condition.

$$\pi(w) = \frac{1}{2}\int_0^1\left(\frac{dw}{dx}\right)^2 dx - \int_0^1 w\,dx - w(x=1) = -1$$

Hence it may be noticed that

$$\pi(u) < \pi(w)$$

Mathematical statement: If u is the exact solution and w is the admissible displacement, then

$$\pi(u) < \pi(w) \text{ unless } w = u$$

i.e., the exact solution minimizes the total potential energy and hence proves the validity of principle of minimum potential energy.

Example 3.1. *Discuss the stability of a spring of stiffness k acted upon by the force F.*

Solution. A spring of stiffness k, when acted upon by the force F, gets displaced by the distance x (deflection). Then we have

$$\text{Strain energy} = U = \frac{1}{2}kx^2$$

Potential energy due to displacement of the load $F = W = Fx$.

The potential energy function, $\pi = U - W$

$$= \frac{1}{2}kx^2 - Fx$$

To derive the equilibrium equation applying minimum potential energy principle

$$\delta\pi = 0$$

$$\frac{\partial \pi}{\partial x} = kx - F = 0 \quad ...(a)$$

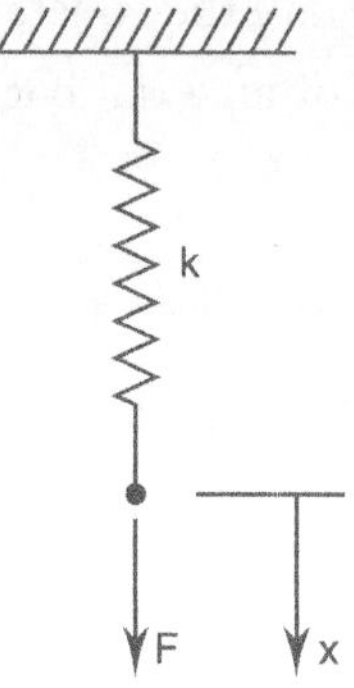

Fig. 3.4. Spring deflection.

Hence the equilibrium equation is given by equation (*b*)

$$F = kx \quad ...(b)$$

Further differentiating equation (*a*) with respect to *x*.

$$\frac{\partial^2 \pi}{\partial x^2} = k > 0 \text{ hence the } \pi \text{ is at stable equilibrium of minimization.}$$

Boundary conditions at $x = 0$, $k = \frac{F}{x} = \infty$ representing the rigid support, and at $x = L$, $k = \frac{F}{L}$ representing linear variation of stiffness with the load.

From equation it may be observed, that at equilibrium, from equation (*b*)

$$W = kx^2 = 2U$$

Example 3.2. *Derive the potential energy functional for an axial member subjected to traction force, q.*

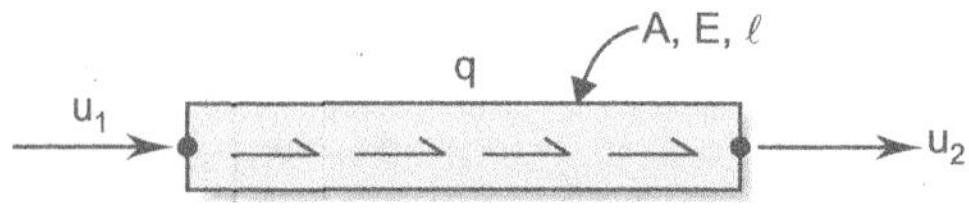

Fig. 3.5. Axial member.

Solution. An axial member with area of cross-section *A*, modulus elasticity *E* is as shown in Fig. 3.5. It is subjected to an axial traction force *q*, along the axis of the member of length *l*.

The total potential energy functional is given by the

$$\pi = U - W$$

Considering the differential element, *dx* (small length) the strain energy is given by

$$U = \int_0^l \frac{F^2}{2AE} dx \quad ...(a)$$

But $F = \sigma A = \in AE$ ...(*b*)

Substituting (*b*) in (*a*) we have

$$U = \int_0^i \left(\frac{AE}{2}\right) \in^2 dx$$

But $\in^2 = \left(\frac{du}{dx}\right)^2$ where du is small change in length of element dx. Hence

$$U = \int_0^l \left(\frac{AE}{2}\right)\left(\frac{du}{dx}\right)^2 dx \qquad ...(c)$$

If the change in length of the rod due to traction force q, is u, then the work done

$$W = \int_0^l (q\,dx)\,u \qquad ...(d)$$

From (c) and (d) we obtain the functional as

$$\pi = \int_0^l \left[\frac{AE}{2}\left(\frac{du}{dx}\right)^2 dx - qu\,dx\right] \qquad ...(3.10)$$

Example 3.3. *Derive the expression for the functional of a cantilever beam subjected to point load F at the free end.*

Fig. 3.6. A cantilever beam with point load.

Solution. A cantilever beam of length l, sectional property I, and the material property (Young's modulus) E, is as shown in Fig. 3.6. The point load F at the free end result in the deflections δ.

The elastic energy in bending,

$$U = \int_0^l \frac{M^2}{2EI} dx \qquad ...(a)$$

From the deflection equation in strength of materials

$$M = EI\frac{d^2y}{dx^2} \qquad ...(b)$$

Hence the strain energy from equations (a) and (b), is

$$U = \frac{EI}{2}\int_0^l \left(\frac{d^2y}{dx^2}\right)^2 dx \qquad ...(c)$$

To determine the work done due to deflection, by the force F.

$$W = F.\delta \qquad ...(d)$$

From equations (c) and (d) the total potential (functional is formulated as)

$$\pi = U - W = \frac{EI}{2}\int_0^l \left(\frac{d^2y}{dx^2}\right)^2 dx - F\delta \qquad ...(3.11)$$

Example 3.4. *Obtain the total potential energy for a beam column shown in Fig. 3.7, subjected to an UDL and an axial force.*

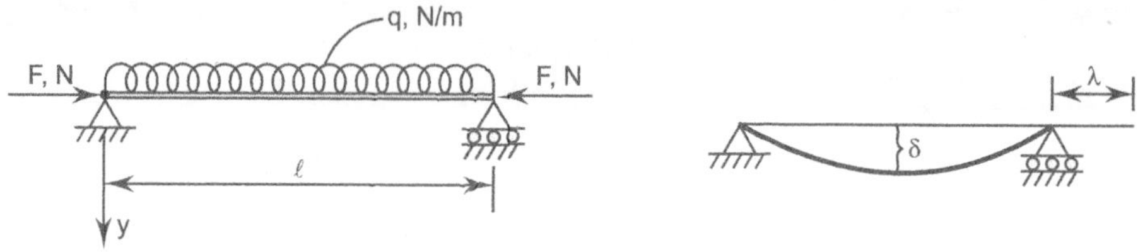

Fig. 3.7. Beam column.

Solution. The beam column subjected to transverse UDL and axial load F as shown in Fig. 3.7 has the sectional property I and the material property E. The maximum transverse deflection is δ and the axial deflection is λ.

The strain energy of the set up is

$$U = \frac{EI}{2}\int_0^l \left(\frac{d^2y}{dx^2}\right)^2 dx \qquad ...(a)$$

The work potential is

$$W = F\lambda + \int_0^l qy\, dx \qquad ...(b)$$

To determine λ we consider a differential element ds as shown in Fig. 3.8

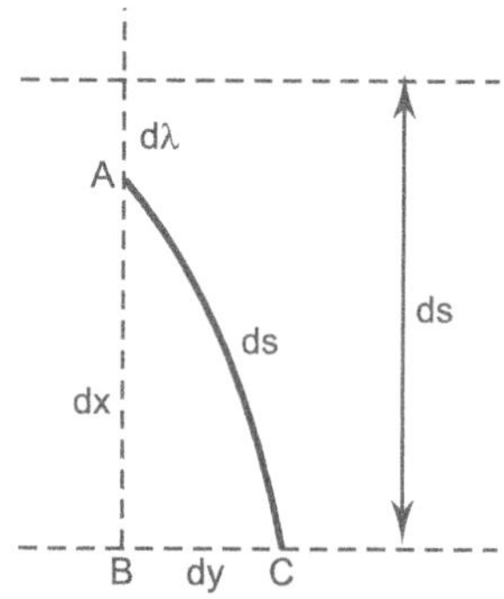

Fig. 3.8. Differential element.

$$d\lambda = ds - dx.$$

From Δ^{le} ABC

$$ds = \sqrt{dx^2 + dy^2}$$

$$= dx\left[1 + \left(\frac{dy}{dx}\right)^2\right]^{\frac{1}{2}}$$

$$= dx\left[1 + \frac{1}{2}\left(\frac{dy}{dx}\right)^2\right]$$

Hence

$$d\lambda = \frac{1}{2}\left(\frac{dy}{dx}\right)^2 \text{ and } \lambda = \int_0^l \frac{1}{2}\left(\frac{dy}{dx}\right)^2 dx$$

From equations (a) and (b)

$$\pi = \int_0^l \frac{EI}{2}\left(\frac{d^2y}{dx^2}\right) dx - \frac{F}{2}\int_0^l \left(\frac{dy}{dx}\right)^2 dx - \int_0^l qy\, dx \qquad ...(3.12)$$

Example 3.5. *Derive the functional for the cantilever column beam with spring attached at the free end as shown in Fig. 3.9.*

Solution. To determine the total strain energy

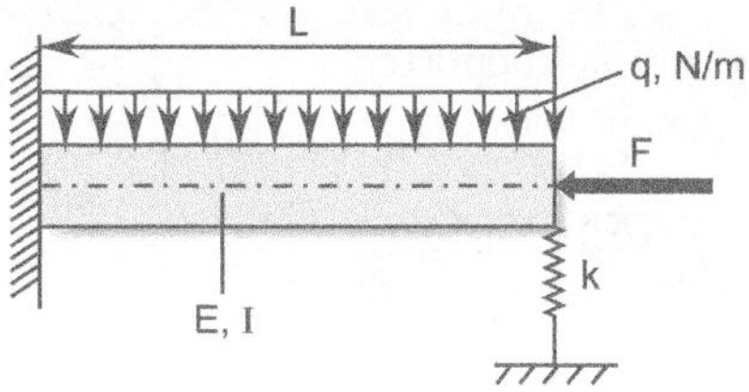

Fig. 3.9. Cantilever beam column.

Strain energy due to bending moment is given by

$$U_b = \frac{EI}{2}\int_0^L \left(\frac{d^2y}{dx^2}\right)^2 dx$$

Strain energy due to spring deflection

$$U_s = \frac{1}{2}ky_L^2$$

To determine the work potential due to q and F.

$$W = \int_0^L (qdx)y + \frac{F}{2}\int_0^L \left(\frac{dy}{dx}\right)^2 dx$$

Hence the total energy potential is written as

$$\pi = U - W = (U_b + U_s) - W$$

$$\pi = \frac{EI}{2}\int_0^L \left(\frac{d^2y}{dx^2}\right)^2 dx + \frac{1}{2}ky_L^2 - \int_0^L qy\,dx - \frac{E}{2}\int_0^L \left(\frac{dy}{dx}\right)^2 dx \quad ...(3.13)$$

The functional (3.13) has the contribution of bending and spring deflection, and work potential lost due to q and F.

TABLE 3.1. Potential Energy Functionals

Figure	Strain energy (U)	Work potential (W)	Functional $\pi = U - W$
• Axial Bar with traction force 	$\frac{AE}{2}\int_0^L \left(\frac{du}{dx}\right)^2 dx$	$\int_0^L qu\,dx$	$\pi = \int_0^L \frac{AE}{2}\left(\frac{du}{dx}\right)^2 dx - \int_0^L qu\,dx$
• Cantilever Beam with point load 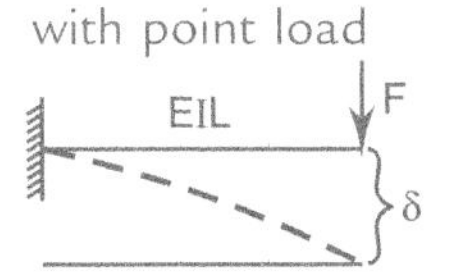	$\frac{EI}{2}\int_0^L \left(\frac{d^2y}{dx^2}\right)^2 dx$	$F\delta$	$\pi = \int_0^L \frac{EI}{2}\left(\frac{d^2y}{dx^2}\right)^2 dx - F\delta$
• Cantilever beam with UDL 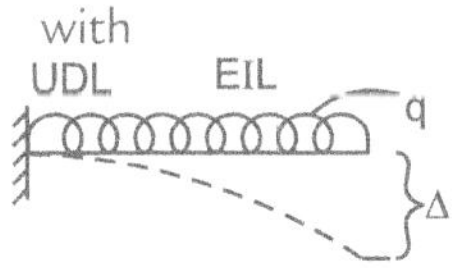	$\frac{EI}{2}\int_0^L \left(\frac{d^2y}{dx^2}\right)^2 dx$	$\int_0^L qy\,dx$	$\pi = \int_0^L \left[\frac{EI}{2}\left(\frac{d^2y}{dx^2} - qy\right)\right] dx$
• Simply supported beam with point load. 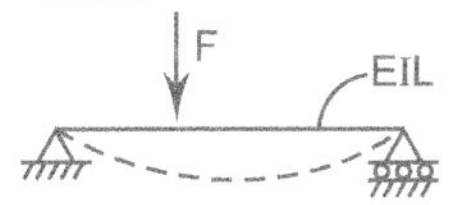	$\frac{EI}{2}\int_0^L \left(\frac{d^2y}{dx^2}\right)^2 dx$	FY	$\pi = \int_0^L \left[\frac{EI}{2}\left(\frac{d^2y}{dx^2}\right)^2\right] dx - Fy$

• Simply supported beam with UDL. 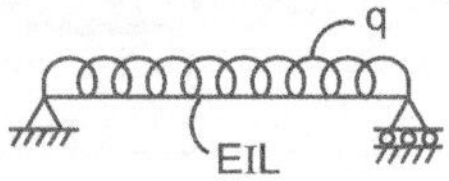	$\frac{EI}{2}\int_0^L\left(\frac{d^2y}{dx^2}\right)^2 dx$	$\int_0^L qy\,dx$	$\pi = \int_0^L\left[\frac{EI}{2}\left(\frac{d^2y}{dx^2}\right)^2 - qy\right]dx$
• Beam column 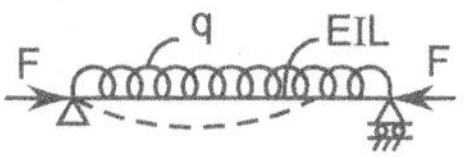	$\frac{EI}{2}\int_0^L\left(\frac{d^2y}{dx^2}\right)^2 dx$	$\frac{F}{2}\int_0^L\left(\frac{dy}{dx}\right)^2 dx$ $-\int qy\,dx$	$\pi = \int_0^L\left[\frac{EI}{2}\left(\frac{d^2y}{dx^2}\right) - \frac{F}{2}\left(\frac{dy}{dx}\right)^2 - qy\right]dx$

3.3 LAGRANGE EULER EQUATION

Consider a scalar quantity like kinetic energy in a system to represent a functional. A small change in the functional corresponds to variation in the solution domain. The analogy of extremization of simple function in ordinary calculus, the vanishing of first derivative of the functional would provide the stationarity condition for the variables of the functional which has been discussed in section (3.2) in the variational approach.

Let the displacement vector, X, be the function of the generalized co-ordinates q_i and time t. That is

$$X = [(q_1, q_2, q_3, ..., q_n), t]$$

The scalar quantity considered in the derivation of L-E (Lagrange-Euler) formulation, be the kinetic energy k given by

$$K = \frac{1}{2}mV^2 \qquad ...(3.14)$$

Since V is the first time derivative of displacement vector X

$$K = \frac{1}{2}m(\dot{X})^2 \qquad ...(3.15)$$

Differentiating k with respect to $\dot{q}$ we have

$$\frac{\partial k}{\partial \dot{q}} = m\dot{X}\frac{\partial \dot{X}}{\partial \dot{q}} \qquad ...(3.16)$$

Since $$\frac{\partial \dot{X}}{\partial \dot{q}} = \frac{\partial X}{\partial t}\cdot\frac{\partial t}{\partial q} = \frac{\partial X}{\partial q} \qquad ...(3.17)$$

Form equations (3.16) and (3.17)

$$\frac{\partial k}{\partial \dot{q}} = m\dot{X}\frac{\partial X}{\partial q} \qquad ...(3.18)$$

Now differentiate partially, the equation (3.15) with respect to generalized co-ordinates q

$$\frac{\partial k}{\partial q} = m\dot{X}\frac{\partial \dot{X}}{\partial q} \qquad ...(3.19)$$

or the equation (3.19) can be in the form

$$\frac{\partial k}{\partial q} = m\dot{X}\frac{d}{dt}\left(\frac{\partial X}{\partial q}\right) \qquad ...(3.20)$$

From equation (3.18), the first variation

$$\frac{d}{dt}\left[\frac{\partial k}{\partial \dot{q}}\right] = \frac{d}{dt}\left[m\dot{X}\left(\frac{\partial X}{\partial q}\right)\right]$$

$$= m\ddot{X}\frac{\partial X}{\partial q} + m\dot{X}\frac{d}{dt}\left(\frac{\partial X}{\partial q}\right) \qquad ...(3.21)$$

From equations (3.20) and (3.21)

$$\frac{d}{dt}\left[\frac{\partial k}{\partial \dot{q}}\right] - \frac{\partial k}{\partial q} = m\ddot{X}\frac{\partial X}{\partial q} \qquad ...(3.22)$$

In the equation (3.22), the factor $m\ddot{X}$ is the force vector = mass × acceleration vector

$$\frac{FdX}{dq} = \frac{d}{dt}\left[\frac{\partial k}{\partial \dot{q}}\right] - \frac{\partial k}{\partial q}$$

$$FdX = \left[\frac{d}{dt}\left(\frac{\partial k}{\partial \dot{q}}\right) - \frac{\partial k}{\partial q}\right]dq$$

The functional

$$\pi = \int FdX = \left[\frac{d}{dt}\left(\frac{\partial k}{\partial \dot{q}}\right) - \frac{\partial k}{\partial q}\right]q \qquad ...(3.23)$$

$$\delta\pi = \int \delta FdX = 0 \text{ for extremization.}$$

But δq being arbitrary

$$\frac{d}{dt}\left(\frac{\partial k}{\partial \dot{q}}\right) - \frac{\partial k}{\partial q} = 0$$

or

$$\frac{\partial k}{\partial q} - \frac{d}{dt}\left(\frac{\partial k}{\partial \dot{q}}\right) = 0 \qquad ...(3.24)$$

In the general form the L-E equation can be presented as

$$\frac{\partial f}{\partial y} - \frac{d}{dx}\left(\frac{\partial f}{\partial \dot{y}}\right) + \frac{d^2}{\partial x^2}\left(\frac{\partial f}{\partial \ddot{y}}\right) = 0 \qquad ...(3.25)$$

Where

The potential energy functional

$$\pi = \int f(x, y, \dot{y}, \dot{y})\, dx$$

with

$$\dot{y} = \frac{dy}{dx} \qquad \ddot{y} = \frac{d^2y}{dx^2}$$

The boundary condition for the governing equation (3.25) for the problem are

$$\left[\frac{\partial f}{\partial \dot{y}} - \frac{d}{dx}\left(\frac{\partial f}{\partial \ddot{y}}\right)\right]\delta y\Bigg|_{x_1}^{x_2} = 0 \qquad ...(3.26)$$

and $$\left[\left(\frac{\partial f}{\partial \ddot{y}}\right)\delta \dot{y}\right]\Bigg|_{x_1}^{x_2} = 0 \qquad ...(3.27)$$

Since δy and $\delta \dot{y}$ are arbitrary the conditions

$$\left[\frac{\partial f}{\partial \dot{y}} - \frac{d}{dx}\left(\frac{\partial f}{\partial \ddot{y}}\right)\right]\Bigg|_{x_1}^{x_2} = 0 \qquad \frac{\partial f}{\partial \ddot{y}}\Bigg|_{x_1}^{x_2} = 0 \text{ are known as}$$

Natural boundary conditions. The satisfied natural boundary conditions are called the free boundary conditions. If of the two natural boundary conditions are not satisfied the arbitrary variables.

The arbitrary variables.

$$\delta y\,(x_1) = 0\,; \qquad \delta y\,(x_2) = 0 \qquad ...(3.28)$$

And $$\delta y(x_1) = 0\,; \qquad \delta y(x_2) = 0$$

The conditions (3.28) are known as geometric or forced boundary conditions. In the L-E formulation of calculus of variations has led to one governing differential equation and four boundary conditions-two free boundary conditions and two forced boundary conditions. The L-E formulation provides a means of obtaining the governing equation for the physical phenomenon describing the problems in engineering, and establishing the boundary conditions that aid in solving the equation for the state variables. This variational principle can be successfully employed for simple continuous problems and the complicated descretrized problems in finite element method. Some of the problems in structural mechanics are dealt in the following examples. A tension rod, a cantilever beam, simple supported beam and a beam column examples are solved to demonstrate the importance of Lagrange–Euler formulations in structural mechanics problem.

Example 3.6. *Using the example (3.2) obtain the governing differential equation by L-E formulation.*

Solution. The functional for equation (3.10) of example (3.2), is

$$\pi = \int_0^l \left[\frac{AE}{2}\left(\frac{du}{dx}\right)^2 - qu\right] dx$$

Hence $$f = \frac{AE}{2}\left(\frac{du}{dx}\right)^2 - qu$$

We obtain $$\frac{\partial f}{\partial u} = -q, \quad \frac{\partial f}{\partial \dot{u}} = AE\frac{du}{dx} \qquad ...(a)$$

From *L-E* formulation in equation (3.25)

$$\frac{\partial f}{\partial u} - \frac{d}{dx}\left(\frac{\partial f}{\partial \dot{u}}\right) = 0 \qquad ...(b)$$

Substituting (*a*) in (*b*)

i.e., $$-q + AE\frac{du}{dx} = 0$$

or $$AE\frac{du}{dx} = q \qquad ...(3.29)$$

The equation (3.29) is the governing equation for the axial bar under tensile traction force.

Since $$AE\frac{du}{dx} \neq 0$$

The geometric boundary conditions are

$$\delta u(x_1) = 0$$
$$\delta \dot{u}(x_1) = 0$$

Example 3.7. *With help of the functional given in Table (3.1) obtain the governing differential equation and the associated boundary condition, for the simple supported beam with point load, in the mid span.*

Solution. The functional from Table 3.1 is given in the form

$$\pi = \int_0^l \frac{EI}{2}\left(\frac{d^2y}{dx^2}\right)^2 dx - FY_L$$

The function f is

$$f = \frac{EI}{2}\left(\frac{d^2y}{dx^2}\right)^2 - F_x Y \qquad \text{since } \frac{F_x}{Y} = \frac{F}{Y_L}$$

$$= \frac{EI}{2}\left(\frac{d^2y}{dx^2}\right)^2 - \frac{FY^2}{Y_L} \qquad F_x = \frac{FY}{Y_L}$$

We obtain $$\frac{\partial f}{\partial y} = \frac{2Fy}{Y_L} \text{ and } \frac{\partial f}{\partial \dot{y}} = 0 \qquad \frac{\partial f}{\partial \ddot{y}} = EI\frac{d^2y}{dx^2}$$

Substituting in L-E equation (3.25)

$$\frac{2Fy}{y_L} + EI\frac{d^2}{dx^2}\left(\frac{d^2y}{dx^2}\right) = 0$$

i.e., $$EI\frac{d^4y}{dx^4} + \frac{2Fy}{y_L} = 0 \qquad ...(3.30)$$

Since $$\frac{\partial f}{\partial \ddot{y}} = EI\frac{d^2y}{dx^2} \neq 0 \quad \text{and} \quad EI\frac{d^3y}{dx^3} = 0$$

The associated boundary conditions are

$$\delta y(x_1) = 0; \qquad \delta y(x_2) = 0$$

Example 3.8. *Solve the functional in equation (3.12) of example (3.4) for the governing differential equation and associated boundary condition by L-E formulation.*

Solution. The functional is given by

$$\pi = \int_0^l \frac{EI}{2}\left(\frac{d^2y}{dx^2}\right)^2 dx - \frac{F}{2}\int_0^l \left(\frac{dy}{dx}\right)^2 dx - \int_0^l qy\, dx$$

Hence the function

$$f = \frac{EI}{2}\left(\frac{d^2y}{dx^2}\right)^2 - \frac{F}{2}\left(\frac{dy}{dx}\right)^2 - qy$$

We obtain $\dfrac{\partial f}{\partial y} = -q, \; \dfrac{\partial f}{\partial \dot{y}} = \dfrac{2F}{2}\left(\dfrac{dy}{dx}\right)$ and $\dfrac{\partial f}{\partial \ddot{y}} = EI\dfrac{d^2y}{dx^2}$

Substituting in equation (3.25)

$$-q - \frac{d}{dx}\left[F\left(\frac{dy}{dx}\right)\right] + \frac{d^2}{dx^2}\left[EI\frac{d^2y}{dx^2}\right] = 0$$

or
$$EI\frac{d^4y}{dx^4} - \frac{Fd^2y}{dx^2} = q \qquad \text{...(3.31)}$$

Since $EI\dfrac{d^2y}{dx^2} \neq 0$ and $F\dfrac{dy}{dx} - EI\dfrac{d^3y}{dx^3} = 0$

The forced boundary conditions are

$\delta y(x_1) = 0$ and $\delta y(x_2) = 0$. Also $EI\dfrac{d^2y}{dx^2} = F$ is the free boundary condition.

The equation (3.31) gives the governing differential equation for the beam column with transverse *UDL* and axial compressive force.

Example 3.9. *Find the governing differential equation for a cantilever beam with UDL using L-E formulation.*

Solution. As given in Table (3.1) the total potential energy functional is

$$\pi = \frac{EI}{2}\int_0^L \left(\frac{d^2y}{dx^2}\right)^2 dx - \int_0^L qy\,dx$$

The function
$$f = \frac{EI}{2}\left(\frac{d^2y}{dx^2}\right)^2 - qy$$

The derivatives are obtained as

$$\frac{\partial f}{\partial y} = -q \qquad \frac{\partial f}{\partial \dot{y}} = 0 \qquad \frac{\partial f}{\partial \ddot{y}} = EI\frac{d^2y}{dx^2}$$

Form L-E formulation in equation (3.25)

$$-q - \frac{d}{dx}(0) + \frac{d^2}{dx^2}\left(EI\frac{d^2y}{dx^2}\right) = 0$$

or
$$EI\frac{d^4y}{dx^4} = q \qquad \text{...(3.32)}$$

Associated boundary conditions are $y(0) = 0$, $\dot{y}(0) = 0$.

3.4 RAYLEIGH-RITZ METHOD

Consider the functional

$$\pi = \int f(x, y, \dot{y}, \ddot{y})\, dx \tag{3.33}$$

to represent total potential energy of a system. In many of the situations in physical real world problems, the exact solution to π does not exist. In such cases the Rayleigh-Ritz method of variational principle gives the approximate solution. The exaction solution to equation (3.33) should satisfy the compatibility conditions and equilibrium requirement. The approximate solution y obtained from Rayleigh-Ritz method, is found to satisfy

- Compatible condition (Geometric boundary condition) perfectly.
- Equilibrium condition approximately.

Rayleigh-Ritz method is signified by the assumption of a trial function for y that approximately satisfies the functional π in equation (3.33). The procedure of Rayleigh-Ritz is highlighted by the following steps:

1. Derivation of potential energy functional.
2. Assuming a trial function (a polynomial) for the displacement field and applying boundary condition.
3. Obtaining the unknown co-efficients a_i of the trial function by exercising the principle of stationarity $\left(i.e., \dfrac{\partial \pi}{\partial a_i} = 0\right)$.
4. Establishing the relation for displacement, strain and stress.

It may be observed that the geometric boundary conditions are to be satisfied by the approximate trial function. Higher the order of the polynomial assumed for the trial function, higher is the accuracy of the solution obtained. The stability of equilibrium is checked by the satisfaction of the condition $\delta^2 \pi > 0$.

Example 3.10. A tension bar fixed at both the ends, is acted upon by an axial force, P in the mid point of length L. Find by Rayleigh-Ritz method the expression for stress distribution. Sketch the plot of displacement and the stress along the length of the rod.

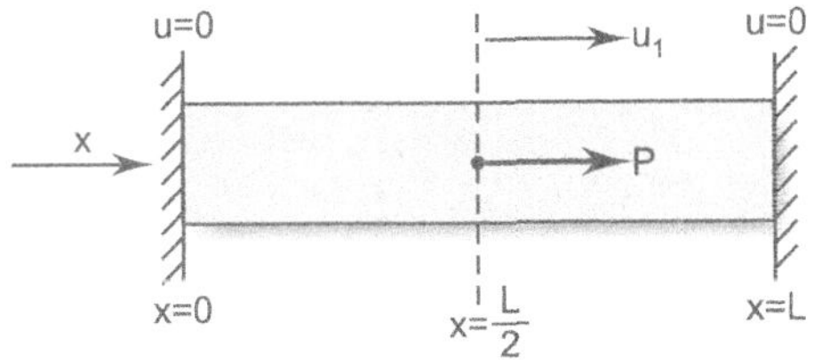

Fig. 3.10. Axial tension bar.

The diagrammatic represention of the given problem is shown in Fig. 3.10. As the bar is fixed at both ends the displacements are zero at $x = 0$ and at $x = L$. Let the displacement at mid-point $\left(i.e., \text{at } x = \dfrac{L}{2}\right)$ be u_1.

The strain energy or the elastic potential is given by

$$U = \int_0^L \frac{1}{2} \in \sigma \, dv$$

$$= \frac{1}{2}\int_0^L \left(\frac{du}{dx}\right) . E\left(\frac{du}{dx} . (A\,dx)\right)$$

$$= \frac{AE}{2}\int_0^L \left(\frac{du}{dx}\right)^2 dx \quad ...(a)$$

The work potential is given by

$$W = Pu_1.$$

The functional is formed as

$$\pi = U - W = \frac{AE}{2}\int_0^L \left(\frac{du}{dx}\right)^2 dx - Pu_1 \quad ...(b)$$

The trial function is assumed with three Ritz constants as

$$u = a_0 + a_1 x + a_2 x^2 \quad ...(c)$$

The geometric boundary condition for compatibility are

at $x = 1, u = 0$, hence $a_0 = 0$

$x = L, u = 0$

Hence $$a_1 L + a_2 L^2 = 0 \quad ...(d)$$

But it is assumed that at $x = \frac{L}{2}$, $u = u_1$

hence $$a_1 L/2 + a_2 L^2/4 = u_1 \quad ...(e)$$

From relations (*d*) and (*e*)

$$a_1 = -a_2 L$$

$$u_1 = -a_2 L^2/4$$

and $$u = a_2 x^2 - a_2 Lx \quad ...(f)$$

Differentiating relation (*f*) for *u*, with respect to *x*

$$\frac{du}{dx} = 2a_2 x - a_2 L \quad ...(g)$$

Substituting the relations for u_1 and $\frac{du}{dx}$ in equation (*b*) we have

$$\pi = \frac{1}{2} AE \int_0^L (2a_2 x - a_2 L)^2 dx + Pa_2 L^2/4$$

$$= \frac{AE}{2}\int_0^L (4a_2^2 x^2 - 4a_2^2 xL + a_2^2 L^2)\, dx + Pa_2 L^2/4$$

$$= \frac{AE}{2}\left[\frac{4}{3}a_2^2L^3 - 2a_2^2L^3 + a_2^2L^3\right] + Pa_2L^2/4$$

$$\pi = \frac{AE}{6}a_2^2L^3 + Pa_2L^2/4 \qquad ...(h)$$

Exercising the stationarity condition on relation (h)

$$\frac{\partial \pi}{\partial a_2} = \frac{AE}{3}a_2L^3 + PL^2/4 = 0$$

Hence

$$a_2 = -\frac{3}{4}\frac{P}{AEL} \qquad ...(i)$$

Substituting the value of a_2 in (f) we have

$$u = -\frac{3}{4}(x^2 - Lx)\frac{P}{AEL} \qquad ...(j)$$

and

$$u_1 = \frac{-a_2L^2}{4} = \frac{3}{16}\frac{PL}{AE}$$

To obtain the expression for stress distribution the Hooke's Law is made use. That is

$$\sigma = E \in = E\left(\frac{du}{dx}\right)$$

$$= E(2a_2x - a_2L)$$

$$= E\left(-\frac{3}{4}\frac{P}{AEL}\right)(2x - L)$$

$$\sigma = -\frac{3}{4}\frac{P}{AL}(2x - L) \qquad ...(k)$$

- **Displacement plot:** From the relation (j) it is evident that the displacement follows second order curve in x. with zero value at $x = 0$ and $x = L$. The maximum displacement $\left(u_1 = \frac{3}{16}\frac{PL}{AE}\right)$ is observed at $x = L/2$.

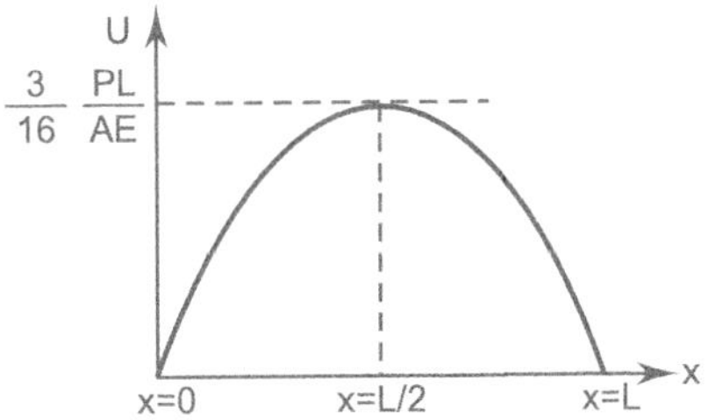

Fig. 3.11. Displacement plot.

- **Stress plot:**

In the relation (k) at $x = 0$; $\sigma_0 = \frac{3}{4}\frac{P}{A}$

at $x = L/2$; $\sigma_1 = 0$

and at $x = L$, $\sigma_L = -\dfrac{3}{4}\dfrac{P}{A}$ (the negative sign is signified by the compressive stress at right support fixation).

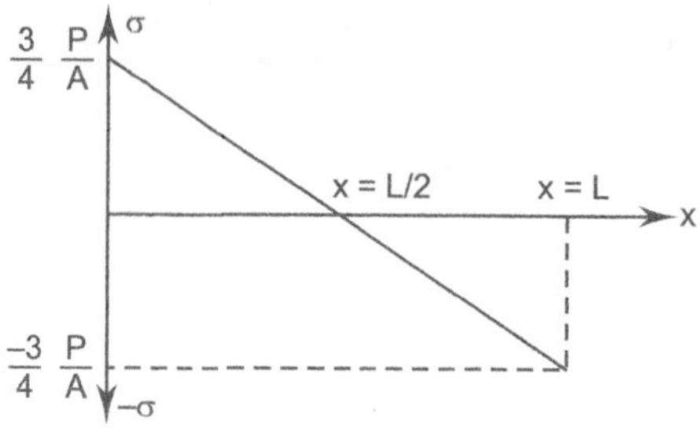

Fig 3.12. The stress plot.

The equation in (k) is linear in x hence the stress plot shows a linear variation along the length of the rod.

Example 3.11. *A vertically suspended bar fixed at one end having uniform cross-section A and length L, is having properties E as Young's modulus and ρ as density. Using three Ritz's constants for the trial function derive the expression for stress distribution along the length of the bar.*

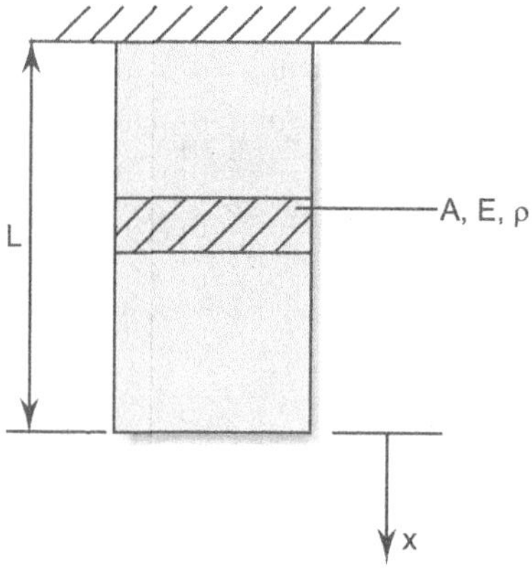

Fig. 3.13. Vertical bar with self weight.

Solution. The problem is transformed to a diagramatic representation as in Fig. 3.13. The bar that is vertical is fixed at the top and the body force due to selfweight act downward. The elastic potential for an axial member is

$$U = \frac{AE}{2}\int_0^L \left(\frac{du}{dx}\right)^2 dx \qquad ...(a)$$

The work potential due to body force is

$$W = \int_0^L F_b u \, dV = \int_0^L (\rho g) u_1 A dx \qquad ...(b)$$

The functional is expressed using the relations (a) and (b) as

$$\pi = U - W = \int_0^L \left[\frac{AE}{2}\left(\frac{du}{dx}\right)^2 - \rho A_g u\right] dx \qquad ...(3.34)$$

It is suggested use a trial function for the displacement field using three Ritz's constants. Hence

$$u = a_0 + a_1 x + a_2 x^2 \qquad \text{where } [a_0, a_1, a_2] \text{ are Ritz's constants.}$$

Applying the geometric boundary conditions

at $x = 0$; $u = 0$ leads to $a_0 = 0$

Hence, $u = a_1 x + a_2 x^2$

The derivative (first) with respect to x is

$$\frac{du}{dx} = a_1 + 2a_2 x$$

Substituting the expression for u and $\frac{du}{dx}$ in equation (3.34) and integrating yields.

$$\pi = \int_0^L \left[\frac{AE}{2}(a_1 + 2a_2 x)^2 - \rho A g\,(a_1 x + a_2 x^2)\right] dx$$

$$= \frac{AE}{2}(a_1^2 L) + \frac{AE}{2}(4a_1 a_2 L^2/2) + \frac{AE}{2}\left(4a_2^2 \frac{L^3}{3}\right) - \rho A g\left[\left(a_1 \frac{L^2}{2}\right) + a_2 L^3/3\right]$$

From the principle of minimum potential energy

$$\frac{\partial \pi}{\partial a_i} = 0 \qquad i.e. \qquad \frac{\partial \pi}{\partial a_1} = 0; \qquad \frac{\partial \pi}{\partial a_2} = 0$$

Hence we have

$$\frac{AE}{2}[2a_1 L + 2a_2 L^2] - \rho A g \frac{L^2}{2} = 0 \qquad ...(c)$$

$$\frac{AE}{2}\left[2a_1 L^2 + \frac{8}{3}a_2 L^3\right] - \rho A g L^3/3 = 0 \qquad ...(d)$$

Solving the relations (c) and (d) simultaneously

$$\frac{AE}{2}\left(\frac{2}{3}a_2 L^3\right) + \rho A g L^3/6 = 0$$

$$a_2 = -\frac{\rho A g}{4}\left(\frac{2}{AE}\right) = -\frac{\rho g}{2E}$$

$$a_1 = \frac{\rho g L}{E}$$

Substituting the constants a_1 and a_2 in u and $\frac{du}{dx}$

$$u = \frac{\rho g L}{E}x - \frac{\rho g}{2E}x^2 = \frac{\rho g}{E}\left(Lx - \frac{x^2}{2}\right)$$

$$\in = \frac{du}{dx} = \frac{\rho g L}{E} - \frac{\rho g x}{E} = \frac{\rho g}{E}(L - x)$$

The expression for stress distribution works out to

$$\sigma = E\in = \rho g(L - x)$$

The displacement plot and stress plot are given in Fig. 3.14 and Fig. 3.15 respectively.

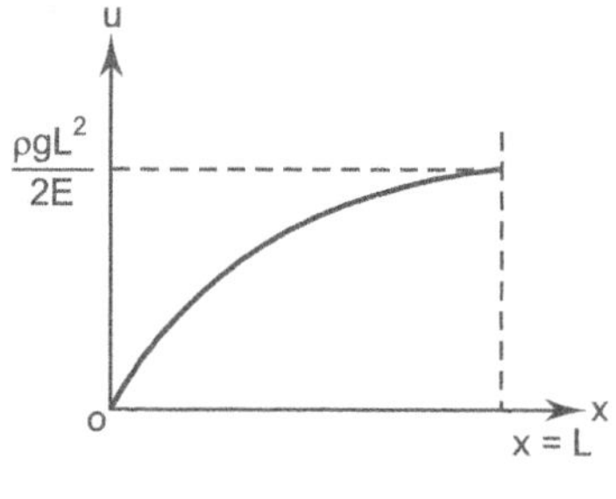

Fig. 3.14. Displacement plot.

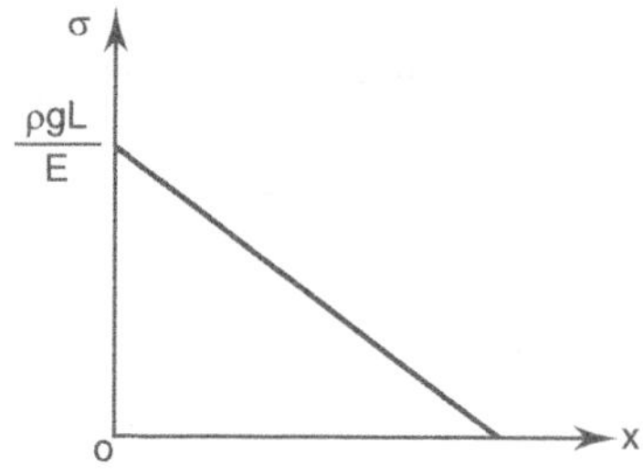

Fig. 3.15. Stress $x = L$ plot.

Example 3.12. *Find the deflection at the free end of a cantilever beam with point load at free end using Rayleigh-Ritz method.*

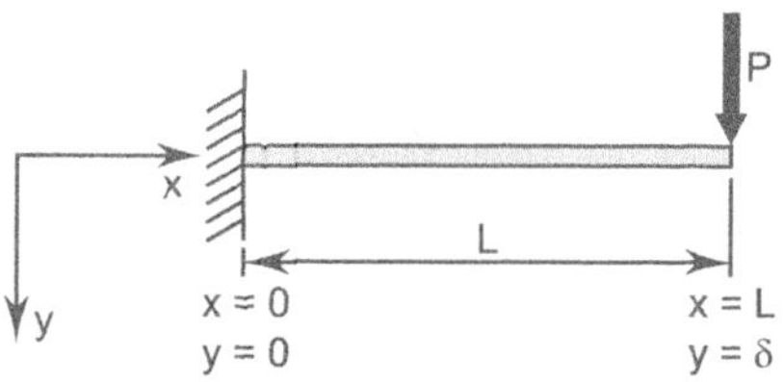

Fig. 3.16. Cantilever beam.

Solution. Let a point load P act at the free end of a cantilever beam of length L as shown in Fig. 3.16. Let the deflection at free end be δ.

The functional for the arrangement is written as

$$\pi = \frac{EI}{2}\int_0^L \left(\frac{d^2y}{dx^2}\right)^2 dx - P\delta \qquad ...(a)$$

Assumed trial function is

$$y = a_0 + a_1x + a_2\,x^2 \qquad ...(b)$$

For the boundary conditions

at $x = 0$ gives $y = 0$, hence $a_0 = 0$

at $x = L$ gives $y = \delta$, hence $\delta = a_1\,L + a_2\,L^2$

at $x = 0$ gives $\frac{dy}{dx} = 0$, hence $a_1 = 0$.

From the above analysis

$$y = a_2x^2, \quad \frac{d^2y}{dx^2} = 2a_2$$

which is substituted in relation (a) and integrated to obtain

$$\pi = 2EIa_2^2L - Pa_2L^2 \qquad ...(c)$$

Exercising stationarity condition on π in relation (*c*)

$$\frac{\partial \pi}{\partial a_2} = 4EIa_2L - PL^2 = 0$$

This results in

$$a_2 = \frac{PL}{4EI}$$

$$\delta = \frac{PL^3}{4EI}$$

The expression for the displacement field is

$$y = \frac{PL}{4EI}x^2$$

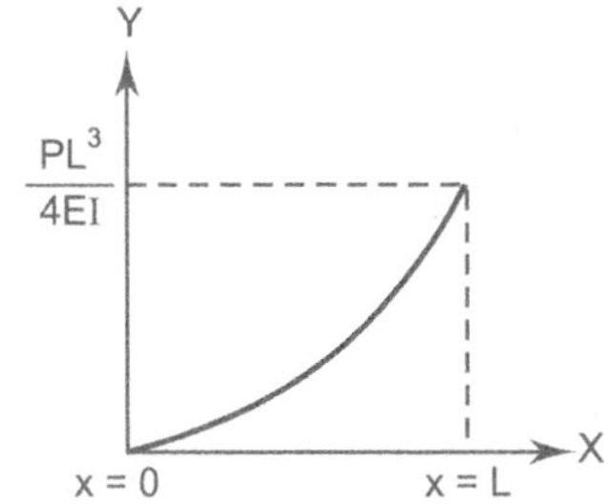

Fig. 3.17

The displacement plot for the cantilever beam is given in Fig. 3.17.

The expression for stress is obtained by flexure equation and deflection equation as

$$\sigma_b = Y.\frac{M}{I}, \quad \text{but } M = EI\frac{d^2y}{dx^2}$$

$$= YE\frac{d^2y}{dx^2} = \left(\frac{PL}{4EI}x^2\right).E.\left(\frac{2PL}{4EI}\right) = \left(\frac{PL}{4EI}\right)^2 2Ex^2$$

Example 3.13. *A simply supported beam with uniformly distributed load is as shown in Fig. 3.18. Using Rayleigh Ritz method obtain the expression for stress distribution.*

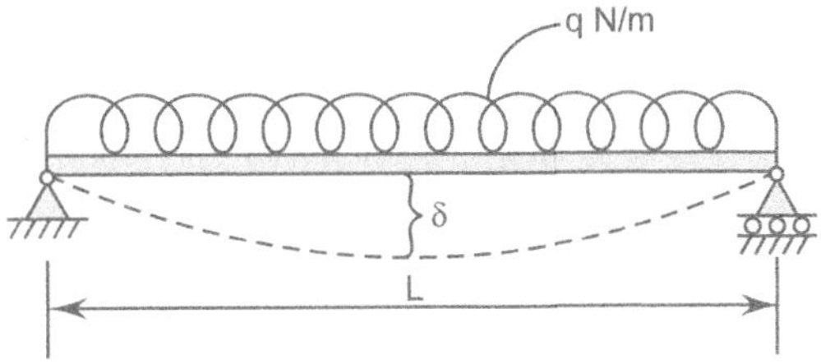

Fig. 3.18. Simply supported beam with UDL.

Solution. The total potential energy (functional), π is expressed as the difference of elastic potential and work potential in the form.

$$\pi = \frac{EI}{2}\int_0^L \left(\frac{d^2y}{dx^2}\right)^2 dx - \int_0^L qy\,dx \qquad ...(a)$$

The trial function is the arbitrary polynomial of second degree with three Ritz constants.

$$y = a_0 + a_1x + a_2x^2 \qquad ...(b)$$

The boundary conditions

at $x = 0$ implies $y = 0$; gives $a_0 = 0$.

at $x = L$ implies $y = 0$; gives $a_1L + a_2L^2 = 0$.

Hence we have $a_1 = -a_2L$

The displacement function

$$y = -a_2Lx + a_2x^2 \quad ...(c)$$

and

$$\frac{d^2y}{dx^2} = 2a_2 \quad ...(d)$$

Substituting (*c*) and (*d*) in (*a*) we have

$$\pi = \frac{EI}{2}\int_0^L 4a_2^2 dx - \int_0^L q(a_2x^2 - a_2Lx)\, dx$$

or

$$\pi = 2EI\, a_2^2 L - q\left(\frac{a_2L^3}{3} - a_2\frac{L^3}{2}\right)$$

$$= 2EIa_2^2L + qa_2L^3/6$$

By invoking stationarity condition

$$\frac{\partial \pi}{\partial a_2} = 4EIa_2L + qL^3/6 = 0$$

Hence

$$a_2 = -\frac{qL^2}{24EI}$$

Substituting a_2 in relations (*c*) and (*d*)

$$y = \frac{qL^2}{24EI}(Lx - x^2) \quad ...(e)$$

$$\frac{d^2y}{dx^2} = \frac{-qL^2}{12EI}$$

To express the bending stress

$$\sigma_b = -yE\frac{d^2y}{dx^2}$$

$$= \frac{qL^2}{12I}\left[\frac{qL^2}{24EI}(Lx - x^2)\right]$$

The displacement plot

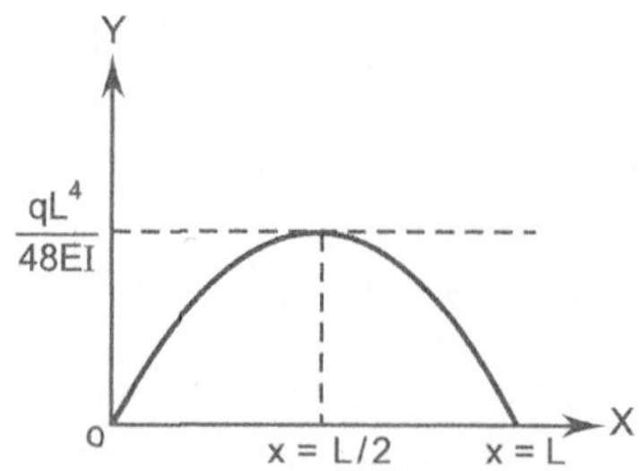

Fig 3.19. Displacement plot.

Example 3.14. *A cantilever beam column is as shown in Fig. 3.20. Obtain the expression for beam deformation and spring deflection.*

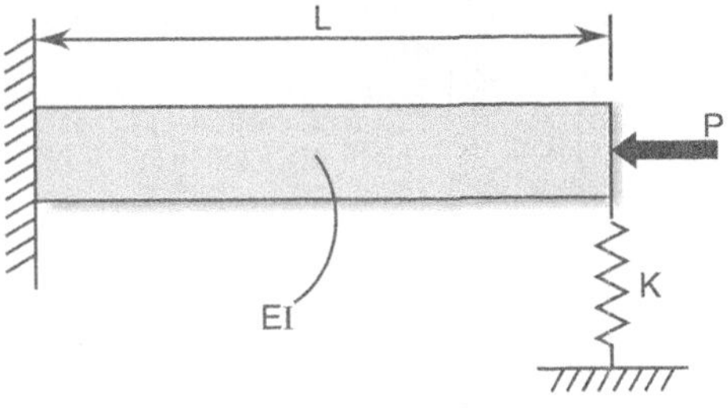

Fig. 3.20. Beam column problem.

Solution. The deflection in the beam column is inclusive of lateral and buckling. The buckling deformation is made dependent on lateral deflection by approximation as analysed in example (3.4). The attached spring as shown in Fig. 3.20 adds strain energy component to the functional. The functional can be directly written as

$$\pi = \int_0^L \frac{EI}{2}\left(\frac{d^2y}{dx^2}\right)^2 dx + \frac{1}{2}ky_L^2 - \frac{P}{2}\int_0^L \left(\frac{dy}{dx}\right)^2 dx \qquad ...(a)$$

The polynomial of second order with three Ritz constants is assumed as trial function for approximation. For the relation (*a*) we need to find the expressions for $\frac{d^2y}{dx^2}, \frac{dy}{dx}$ and y_L before exercising the minimum potential energy principle.

The polynomials is

$$y = a_0 + a_1x + a_2x^2 \qquad ...(b)$$

The boundary conditions

at $x = 0$ leads to $y = 0$ and hence $a_0 = 0$

at $x = 0$ gives $dy/dx = 0$ and hence $a_1 = 0$

at $x = L$ gives $y = y_L$ results in $y_L = a_2 L^2$

Substituting $\frac{dy}{dx}$ and y_L also $\frac{d^2y}{dx^2} = 2a_2$ in relation (*a*) we arrive at

$$\pi = \int_0^L \frac{EI}{2}(2a_2)^2 dx + \frac{1}{2}k(a_2L^2)^2 - \int_0^L \frac{P}{2}(2a_2x)^2 dx$$

After integration

$$\pi = 2EIa_2^2L + \frac{1}{2}ka_2^2L^4 - \frac{2}{3}Pa_2L^3 \qquad ...(c)$$

Using the principle of stationarity

$$\frac{\partial \pi}{\partial a_2} = 4EIa_2L + ka_2L^4 - \frac{2P}{3}L^3 = 0$$

Hence

$$a_2 = \frac{2/3PL^2}{(4EI + kL^3)}$$

The deflection of the spring, $$y_L = \frac{2PL^4}{3(4EI + kL^3)}$$

The axial deflection, $$\lambda = \frac{1}{2}\left(\frac{dy}{dx}\right)^2$$

$$= \frac{1}{2}\left[\frac{4}{3}\frac{PL^3}{(4EI + kL^3)}\right]^2$$

The lateral deflection $$y = \frac{2PL^2x^2}{(4EI + kL^3)} \quad ...(d)$$

The lateral deflection $y = 0$ at $x = 0$

and $$y = \frac{2PL^4}{(4EI + kL^3)} \quad \text{at } x = L.$$

The relation (*d*) signifies that the transverse deflection is parabolic in distribution along the length x.

Example 3.15. *A physical phenomena in engineering is described by the differential equation*

$$\frac{d^2y}{dx^2} + 8y^2 - 4x^3 = 0 \quad \text{in the range } 0 \le x \le L.$$

Subjected to the boundary conditions $y(0) = 0$, $y(L) = 0$. *Obtain the functional and using Rayleigh-Ritz method derive the expression for* y.

Solution. The Lagrange-Euler equation is given by

$$\frac{\partial f}{\partial y} - \frac{d}{dx}\left(\frac{\partial f}{\partial \dot{y}}\right) = 0 \quad ...(a)$$

Comparing the relation (*a*) with that of the give *DE*

$$\frac{\partial f}{\partial y} = 8y^2 - 4x^3 \quad ...(b)$$

and $$-\frac{d}{dx}\left(\frac{\partial f}{\partial \dot{y}}\right) = \frac{d^2y}{dx^2}$$

i.e., $$\frac{\partial f}{\partial \dot{y}} = -\frac{dy}{dx} = -\dot{y} \quad ...(c)$$

Integrating relations (*b*) and (*c*)

$$f = \frac{8}{3}y^3 - 4x^3y + c_1(\dot{y}) \quad ...(d)$$

$$f = -\frac{\dot{y}^2}{2} + c_2(y) \quad ...(e)$$

Comparing relations (*d*) and (*e*)

$$c_1(\dot{y}) = -\frac{\dot{y}^2}{2}$$

$$c_2(y) = \frac{8}{3}y^3 - 4x^3y$$

Therefore $$f = \frac{8}{3}y^3 - 4x^3y - \frac{\dot{y}^2}{2}$$

Hence the functional becomes

$$\pi = -\frac{1}{2}\int_0^L \left(\frac{dy}{dx}\right)^2 dx + \int_0^L y\left(\frac{8}{3}y^2 - 4x^3\right)dx \qquad ...(f)$$

The trial function assumed is

$$y = a_0 + a_1x + a_2x^2$$

From the boundary conditions $y(0) = 0$ and $y(L) = 0$

$$a_0 = 0$$

$$a_1L + a_2L^2 = 0$$

$$a_1 = -a_2L$$

Hence $$y = -a_2Lx + a_2x^2$$

$$\frac{dy}{dx} = -a_2L + 2a_2x$$

Substituting these in relation (f)

$$\pi = -\frac{1}{2}\int_0^L (2a_2x - a_2L)^2 dx + \int_0^L \left[\frac{8}{3}(a_2x^2 - a_2Lx)^3 - 4x^3(a_2x^2 - a_2Lx)\right]dx$$

$$= -\frac{1}{2}\left[\left(\frac{4}{3}a_2^2L^3\right) - \frac{2}{2}a_2^2L^3 + a_2^2L^3\right] + \frac{8}{3}\left[\frac{a_2^3}{7}L^7 - \frac{2a_2^3}{6}L^7 + \frac{a_2^3L^7}{5}\right.$$

$$\left. - \frac{a_2^3L^7}{6} + \frac{2a_2^3L^7}{5} - \frac{a_2^3L^7}{4}\right] - \frac{4}{6}a_2L^6 - \frac{4}{5}a_2L^6$$

$$= -\frac{2}{3}a_2^2L^3 - 0.019a_2^3L^7 - 1.467a_2L^6$$

By the principle of stationarity

$$\frac{\partial \neq}{\partial a_2} = 0, \quad \text{i.e.,} \quad -\frac{4}{3}a_2L^3 - 0.57a_2^2L^7 - 1.467L^6 = 0$$

$$+\frac{4}{3}a_2 + 0.57a_2^2L^4 + 1.467L^3 = 0.$$

This is a quadratic equation in a_2 which has the solution

as $$a_2 = \frac{1.33 \pm \sqrt{(1.33)^2 - 4(0.57)L^4(1.467L^3)}}{2(0.57L^4)}$$

Hence the equation for deflection, y is

$$y = a_2(x^2 - Lx)$$

By substituting the value for L we get a_2 and substituting a_2 we arrive at the expression for y.

Example 3.16. *The equilibrium equation of a structural system is given by*

$$100\frac{d^4y}{dx^4} + 50\frac{d^2y}{dx^2} - 10 = 0$$

Determine the type of the system and the system behaviour using Rayleigh-Ritz method.

Solution. The Lagrange-Euler equation for a system in variational approach is

$$\frac{\partial f}{\partial y} - \frac{d}{dx}\left(\frac{\partial f}{\partial \dot{y}}\right) + \frac{d^2}{dx^2}\left(\frac{\partial f}{\partial \ddot{y}}\right) = 0 \qquad ...(a)$$

Comparing the terms of given equilibrium equation with that of L-E equation in (*a*)

$$\frac{\partial f}{\partial y} = -10; \qquad -\frac{d}{dx}\left(\frac{\partial f}{\partial \dot{y}}\right) = 50\frac{d^2y}{dx^2} \quad \text{and} \quad \frac{d^2}{dx^2}\left(\frac{\partial f}{\partial \ddot{y}}\right) = 100\frac{d^4y}{dx^4}$$

Hence

$$f = -10y + f(\dot{y}) \qquad ...(b)$$

$$\frac{\partial f}{\partial \dot{y}} = -50\dot{y} \quad \text{or} \quad f = -\frac{50}{2}\dot{y}^2 + g(y) \qquad ...(c)$$

$$\frac{\partial f}{\partial \ddot{y}} = 100\ddot{y}$$

Integrating

$$f = \frac{100}{2}\ddot{y}^2 + k \qquad ...(d)$$

Comparing relations (*b*) and (*c*)

$$f = -25\dot{y}^2 - 10y \qquad ...(e)$$

Comparing relations (*d*) and (*e*)

$$k = -25\dot{y}^2 - 10y$$

Hence

$$f = 50\ddot{y}^2 - 25\dot{y}^2 - 10y$$

Hence the functional is

$$\pi = \int_0^L f(y, \dot{y}, \ddot{y})\,dx$$

$$= \int_0^L 50\left(\frac{d^2y}{dx^2}\right)^2 dx - 25\int_0^L \left(\frac{dy}{dx}\right)^2 dx - 10\int_0^L y\,dx \qquad ...(f)$$

This compares with the potential energy functional of the type

$$\pi = \frac{EI}{2}\int_0^L \left(\frac{d^2y}{dx^2}\right)^2 dx - \frac{P}{2}\int_0^L \left(\frac{dy}{dx}\right)^2 dx - \int_0^L qy\,dx$$

Which represents a beam column simply supported with a *UDL* as shown by the Fig. 3.21.

To predict system behaviour the deflection pattern, y is to be determined using Rayleigh–Ritz method.

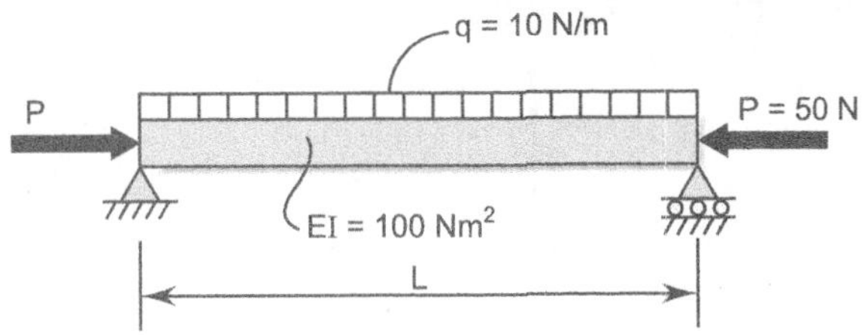

Fig. 3.21. Beam column.

Assume a trial function for y with three Ritz's constant

as $$y = a_0 + a_1x + a_2x^2 \qquad ...(g)$$

The boundary conditions are

$$y(0) = 0 \text{ and } y(L) = 0$$

Hence $$a_0 = 0$$

and $$a_1L + a_2L^2 = 0$$

implies $$a_1 = -a_2L$$

This results in $$y = a_2x^2 - a_2Lx \qquad ...(h)$$

$$\frac{dy}{dx} = 2a_2x - a_2L$$

$$\frac{d^2y}{dx^2} = 2a_2$$

Substituting these values in relation (f) we arrive at

$$\pi = 50\int_0^L (2a_2)^2dx - 25\int_0^L (2a_2x - a_2L)^2dx - 10\int_0^L (a_2x^2 - a_2Lx)dx$$

$$= 200a_2^2L - 25\left(\frac{4}{3}a_2^2L^3 - 2a_2^2L^3 + a_2^2L^3\right) - 10\left(\frac{a_2}{3}L^3 + - \frac{a_2L^3}{2}\right)$$

$$= 200a_2^2L - \frac{25}{3}a_2^2L^3 - \frac{17}{6}a_2L^3$$

By invoking the condition of stationarity

$$\frac{\partial\pi}{\partial a_2} = 400a_2L - \frac{50}{3}a_2L^3 - \frac{17}{6}L^3 = 0$$

Hence $$a_2 = \frac{17}{6}\frac{L^2}{\left(400 - \frac{50}{3}L^2\right)}$$

Substituting the value of a_2 in relation (h)

$$y = \frac{17L^2}{6\left(400 - \frac{50}{3}L^2\right)}(x^2 - Lx).$$

Deflection plot is as shown by the Fig. 3.22

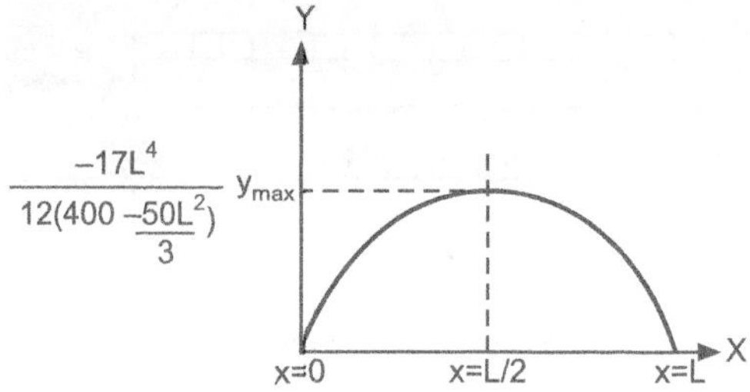

Fig. 3.22. Deflection plot of beam column.

The negative y_{max} indicates that the deflection is downwards.

Example 3.17. *Compare the deflection pattern and the stress distribution in an axial bar with traction force using three and four Ritz's constants in the trial function.*

Solution. The physical representation of the given system (an axial bar under tension) is as shown by the Fig. 3.23.

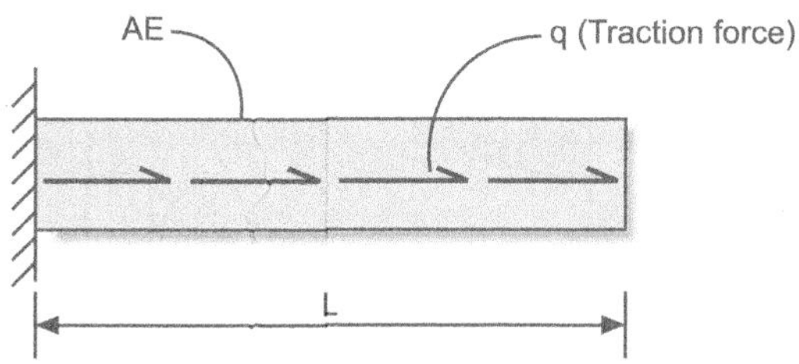

Fig. 3.23. Axial Bar under traction.

The total potential energy (functional, π) given by the difference in the elastic and work potential is,

$$\pi = \frac{AE}{2}\int_0^L \left(\frac{du}{dx}\right)^2 dx - \int_0^L qu\, dx \qquad ...(a)$$

where A is the area of cross-section and E is the modulus of elasticity.

(*a*) Polynomial trial function with three Ritz constants

$$u = a_0 + a_1 x + a_2 x^2$$

Applying boundary conditions

$$u(0) = 0 \text{ and } \left.\frac{du}{dx}\right|_{x=0} = 0$$

Hence
$$a_0 = 0 \text{ and } a_1 = 0.$$

also
$$u = a_2\, x^2$$

and
$$\frac{du}{dx} = 2a_2 x$$

Substituting u and $\frac{du}{dx}$ in relation (*a*)

$$\pi = \frac{AE}{2}\int_0^L (2a_2x)^2 dx - \int_0^L q\, a_2x^2 dx$$

$$= \frac{2AE}{3}a_2^2L^3 - qa_2\frac{L^3}{3}$$

Exercising the stationarity condition

$$\frac{\partial \pi}{\partial a_2} = \frac{4}{3}AEa_2L^3 - qL^3/3 = 0$$

Hence $$a_2 = \frac{q}{4AE}$$

Deflection and stress pattern are given by

$$u = \frac{q}{4AE}x^2 \quad \text{and} \quad \sigma = \frac{q}{2A}x \qquad \left(\because \sigma = E\frac{du}{dx}\right)$$

(*b*) Polynomial trial function with four Ritz constants

$$u = a_0 + a_1x + a_2x^2 + a_3x^3$$

Exercising boundary conditions

$$u\,(0) = 0 \text{ implies that } a_0 = 0$$

$$\left.\frac{du}{dx}\right|_{x=0} = 0 \text{ implies that } a_1 = 0$$

Hence the trial function and its first derivatives are

$$u = a_2x^2 + a_3x^3$$

$$\frac{du}{dx} = 2a_2x + 3a_3x^2$$

Substituting these values in relation (*a*)

$$\pi = \int_0^L \frac{AE}{2}(2a_2x + 3a_3x^2)^2 dx - q\int_0^L (a_2x^2 + a_3x^3)dx$$

$$= \frac{AE}{2}\left[4a_2^2\left(\frac{L^3}{3}\right) + \frac{12}{4}a_2a_3L^4 + \frac{9}{5}a_3^2L^5\right] - q\left(\frac{a_2}{3}L^3 + \frac{a_3}{4}L^4\right)$$

By invoking condition of stationarity

$$\frac{\partial \pi}{\partial a_2} = 0 \quad \Rightarrow \quad \frac{AE}{2}\left[\frac{4}{3}\left(\frac{L^3}{3}\right)2a_2 + 3a_3L^4\right] - \frac{q}{3}L^3 = 0$$

$$\frac{\partial \pi}{\partial a_3} = 0 \quad \Rightarrow \quad \frac{AE}{2}\left[3a_2L^4 + \frac{18}{5}a_3L^5\right] - \frac{q}{4}L^4 = 0$$

Solving the above two equations simultaneously

$$a_2 = -\frac{9}{116}q\left(\frac{2}{AE}\right) \quad \text{and} \quad a_3 = \frac{35}{261}\frac{q}{L}\left(\frac{2}{AE}\right)$$

Hence, $$u = \left[-\frac{9}{116}qx^2 + \frac{35}{261}\frac{q}{L}x^3\right]\frac{2}{AE}$$

$$\sigma = E\frac{du}{dx} = \frac{2}{A}\left[-\frac{18}{116}qx + \frac{35}{87}\frac{q}{L}x^2\right]$$

- **The displacement plot**

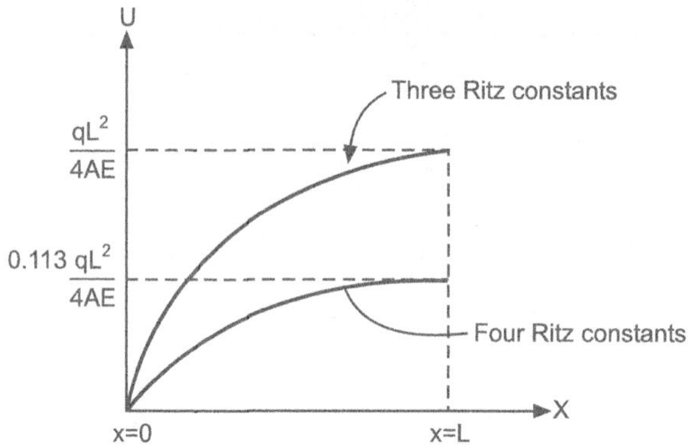

Fig. 3.24. Deflection plot.

- **Stress plot**

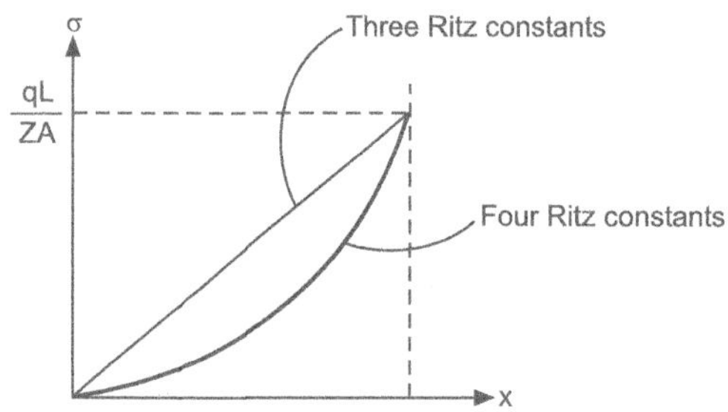

Fig. 3.25. Stress plot.

3.5 WEIGHTED RESIDUAL METHODS

In the previous sections the structural mechanics problems were given approximate solution by direct integration and variational approaches. Of these methods it is Rayleigh-Ritz method, was popularly employed to arrive at the solution to approximate equilibrium. But for the non-structural problems use of variational approach can not be successful because of the difference in behaviour of the system. Hence different methods known as weighted residual methods are to be used effectively. The following sections deal with the weighted residual methods in brief. The weighted residual methods are extensively used in the finite element method. They are the extensions of the solution techniques in FEM. For example, one dimensional heat conduction, steady state conduction and convection problems in FEM are successfully exercised using weighted residual methods.

Formulation of Method

The system is governed and defined by the equilibrium equations in the form of set of differential equation. By the assumption of an approximate trial function for the governing differential equation, as the solution results in a gross error, known as the residual.

Residual is the value obtained by substituting the trial function and its derivatives in the differential equations of equilibrium. For the solutions to be close the residual has to be zero when summed up.

Let the trial function assumed be

$y = f(x_i, a_i)$ that would satisfy the boundary conditions. By substituting y in the equilibrium equation results in the residual.

$$R(x_i, a_1, a_2, a_3, \ldots a_n)$$

where a_i $(= a_1, a_2 \ldots a_n)$ are the unknowns that are determined such that

$$\int_D w_i R(x_i, a_i)\,dx = 0 \qquad \ldots(3.35)$$

where $w_i = w(x)$ is the weighting function. Several methods used to solve equation (3.35) are

- Point Collocation Method.
- Subdomain Collocation Method.
- Least Square Method.
- Galerkin's Method.

Of these four methods Galerkin's Method has wide application in finite element method. All the four methods are tackled in brief in the following paragraphs.

3.5.1 Point Collocation Method

In this method the weigting function W_i can be denoted by $f(x - x_i)$ which leads to the definition that at collocation points, *i.e.*, at $x = x_i$ results in

$$W_i = \int_a^b f(x - x_i)\,dx = 1 \qquad \ldots(3.36)$$

and at non-collocation points $\int_a^b f(x - x_i)\,dx = 0$ for $x \neq x_i$

The selection of collocation point is dependent on the analyzer. The substitution of equation (3.36) in equation (3.35) results in 'n' equations to be solved for a_i (for $i = 1, 2, \ldots n$). Hence the equation (3.35) takes the form.

$$\int_D f(x - x_i) R(x_i, a_i)\,dx - 0$$

At collocation point $\int f(x - x_i)\,dx = 1$ for $x = x_i$

Hence we have the set of equations

$$\int_D R(x_i, a_i)\,dx = 0 \qquad \ldots(3.37)$$

which is to be solved for a_i at the collocation points $x = (x_1, x_2, \ldots, x_n)$.

Example 3.18. *The physical phenomenon is described by the differential equation* $\frac{d^2y}{dx^2} + 50 = 0$, *for which the boundary conditions are* $y(0) = 0$, $y(8) = 0$. *Using suitable trial function obtain the solution for the DE using point collocation method.*

Solution. Trial function assumed is, $y = a_0 + a_1x + a_2x^2$, which on satisfaction of boundary conditions

$$y(0) = 0 \text{ gives } a_0 = 0.$$

$$y(8) = 0 \text{ gives } 8a_1 + 64a_2 = 0 \Rightarrow a_1 = -8a_2$$

and
$$\frac{d^2y}{dx^2} = 2a_2$$

Substituting this in the given differential equation residual,

$$R = 2a_2 + 50 \qquad ...(a)$$

The residual (*a*) is independent of the collocation point in the given problem.

Hence $\qquad 2a_2 + 50 = 0$

$$a_2 = -25 \text{ and } a_1 = 200$$

Therefore the solution is

$$y = 200x - 25x^2 \qquad ...(b)$$

It may be observed that by direct integration of given *DE* results in the same equation as that of the relation (*b*).

Example 3.19. *The governing differential for one dimensional heat conduction given by*

$$A(\phi) = \frac{d^2\phi}{dx^2} + Q(x) = 0$$

Assuming the trial function $\phi = \Sigma a_i \sin\frac{\pi x i}{L}$. *Obtain the solution* ϕ *for the given problem, using point collocation method.*

Solution. It is given that

$$\frac{d^2\phi}{dx^2} + Q(x) = 0 \quad (0 \le x < L) \qquad ...(a)$$

It is assumed that $\qquad Q(x) = 1$ for $0 \le x < L/2$

and $\qquad Q(x) = 0$ for $L/2 < x \le L.$

The boundary conditions are

$$\phi(0) = 0 \text{ and } \phi(L) = 0. \qquad ...(b)$$

in the given trial function

$$\phi = \sum a_i \sin\frac{\pi x_i}{L}$$

use $i = 1$ and $x_i = L/2$.

Hence,
$$\phi = a_i \sin\frac{\pi x}{L}$$

$$\frac{d^2\phi}{dx^2} = -a_1\frac{\pi^2}{L^2}\sin\frac{\pi x}{L}.$$

Since for $x = L/2$, Q is either zero or one.

Hence use $Q = \frac{1}{2}$

The residual

$$R = -a_1 \frac{\pi^2}{L^2} \sin \frac{\pi}{2} + \frac{1}{2} = 0$$

$$a_1 = \frac{L^2}{2\pi^2}$$

Hence the solution to the differential equation governing the one dimensional heat conduction is

$$\phi = \frac{L^2}{2\pi^2} \sin \frac{\pi x}{L}$$

It may be easily verified to satisfy the relation (*a*)

as
$$\frac{d^2\phi}{dx^2} = -\frac{1}{2}$$

and
$$\frac{d^2\phi}{dx^2} + Q = -\frac{1}{2} + \frac{1}{2} = 0.$$

Example 3.20. *Solve the example (3.19) by two point collocation. To exercise the point collocation using two points we divide the domain into two parts i.e., between (0, L/2)* and *(L/2, L). The collocation points chosen are* $x_1 = L/3$ *and* $x_2 = \frac{2}{3}L$

at
$$x = x_1;\ Q = 1$$

and
$$x = x_2;\ Q = 0.$$

Solution. The trial function for $i = 2$ is given by

$$\phi = a_1 \sin \frac{\pi x}{L} + a_2 \sin \frac{2\pi x}{L} \qquad ...(a)$$

The boundary conditions are $\phi(0) = \phi(L) = 0$.

Differentiating relation (*a*) w.r.t. x for the second derivative

$$\frac{d^2\phi}{dx^2} = -a_1 \frac{\pi^2}{L^2} \sin \frac{\pi x}{L} - a_2 \frac{4\pi^2}{L^2} \sin \frac{2\pi x}{L}$$

at $x = L/3$
$$-\frac{\pi^2}{L^2} a_1 \left(\sin \frac{\pi}{3}\right) - \frac{4\pi^2}{L^2} a_2 \left(\sin \frac{2\pi}{3}\right) + 1 = 0$$

at $x = \frac{2L}{3}$,
$$-\frac{\pi^2}{L^2} a_1 \sin \frac{2\pi}{3} - \frac{4\pi^2}{L^2} a_2 \sin \frac{4\pi}{3} + 0 = 0$$

or on simplification

$$-8.547 \frac{a_1}{L^2} - \frac{34.19}{L^2} a_2 + 1 = 0 \qquad ...(b)$$

$$-8.547 \frac{a_1}{L^2} + \frac{34.19}{L^2} a_2 = 0 \qquad ...(c)$$

Solving (*b*) and (*c*) simultaneously

$$a_1 = \frac{L^2}{17.095}$$

$$a_2 = \frac{L^2}{68.379}$$

Hence the solution to trial function is

$$\phi = \frac{L^2}{17.095}\sin\left(\frac{\pi x}{L}\right) + \frac{L^2}{68.379}\sin\left(\frac{2\pi x}{L}\right)$$

It may be verified at $x = \frac{L}{3}$; $\frac{d^2\phi}{dx^2} = -1$

at $x = \frac{2L}{3}$; $\frac{d^2\phi}{dx^2} = 0.$

3.5.2 Subdomain Collocation Method

The weighting function are assumed to be unity in the portion of the domain and to be zero else where in the domain. To define the procedure mathematically

$$w_1 = \begin{cases} 1 \text{ for } x \text{ in domain } D_1 \\ 0 \text{ for } x \text{ not in domain } D_1 \end{cases}$$

$$w_2 = \begin{cases} 1 \text{ for } x \text{ in the domain } D_2 \\ 0 \text{ for } x \text{ not in the domain } D_2 \end{cases} \qquad \text{...(3.38)}$$

$$\vdots \qquad \vdots$$

$$w_n = \begin{cases} 1 \text{ for } x \text{ in the domain } D_n \\ 0 \text{ for } x \text{ not in the domain } D_n \end{cases}$$

The weighted function

$$\int_0^L f(x_i, a_i)\, R(x_i, a_i)\, dx = 0$$

or

$$\int_0^L w_i R(x_i, a_i)\, dx = 0 \qquad \text{...(3.39)}$$

In the domain (0, L/2) in which x_i lies

$$w_i = 1 \text{ and } \int_0^{L/2} R(x_i, a_i) = 0$$

$$w_i = 0 \text{ when } x_i \text{ is in } (L/2, L)$$

Hence

$$\int_0^{L/2} R(x_i, a_i) \neq 0$$

Example 3.21. *A physical phenomenon be described by the differential equilibrium equation.*

$$\frac{d^2y}{dx^2} + 100\,x^2 = 0 \text{ for } 0 \le x \le 1$$

With the boundary condition $y(0) = 0$ and $y(1) = 0$ and using the trial function $y = a_1 (x - x^2) + a_2 (x - x^3)$. Find the constants a_1 and a_2 by subdomain collocation method.

Solution. Differentiating the trial function given

$$y = a_1 (x - x^2) + a_2 (x - x^3)$$

$$\frac{dy}{dx} = a_1 (1 - 2x) + a_2 (1 - 3x^2)$$

$$\frac{d^2y}{dx^2} = -2a_1 - 6a_2x$$

The residual function

$$R = -2a_1 - 6a_2x + 100x^2$$

In the Domain $D_1 = (0, 1/2)$, $w_1 = 1$ and $\int_0^{1/2} R_1 = 0$

In the Domain $D_2 = \left(\frac{1}{2}, 1\right)$, $w_2 = 1$ $\int_{1/2}^{1} R_2 = 0$

Hence in the two domains

$$\int_0^{1/2} (-2a_1 - 6a_2x + 100x^2)\,dx = 0$$

$$-2a_1x \;\; 3a_2x^2 + \frac{100}{3}x^3\Big|_0^{1/2} = 0$$

$$-2a_1 - \frac{3}{2}a_2 + \frac{100}{12} = 0$$

or

$$4a_1 + 3a_2 = \frac{50}{3} \qquad ...(a)$$

In the domain $D_2 = \left(\frac{1}{2}, 1\right)$

$$\int_{1/2}^{1} (-2a_1 - 6a_2x + 100x^2)\,dx = 0$$

$$-2a_1\left(\frac{1}{2}\right) - 3a_2\left(1 - \frac{1}{4}\right) + \frac{100}{3}\left(1 - \frac{1}{8}\right) = 0$$

$$-a_1 - \frac{9}{4}a_2 + \frac{700}{24} = 0$$

$$4a_1 + 9a_2 = \frac{700}{6} \qquad ...(b)$$

Solving relations (*a*) and (*b*)

$$a_1 = -\frac{25}{3}$$

$$a_2 = 9.26$$

Hence the trial function is $y = \frac{25}{3}(x^2 - x) + 9.26\,(x - x^3)$.

Example 3.22. *Solve the example of one dimensional heat conduction by subdomain collocation.*

Solution. The governing equation for one dimensional conduction phenomenon is

$$\frac{d^2\phi}{dx^2} + Q = 0.$$

Given that

$$Q = 1 \text{ for } 0 \le x < L/2$$

$$Q = 0 \text{ for } L/2 < x \le L.$$

The boundary conditions assumed will be

$$\phi(0) = 0; \quad \phi(L) = 0.$$

Consider two term approximation of the Fourier series as the trial function.

$$\phi = -\,a_1 \sin\frac{\pi x}{L} + a_2 \sin\frac{2\pi x}{L}$$

Hence by differentiation

$$\frac{d^2\phi}{dx^2} = a_1\frac{\pi^2}{L^2}\sin\frac{\pi x}{L} - a_2\frac{4\pi^2}{L^2}\sin\frac{2\pi x}{L}$$

In the domain (0, $L/2$)

The Residual

$$R = -\,a_1\frac{\pi^2}{L}\sin\frac{\pi x}{L} - a_2\frac{4\pi^2}{L^2}\sin\frac{2\pi x}{L} + 1 = 0$$

$$\int_0^{L/2} R\,dx = +\,a_1\frac{\pi^2}{L^2}\cos\frac{\pi x}{L} + a_2\frac{4\pi^2}{L^2}\cos\frac{2\pi x}{L} + x\Big|_0^{L/2} = 0$$

$$a_1\frac{\pi^2}{L^2}(-1) + a_2\left(\frac{4\pi^2}{L^2}\right)(-1-1) + \frac{L}{2} = 0$$

$$a_1 + 8a_2 = \frac{L^3}{2\pi^2} \qquad \text{...}(a)$$

In the domain ($L/2$, L), $\quad Q = 0$

$$\int_{L/2}^{L} R\,dx = a_1\left(\frac{\pi^2}{L^2}\right)\cos\frac{\pi x}{L} + a_2\left(\frac{4\pi^2}{L^2}\right)\cos\frac{(2\pi x)}{L}\Big|_{L/2}^{L} = 0$$

$$a_1\left(\frac{\pi^2}{L^2}\right)(-1) + a_2\left(\frac{4\pi^2}{2}\right)(1+1) = 0$$

$$a_1 - 8a_2 = 0 \qquad \text{...}(b)$$

Solving the relations (*a*) and (*b*) simultaneously

$$a_1 = \frac{L^3}{4\pi^2}$$

$$a_2 = \frac{L^3}{32\pi^2}$$

Hence the trial function takes the form

$$\phi = \frac{L^3}{4\pi^2}\sin\frac{\pi x}{L} + \frac{L^3}{32\pi^2}\sin\frac{2\pi x}{L}$$

It may be observed that in the domain $D_1 \equiv (0, L/2)$

$$\frac{d^2\phi}{dx^2} = -1$$

and in the domain $D_2 \equiv (L/2, L)$

$$\frac{d^2\phi}{dx^2} = 0.$$

3.5.3 Least Square Method

This method of weighted residuals is consolidated in the following steps:

- Given the differential equation of equilibrium for the phenomenon, the residual is calculated assuming the trial function.
- The square of the residual function is integrated in the given domain D to obtain the integral equation.

$$I = \int_D R(x; a_i)\,dx \qquad ...(3.40)$$

- This leads to a function in a_i as

$$I = f(a_1, a_2, ..., a_n) \qquad ...(3.41)$$

- For minimization of the function f

$$\frac{\partial f}{\partial a_i} = 0 \qquad ...(3.42)$$

- The simultaneous equations (3.42) are solved to obtain a_i which are substituted to get the trial function as an approximate solution to differential equation of equilibrium describing the physical phenomenon.

Example 3.23. *Solve the one dimensional heat conduction equation by least square method.*

Solution. The given differential equation is

$$\frac{d^2\phi}{dx^2} + Q = 0$$

The trial function with on term Fourier series is

$$\phi = a_1 \sin\frac{\pi x}{L} \qquad ...(a)$$

Residual is

$$R = -a_1\frac{\pi^2}{L}\sin\frac{\pi x}{L} + Q$$

$Q(x) = 1$ in $0 \le x < L/2$

$Q(x) = 0$ in $L/2 < x \le L$

Hence it is assumed that $Q = \frac{1}{2}$ in $(0 < x < L)$

$$\frac{\partial f}{\partial a_1} = \int_0^L R\frac{\partial R}{\partial a_1} = 0$$

$$= \int_0^L \left(-a_1\frac{\pi^2}{L}\sin\frac{\pi x}{L} + \frac{L}{2}\right)\left(-\frac{\pi^2}{L}\sin\frac{\pi x}{L}\right)dx = 0$$

$$= \int_0^L a_1\frac{\pi^4}{L^4}\left(1 - \cos\frac{2\pi x}{L}\right)dx - \int_0^L \frac{\pi^2}{2L^2}\sin\frac{\pi x}{L}\,dx = 0$$

or $$a_1\frac{\pi^4}{L^4}(L - 0) = \frac{-\pi^2}{L^2}(-1-1)$$

Hence $$a_1 = \frac{L}{\pi^2}$$

The solution to trial function is

$$\phi = \frac{L}{\pi^2}\sin\frac{\pi x}{L}$$

3.5.4 Galerkin's Method

This method is presented in the following steps:

Step 1: Obtaining differential equation of equilibrium describing the phenomenon.

Step 2: Assuming the general trial function with unknown a_i as the co-efficients of the terms.

Step 3: Substituting derivatives of the trial function in the governing differential equation to obtain the residual function.

Step 4: Making trial functions themselves as the weighting function.

i.e., $$W_i = Y = N_i$$

Step 5: Obtaining the integral of the product of weighting function (trial function) and the residual function (R) and equating it to zero, to obtain the unknowns a_i

$$\int_D N_i R(x_i\,; a_1, a_2, \ldots, a_n)\,dx = 0 \qquad \ldots(3.43)$$

Step 6: Back substitution of a_i s in the trial function to get the solution for the differential equation.

Example 3.24. *A physical phenomenon is governed by the differential equation*

$$\frac{d^2y}{dx^2} + 90x^2 = 0 \quad \textit{in the domain } (0 < x < 1)$$

Assuming a trial function $y = a_1(x - x^3) + a_2(x - x^4)$

Solve the different equation using Galerkin's method, given the boundary conditions $y(0) = 0$ $y(1) = 0$.

Solution. The given differential equation is

$$\frac{d^2y}{dx^2} + 90x^2 = 0$$

The trial function given is $y = a_1 (x - x^3) + a_2 (x - x^4)$

The derivative

$$\frac{d^2y}{dx^2} = -6a_1x - 12a_2x^2$$

Residual function is

$$R = -6a_1x - 12a_2x^2 + 90x^2$$

The weighting function

$$w_i = a_1(x - x^3) + a_2 (x - x^4)$$

The integral equation, $$I = \int_0^1 w_i R(x_i ; a_i) = 0$$

$$w_1 = a_1(x - x^3)$$

i.e., $$I_1 = \int_0^1 (-6a_1x - 12a_2x^2 + 90x^2)(a_1x - a_1x^3)\,dx = 0$$

or $$\int_0^1 (-6a_1x^2 - 12a_2x^3 + 6a_1x^4 + 12a_2x^5)dx - \int 90x^2(x - x^3)\,dx$$

i.e., $$-2a_1 - 3a_2 + \frac{6}{5}a_1 + 2a_2 = -\left(\frac{90}{4} - 6\right) \quad ...(a)$$

$$9a_1 + 5a_2 = -\frac{33}{2}$$

Now taking $w_2 = a_2(x - x^4)$

$$I_2 = \int_0^1 (x - x^4)(-6a_1x - 12a_2x^2 + 90x^2)\,dx = 0$$

or $$\int_0^1 (-6a_1x^2 - 12a_2x^3 + 6a_1x^5 + 12a_2x^6)dx = -\int_0^1 90x^2 (x - x^4)\,dx$$

$$-3a_1 - 3a_2 + a_1 + \frac{12}{7}a_2 = 0 - \left(\frac{90}{4} - \frac{90}{7}\right)$$

$$14a_1 + 9a_2 = 0 - \frac{290}{28} \quad ...(b)$$

Solving the relations (a) and (b) simultaneously

$$a_1 = -0.558$$

$$a_2 = -0.282$$

Hence the trial function on substitution of a_1 and a_2 gives

$$N_i = -0.558 (x - x^3) - 0.282 (x - x^4)$$

Example 3.25. *Solve the one dimensional heat conduction equation by Galerkin's method of weighted residuals.*

Solution. The differential equation of 1D heat conduction is

$$\frac{d^2\phi}{dx^2} + Q = 0$$

where $Q(x) = 1$ in $(0 \le x < L/2)$

$Q(x) = 1$ in $(L/2 < x \le L)$

The trial function is $$y = a_1 \sin\frac{\pi x}{L}$$

$$\frac{d^2\phi}{dx^2} = -a_1\frac{\pi^2}{L^2}\sin\frac{\pi x}{L}$$

Assuming $Q = \frac{1}{2}$

$$R = -a_1\frac{\pi^2}{L^2}\sin\frac{\pi x}{L} + \frac{1}{2}$$

$$I = \int_0^L N_i R(x, a_i)\,dx = 0$$

$$= \int_0^L a_1 \sin\frac{\pi x}{L}\left(\frac{1}{2} - a_1\frac{\pi^2}{L^2}\sin\frac{\pi x}{L}\right)dx = 0$$

$$= \int_0^L \left(\frac{1}{2}\sin\frac{\pi x}{L} - a_1\frac{\pi^2}{L^2}\sin^2\frac{\pi x}{L}\right)dx = 0$$

$$= \int_0^L \left(\frac{1}{2}\sin\frac{\pi x}{L} - \frac{a_1\pi^2}{L^2} + \frac{a_1\pi^2}{L^2}\cos\frac{2\pi x}{L}\right)dx = 0$$

$$= -\frac{1}{2}(-1-1) - \frac{a_1\pi^2}{L^2} + \frac{a_1\pi^2}{L^2}(0) = 0$$

Hence, $$a_1 = +\frac{L^2}{\pi^2}$$

Hence the solution is $y = \frac{L^2}{\pi^2}\sin\frac{\pi x}{L}$.

Galerkin's Method Applied to Elasticity Problems

Some of the examples of simple elasticity problems are an axial member subjected to tensile tranction, a cantilever beam with point or uniformly distributed load, a simply supported beam with point or uniformly distributed load, and beam-column with uniformly distributed and an axial compressive loads. In all these cases the Galerkin's procedure is out lined in following steps.

- Establishing total potential energy function (functional) for the problem.

- Deriving the governing differential equation by L-E formulation.
- Assuming a suitable trial function for the displacement field.
- Obtaining the residual function by substituting the derivatives of trial function in the governing differential equation of equilibrium for elasticity problem.
- Obtaining and solving the integral equation as explained in section (3.5.4) by Galerkin's Method.

Example 3.26. *An axial bar of area of cross-section A and material of Young's Modulus E, fixed at one end is subjected to tensile traction force q. Determine the expression for displacement and stress distribution in the bar by Galerkin's method of weighted residuals.*

Solution. The diagrametic representation of the problem is as shown in Fig. 3.26.

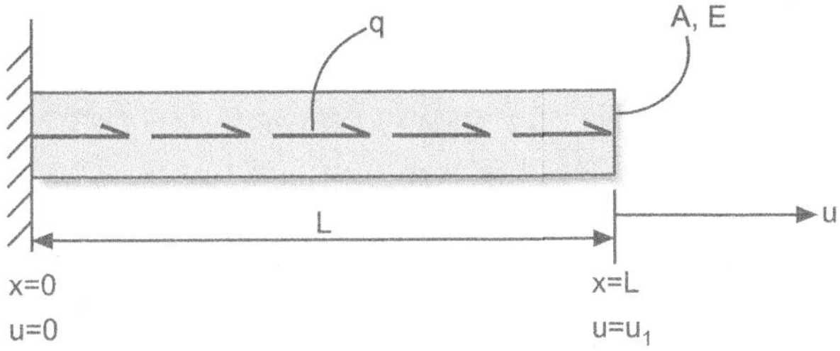

Fig. 3.26. Axial bar.

The potential energy functional for the axial bar is given by

$$\pi = \int_0^L \left[\frac{AE}{2}\left(\frac{du}{dx}\right)^2 + qu\right]dx$$

where $$f = \frac{AE}{2}\left(\frac{du}{dx}\right)^2 + qu$$

The Lagrange-Euler formulation is given by

$$\frac{\partial f}{\partial y} - \frac{d}{dx}\left(\frac{\partial f}{\partial \dot{y}}\right) = 0$$

where $$\frac{\partial f}{\partial y} = +q \qquad \text{(Here } y \text{ is considered to be } u\text{)}$$

$$\frac{\partial f}{\partial \dot{y}} = -AE\frac{dy}{dx} \qquad \left(\text{Here } \dot{y} = \frac{du}{dx}\right)$$

Hence the governing differential equation is

$$-AE\frac{du}{dx} + q = 0$$

Assume a trial function

$$u = a_0 + a_1x + a_2x^2$$

Boundary conditions are

$$u(0) = 0; \qquad \left.\frac{du}{dx}\right|_{x=0} = 0$$

This implies that $a_0 = 0$ and $a_1 = 0$

and $$u = a_2x^2, \text{ also } \frac{du}{dx} = 2a_2x$$

The residual function is $R = -AE(2a_2x) + q$

The integral from Galerkin's method

$$I = \int_0^L w_i R(x, a_i)\, dx = 0$$

$$= \int_0^L a_2x^2(-2AEa_2x + q)\, dx = 0$$

$$-\frac{2}{4}AE\, a_2^2 L^4 + a_2 q \frac{L^3}{3} = 0$$

Hence $$a_2 = +\frac{2}{3}\frac{q}{AE}$$

The displacement function, $$u = +\frac{2}{3}\frac{q}{AE}\cdot x^2$$

The stress distribution is given by

$$\sigma_x = E\epsilon = E\frac{du}{dx} = E(2a_2x)$$

$$= \frac{4}{3}\frac{q}{A}x$$

The displacement plot and the stress plot are as given in Fig. 3.27.

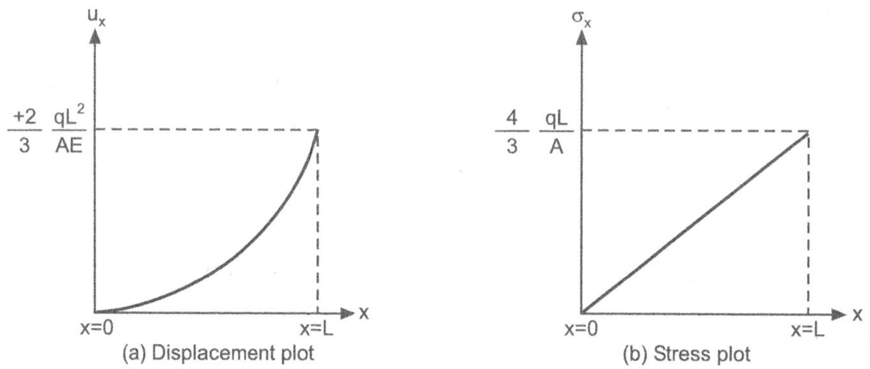

Fig. 3.27. Displacement and stress plot.

Example 3.27. *Consider a simply supported beam with uniformly distributed load. Find the expressions for displacement and stress distribution by Galerkin's Method.*

Solution. The simply supported beam with section property I and material property E is as shown in Fig. 3.28.

The total potential energy functional is given by

$$\pi = \frac{EI}{2}\int_0^L \left(\frac{d^2y}{dx^2}\right)^2 - \int_0^L qy\, dy$$

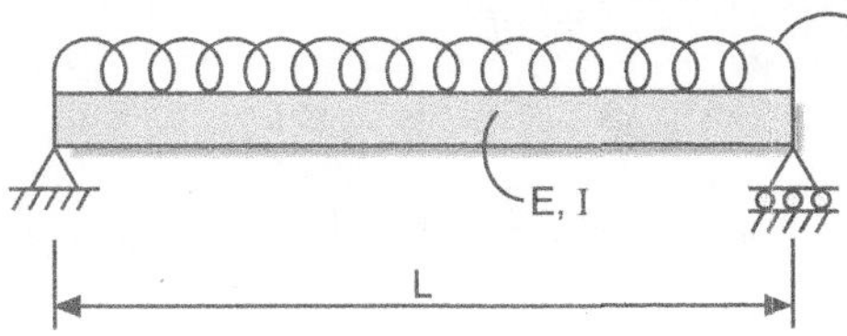

Fig. 3.28. Simply supported beam with UDL.

$$f = \frac{EI}{2}\left(\frac{d^2y}{dx}\right)^2 - qy \quad \text{...(a)}$$

By *L-E* Formulation

$$\frac{\partial f}{\partial y} - \frac{d}{dx}\left(\frac{\partial f}{\partial \dot{y}}\right) + \frac{d^2}{dx^2}\left(\frac{\partial f}{\partial \ddot{y}}\right) = 0 \quad \text{...(b)}$$

In this $\dot{y} = \frac{dy}{dx}$ and $\ddot{y} = \frac{d^2y}{dx^2}$

Assume the displacement function

$$y = a_0 + a_1x + a_2x^2$$

The boundary conditions are $y(0) = 0$, $y(L) = 0$

which implies that $a_0 = 0 \quad a_1 = -a_2L$

Hence trial function has the form

$$y = a_2\,(x^2 - Lx)$$

and

$$\frac{d^2y}{dx^2} = 2a_2$$

Substituting in relation (*b*), the differential function is

$$\frac{EI}{2}\left(2\frac{d^4y}{dx^4}\right) - q = 0$$

$$EI\,\frac{d^4y}{dx^4} - q = 0 \quad \text{...(c)}$$

The residual function from relation (*c*)

$$R = 2EIa_2 - q\frac{x^2}{2}$$

The integral function,

$$I = \int_0^L a_2(x^2 - Lx)\left(2EIa_2 - q\frac{x^2}{2}\right)dx = 0$$

$$\int_0^L \left(2EIa_2x^2 - 2EILa_2x - q\frac{x^4}{2} + q\,\frac{Lx^3}{2}\right)dx = 0$$

$$\frac{2}{3}\,EIa_2L^3 - EIa_2L^3 - \frac{q}{3}\,\frac{L^5}{(5)} + \frac{q}{2}\,\frac{L^5}{(4)} = 0$$

Therefore, $$a_2 = -\frac{7}{40}\frac{qL^2}{EI}$$

The displacement function is $$y = -\frac{7}{40}\frac{qL^2}{EI}(x^2 - Lx)$$

The stress function is $$\sigma_b = -yE\frac{d^2y}{dx^2}$$

$$= \frac{q}{I}y = +\frac{q}{I}\left[\frac{7qL^2}{40EI}(x^2 - Lx)\right]$$

The displacement and stress plot.

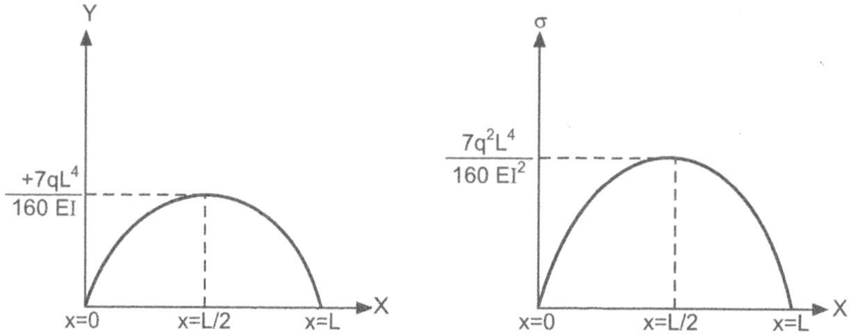

Fig. 3.29. Displacement and stress plot.

Example 3.28. *Derive the expression for stress for a beam fixed at both ends and loaded with UDL, using Galerkin's method.*

Solution. The fixed beam with UDL is as shown in Fig. 3.30

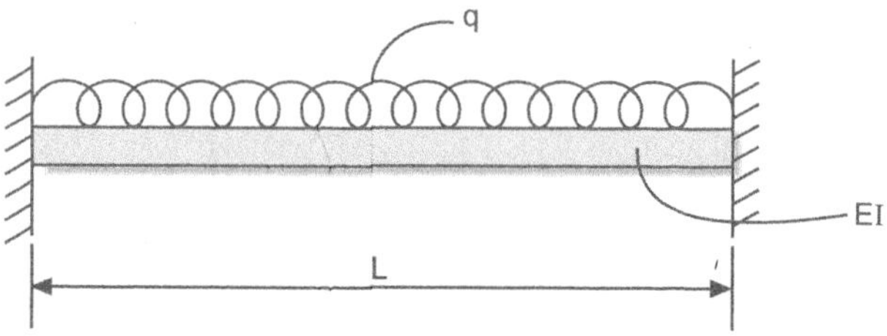

Fig. 3.30. Fixed beam with UDL.

The total potential energy function,

$$\pi = \frac{EI}{2}\int_0^L \left(\frac{d^2y}{dx^2}\right)^2 dx - \int_0^L qy\,dx$$

$$f = \frac{EI}{2}\left(\frac{d^2y}{dx^2}\right)^2 - qy$$

By *L–E* formulation

$$\frac{\partial f}{\partial y} - \frac{d}{dx}\left(\frac{\partial f}{\partial \dot{y}}\right) + \frac{d^2y}{dx^2}\left(\frac{\partial f}{\partial \ddot{y}}\right) = 0$$

$$-q + EI\frac{d^4y}{dx^4} = 0$$

The trial function assumed is

$$y = a_0 + a_1x + a_2x^2 + a_3x^3 + a_4x^4$$

from the boundary conditions

$$y(0) = 0 \qquad y(L) = 0$$

$$\left.\frac{dy}{dx}\right|_{x=0} = 0 \qquad \left.\frac{dy}{dx}\right|_{x=L} = 0$$

implies that

$$a_1 = 0 \qquad a_2 = 0$$

$$a_2L^2 + a_3L^3 + a_4L^4 = 0$$

$$2a_2L + 3a_3L^2 + 4a_4L^3 = 0$$

$$a_3L^3 + 2a_4L^4 = 0 \qquad \ldots(a)$$

$$a_2L^2 - a_4L^4 = 0 \qquad \ldots(b)$$

from (a) and (b)

$$a_3 = -2a_4L$$

$$a_2 = a_4L^2$$

Hence

$$y = a_4L^2x^2 - 2a_4\,Lx^3 + a_4\,x^4$$

$$\frac{d^4y}{dx^4} = 24a_4$$

Residual is

$$R = 24EIa_4 - q$$

Integral is

$$I = \int_0^L yR\,dx = 0$$

$$\int_0^L a_4\,(L^2x^2 - 2Lx^3 + x^4)\,(24EIa_4 - q)\,dx = 0$$

$$\int_0^L 24EIa_4(L^2x^2 - 2Lx^3 + x^4)dx = q\int_0^L (L^2x^2 - 2Lx^3 + x^4)\,dx$$

$$24EIa_4\left(\frac{L^5}{3} - \frac{1}{2}L^5 + \frac{L^5}{5}\right) = q\left(\frac{L^5}{3} - \frac{L^5}{2} + \frac{L^5}{5}\right)$$

Hence

$$a_4 = \frac{q}{24EI}$$

$$y = \frac{q}{24EI}(L^2x^2 - 2Lx^3 + x^4)$$

The bending stress is given by

$$\sigma_b = -yE\frac{d^2y}{dx^2}$$

$$= \frac{-q}{24I}(2L^2 - 12Lx + 12x^2)\,.\,y$$

3.6 PRINCIPLE OF VIRTUAL DISPLACEMENT

The objective of virtual displacement is to determine the nodal force acting on the discrete system by making equivalent statically to the actual boundary stresses and distributed loads. The procedure includes the imposition of virtual (arbitrary) nodal displacement and equating the work done to the work done (internal and external) by various forces and stresses during that displacement.

The principle of virtual displacement is outlined in the following steps:

Step 1: Imposition of virtual displacement. Let δa^e be the virtual displacement at the nodes. The corresponding displacement function and the strains are equal to

$$\delta u = N\delta a^e \qquad \text{...(3.44)}$$

where N is the function describing different shapes of displacement.

and

$$\delta \in = B\delta a^e \qquad \text{...(3.45)}$$

where $B = LN$ since $\in = Lu = LNu$, where L is suitable linear operator. For example, for an axial bar with tensile force the strain $\in = E\frac{du}{dx}$ for which the linear operator $L = \frac{d}{dx}$.

Step 2: Calculation of work done (external), due to virtual displacement.

$$\text{External work done} = \begin{Bmatrix}\text{Work done by individual} \\ \text{force components}\end{Bmatrix} \times \begin{Bmatrix}\text{Corresponding nodal} \\ \text{displacements}\end{Bmatrix}$$

In the matrix language

$$\delta W_e = q^e \delta a^e \qquad \text{...(3.46)}$$

Step 3: Calculation of internal work

The internal work per unit volume done by the stresses and distributed forces

$$\delta W_i = \sigma\delta\in - b\delta u \qquad \text{...(3.47)}$$

where σ is the stress and b is the body force or from relations (3.44) and (3.45)

$$\delta W_i = (\sigma B - bN)\delta a^e \qquad \text{...(3.48)}$$

Step 4: Equating external work with that of internal work done.

The total internal work done is obtained by intergrating equation (3.48) over the volume of the element and is then equated to external work done obtained in equation (3.46)

$$q^e\delta a^e = \left[\int_V \sigma\, BdV - \int_V bN\, dv\right]\delta a^e \qquad \text{...(3.49)}$$

The equation (3.49) holds good for any arbitray displacement δa^e

Hence
$$q^e = \int_V B^T \sigma\, dV - \int_V N^T b\, dV \qquad \text{...(3.50)}$$

The equation (3.50) can be compared with the stress-strain relation in finite element method, given by

$$q^e = K^e a^e + f^e \qquad ...(3.51)$$

Hence it is clear that

$$K^e = \int_V B^T \sigma \, dV \qquad ...(3.52)$$

But it is known for a axial bar $\sigma = E\in$ and $\in = \frac{du}{dx}$. Hence $\sigma = DB$ where D is the elasticity matrix.

Hence the elemental stiffness matrix in relation (3.52) takes the form

$$K^e = \int_V B^T DB \, dV \qquad ...(3.53)$$

And the nodal forces

$$F^e = -\int_V N^T b dV. \qquad ...(3.54)$$

3.7 PRINCIPLE OF VIRTUAL WORK

The principle of virtual work using equilibrium is presented based on Newtons Laws of motion with which the readers are comfortable to apply. The principle makes use of the fact that the internal virtual work is equivalent to the negative of the change in quantity called the strain energy. The location of application of the principle in the symbolic representation of the equations of solid mechanics is as shown in Fig. 3.31.

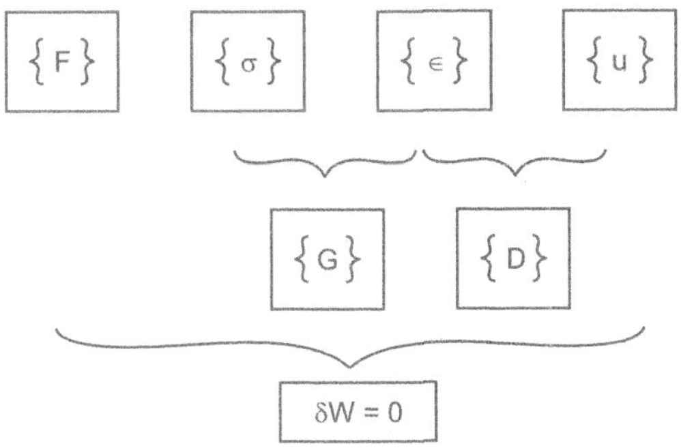

Fig. 3.31. Symbolic representation of virtual work.

Retaining the definition of the stresses, strains and deformations the equilibrium is stated in the form of integral equation, $\delta W = 0$. It may be noted that the applied loads and the geometric constraints are contained explicitly in the integral equation, unlike as boundary conditions in the differential equations of equilibrium.

The equations requiring solutions are

$$\{\sigma\} = [G]\{\in\} \quad \Rightarrow \text{stress-strain equation} \qquad ...(3.55)$$

$$\{\in\} = [D]\{u\} \quad \Rightarrow \text{Strain-displacement relation} \qquad ...(3.56)$$

The equations (3.55) and (3.56) are combined with applied loads, $\{F\}$ and geometric constraints to satisfy the equilibrium in the form

$$\delta W = 0 \qquad ...(3.57)$$

The more useful form of set of linear algebric equations in FEM is

$$[k]\{q\} = \{Q\} \quad ...(3.58)$$

The equation (3.58) is the combined form of equations

- Admissible function equation : $\{u\} = [\Omega]\{q\}$ ...(3.59)
- Generalized force equation : $\{Q\} = [A]\{F\}$ ...(3.60)

where $\{q\}$ = generalized displacements

$\{Q\}$ = generalized forces

$[\Omega]$ = admissible function.

To derive the expression that the virtual work is equal to change in strain energy, the virtual work of the surface forces

$$\delta W_s = \int (X_s \delta u + Y_s \delta V + Z_s \delta W)\, dA \quad ...(3.61)$$

where

$$[X_s, Y_s, Z_s] = [l, m, n]\begin{bmatrix} \sigma_x & \tau_{xy} & \tau_{xz} \\ \tau_{yx} & \sigma_y & \tau_{yz} \\ \tau_{zx} & \tau_{zy} & \sigma_z \end{bmatrix} \quad ...(3.62)$$

By the divergence theorem, the surface forces can be written in terms of surface stresses as

$$X_s = \frac{\partial}{\partial x}(X_s)\, dV$$

$$Y_s = \frac{\partial}{\partial y}(Y_s)\, dV$$

and

$$Z_s = \frac{\partial}{\partial z}(Z_s)\, dV$$

Hence

$$\delta W_s = \int \left[\frac{\partial}{\partial x}(\sigma_x \delta u + \tau_{xy}\delta v + \tau_{xz}\delta w) + \frac{\partial}{\partial y}(\tau_{yx}\delta u + \sigma_y \delta v + \tau_{yz}\delta w) + \frac{\partial}{\partial z}(\tau_{zx}\delta u + \tau_{yz}\delta v + \sigma_z \delta W) \right] dV \quad ...(3.63)$$

We are aware that $\delta \in_x = \frac{\partial \delta u}{\partial x}$ and $\delta \sigma_{xy} = \frac{\partial \delta u}{\partial y} + \frac{\partial \delta v}{\partial x}$ also regrouping the terms of the equation (3.63)

$$\delta W_s = -\int (f_z \delta u + f_y \delta v + f_z \delta w)\, dV + \int [(\sigma_x \delta \in_x + \sigma_y + \sigma_z \delta \in_z) + \tau_{xy}\delta v_{xy} + \tau_{xz}\delta v_{xz} + \tau_{yz}\delta v_{yz}]\, dV \quad ...(3.64)$$

The first integral is the virtual work due to body forces and the second integral in equation (3.64) is virtual work of internal forces, which is defined as the change in strain energy, δu.

Hence

$$\delta W_s = -\delta W_b + \delta U$$

$$\delta W_s + \delta W_b = \delta U \quad ...(3.65)$$

or

$$\delta W_e = \delta U$$

From equation (3.65) it is evident that the external virtual work is equal to change in strain energy.

Example 3.29. *Use principle of virtual work to find deflection, strains and stresses in an axial bar fixed at one end and a concentrated tensile load at the other end.*

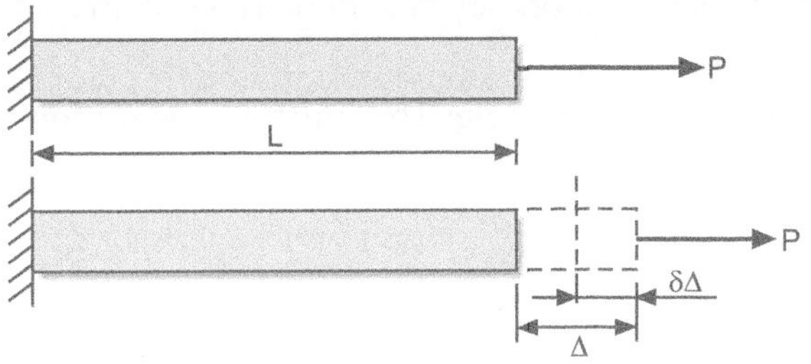

Fig. 3.32. Axially loaded bar.

Solution. A bar of length L, and area of cross-section, A is loaded by tensile concentrated load P, the total deflection due to which is Δ. Fig. 3.32 shows the arangement. The external virtual work is caused by P is

$$\delta W_e = P \,.\, \delta\Delta \qquad ...(a)$$

The strain and the virtual strain can be

$$\in_a = \frac{\Delta}{L} \qquad \delta\in_a = \frac{\delta\Delta}{L} \qquad ...(b)$$

The stress
$$\sigma_a = E \in_a = \frac{E\Delta}{L} \qquad ...(c)$$

Hence the virtual work becomes

$$\delta W = \delta W_e - \delta v$$

$$= P\delta\Delta - \int_0^L \delta \in_a . \sigma_x dV$$

$$= P\delta\Delta - \int_0^L \frac{\delta\Delta}{L} . \left(\frac{E\Delta}{L}\right) . A \,.\, dx$$

But from principle of virtual work $\delta W = 0$.

$$\delta\Delta\left(P - \frac{EA\Delta}{L}\right) = 0$$

Hence
$$P = \frac{EA\Delta}{L}$$

$$\Delta = \frac{PL}{AE} = \text{displacement}$$

$$\in_a = \frac{\Delta}{L} = \frac{P}{AE} = \text{strain}$$

$$\sigma_a = E . \in_a = \frac{P}{A} = \text{stress.}$$

POINTS TO BE REMEMBER

- **Functional:** The total potential energy function which is the difference in elastic-strain energy and work potential.
- **Minimum Potential Energy Principle:** The functional is stationary at the solution point of equilibrium of an elastic system.

$$\delta\pi = 0$$

- **Stability:** Conditions of equilibrium

$$\delta^2\pi > 0 \text{ stable equilibrium}$$
$$\delta^2\pi = 0 \text{ Netural equilibrium}$$
$$\delta^2\pi < 0 \text{ Unstable equilibrium}$$

- **Admissible Function:** The trial function depending on the space variable and generalized co-ordinates. This is the expression for the displacement field.
- **Rayleigh-Ritz Method:** An approximate method to obtain the displacement function as a solution to equation of equilibrium.
- More the number of Ritz constants in the trial function, better is the accuracy of solution.
- **Lagrange-Euler Formulation:** Gives the differential equations of equilibrium and associated boundary conditions.
- **Boundary Value Problem:** The problems depending on space dependent boundary condition.
- **Initial Value Problem:** The problems with time dependent conditions. The dynamic problems are IVP.
- **Natural Boundary Condition:** The boundary conditions associated with L-E formulation.
- The satisfied natural boundary conditions are free boundary condition otherwise forced BCs.
- **Point Collocation:** Depend on equating residual to zero at a particular point in the range.
- **Subdomain Collocation:** Equating the residual to one in the part of the domain and to zero elsewhere in the domain.
- **Least Square Method:** Equating integral of square of the residual to zero.
- **Galerkin's Method:** In this weighting function is made equal to the trial function and integral is equated to zero.
- **Principle of Virtual Displacement:** Obtaining the nodal forces by equating the external work with internal work due to stresses.
- **Principle of Virtual Work:** Establishing relation between stress, strain, displacement by equating virtual work with that of change in strain energy.
- **Galerkin's Method** can be successfully applied to problems in elasticity and fluid flow/heat transfer problems as well, in FEM.
- **Variational Methods:** Can be successfully applied to continuous and discrete systems.

EXERCISE

1. Derive the equation for deflection of a cantilever beam loaded with uniformly distributed load using variational principle.
2. Explain any one weighted residual method for solving solid mechanics problem.
3. Explain the theorem of minimum potential energy and obtain the expression for displacement of system of two springs in parallel with usual notation.
4. Obtain an expression for displacement and stress for the rod given in Fig. 1 by Rayleigh-Ritz method. Neglect body forces.

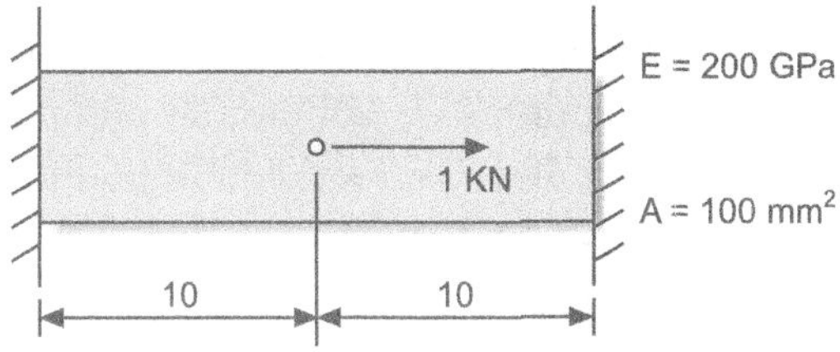

Fig. 1

5. Find the function $y = f(x)$ which extremizes the functional $\int_{x_1}^{x_2} (3y' + 4x)dx$ given $(x_1, y_1) =$ (0, 0) and $(x_2, y_2) = (2, 3)$
6. Using the principle of virtual work and virtual displacement obtain the equilibrium equation and the boundary conditions on the forces of the cantilever of length L, with uniformly distributed load 90% per unit length and a concentrated load P at the free end.
7. State the principle of virtual displacement. Prove that it is an equilibruim requirement and is independent of material behaviour.
8. Obtain the finite formulation of one dimensional problem using virtual displacement method.
9. A cantilever beam of span L is subjected to a point load P at free end. Derive an equation for the deflection at free end by using Rayleigh-Ritz method.
10. Derive and expression for potential energy functional and find the extreme values for the system shown in Fig. 2. Take $E = 200 \times 10^3$ N/mm^2, $P = 3$ kN, $L = 100$ mm, width W = 20 mm and thickness 10 mm.

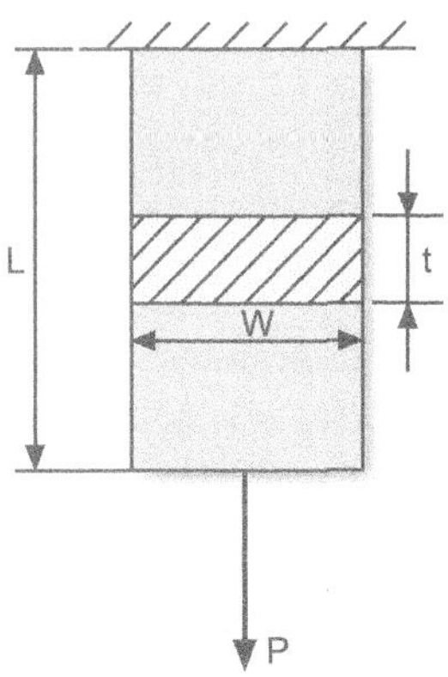

Fig. 2

11. With an example explain Rayleigh-Ritz method.
12. Find the approximate deflection of a simply supported beam with uniformly distributed load of a units/length using Galerkin's method.
13. Define functional. Explain principle of minimum potential energy. Give example.
14. Derive Euler-Lagrange's equation.
15. Discuss FE procedure using Galerkin's method for ideal fluid flow in 1-*D*.
16. Using variational principle, obtain Euler-Lagrange equation and boundary conditions for a cantilever beam by length *L*, modulus of elasticity *E*, area of cross-section A, which is subjected to a uniformly distributed load P_0.
17. A bar of uniform cross-sectional area is subjected to a uniformly distributed load (force/unit length) and a body force (force/unit volume) as shown in Fig. 3. Determine the displacement at mid point of the rod using Rayleigh-Ritz method.

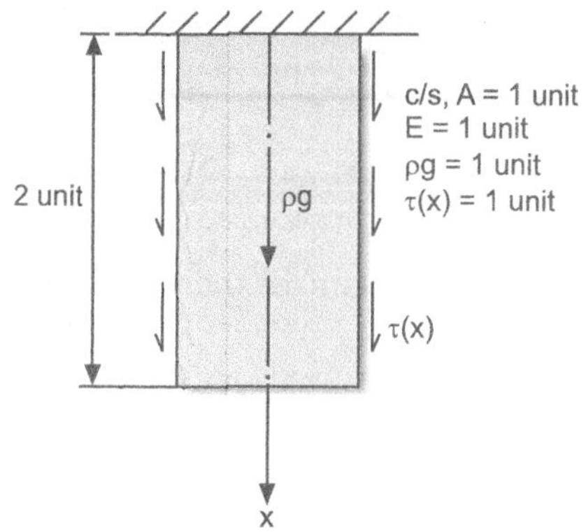

Fig. 3

18. What are the objectives of solution methods for a continuous system?
19. Differentiate with example between boundary value problem and initial value problem.
20. Outline the procedure of variational approach to solve the continuous elasticity problem.
21. Discuss the advantages and disadvantages of variational approach in FEM.
22. What are the characteristics features observed in the variational approach?
23. Distinguish between function and the functional. Give examples.
24. Ennumerate the condition for type of equilibrium in a stationary system.
25. Derive the functionals for a
    (*a*) Cantilever beam and a simply supported beam with a point load.
    (*b*) Simply supported beam and a cantilever beam with UDL.
    (*c*) Cantilever and simply supported beam column.
26. Using energy method, derive the Lagrange-Euler equation of differential equilibrium.
27. Explain free and forced boundary conditions of equilibrium.
28. Outline the Rayleigh-Ritz procedure. What are the conditions to be satisfied?
29. Where does the importance of weighted residual method signified? Give the general formulation of weighted residual methods.
30. List the several weighted residual methods and explain any one with an example.

31. Give the procedure of subdomain collocation. How is it different from point collocation?
32. Write the steps involved in the least square method and demonstrate it on one dimensional heat conduction problem.
33. Explain the steps involved in Galerkin's procedure of weighted residuals.
34. Write the Galerkin's procedure to solve the problems in elasticity.
35. Derive the relation for stiffness matrix by using the principle of virtual displacements.
36. Explain the principle of virtual work by symbolic representation of parameters of elasticity.
37. List the relations required to establish the equation governed by virtual work.
38. Show that virtual work is equal to change in strain energy.
39. Using the principle of virtual work, obtain the expression for deflection of a cantilever beam with a point load at free end.
40. An axial bar of length L and area of cross-section A is connected to a axial spring of stiffness K, as shown in Fig. 4.

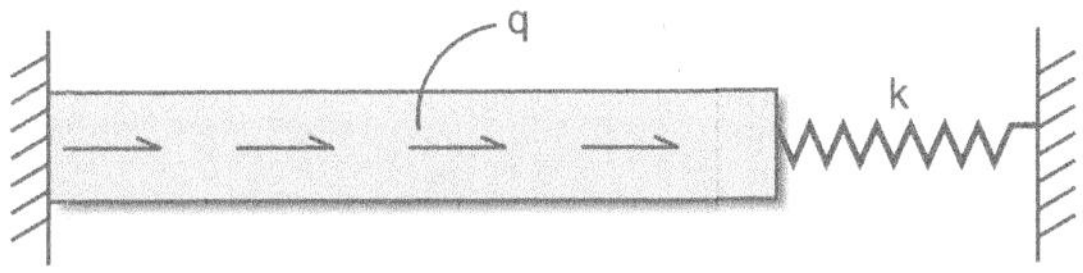

Fig. 4

If the traction force, q acts axially, obtain taking Young's modules, E,

(*a*) The functional.

(*b*) The differential question of equilibruim using Lagrange-Euler Formulation.

(*c*) The displacement and stress function by Rayleigh-Ritz method.

(*d*) Displacement field by Galerkin's method.

(*e*) Compare the result obtained by linear and quadratic trial functions.

41. A cantilever beam column is as shown in Fig. 5. If $EI = 1$ unit, $L = 2$ unit, $q = 1$ unit, $P = 2$ units, obtain the expression for the deflection pattern by Galerkin's method using suitable trial function. Find the deflection at the free end of the beam axially and transversely.

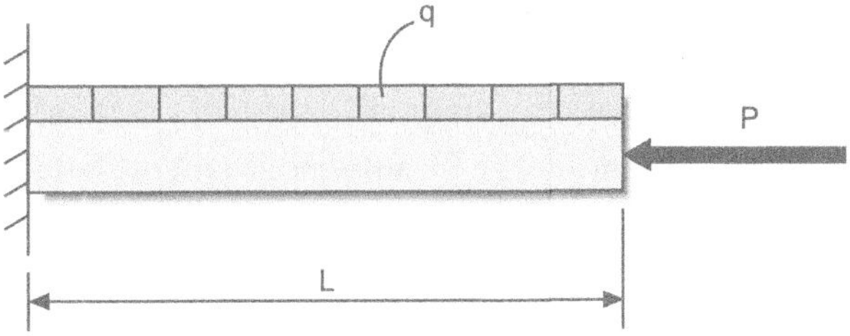

Fig. 5

Chapter 4

INTRODUCTION TO THEORY OF ELASTICITY

न चोरहार्यं न च राजहार्यं।
न भ्रातृ भाज्यं न च भारकारी॥
व्यये कृते वर्धते एव नित्यम्।
विद्या धनं सर्वधन प्रधानम्॥

[The knowledge of study cannot be stolen by others like thieves, king or brothers. It is not even heavy. If you spend it will prosper. It is the basis for all sorts of wealth and fortune].

The theory of elastic structures is the basis for the development of finite element method which is a computational mechanics procedure. This chapter gives an introduction to basic relations like strain-displacement relation, stress-strain equation and the equilibrium equation. The applied forces on a continuous system develops a resistive force internal to the system. For the system to be in equilibrium there has to be balance between these forces which leads to the establishment of equilibrium equations. The external forces applied also results in strains in the system which are related to the deflections of the system by the strain-displacement equation. The resistive force per unit area is the stress induced. By knowing the property of the material used for the system, the relations between the stress-strain are established. If the Newton's Law helps in arriving at the force equilibrium, the Hooke's law is required to derive the stress-strain relation. The above stated three relations provide the strong foundational basis for understanding the concepts of FEM. The components of stress, and the components of strain are given brief thought in this chapter. The plane stress and plane strain approximations based on stress-strain relations with the illustrative examples are also covered. In total this chapter spans the key issues in theory of elasticity useful for the basic knowledge for the concepts in finite element method (FEM).

4.1 INTRODUCTION

The mathematical analysis of solid bodies are understood by the fundamental concepts of stress and strain. The relationship between stress components, body forces and applied loads leads to the formation of equilibrium equations in elasticity. The applied loads with the geometric constraints, given suitable boundary conditions for the elastic structure are defined by the following set of equations.

- Equilibrium equations
- Stress-strain equations
- Strain-displacement equation.

To give a symbolic representation as in Fig. 4.1, for the equilibrium equation, the stress, the sectional property and the forces applied are related.

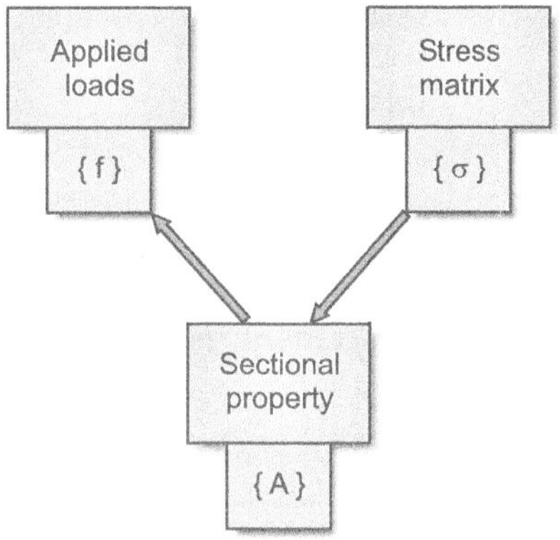

Fig. 4.1. Symbolic representation of equilibrium equation.

The matrix equation of the above representation leading to the equation of equilibrium, is

$$\{A\}\,[\sigma] + \{f\} = 0 \qquad \text{...(4.1)}$$

Hence the equilibrium equation is defined by the equation (4.1) states that there exists an equilibrium between the resistive force in the body due to stress distribution and the applied load, and body forces.

To give a symbolic representation of the stress-strain relation, the Fig. 4.2 is referred.

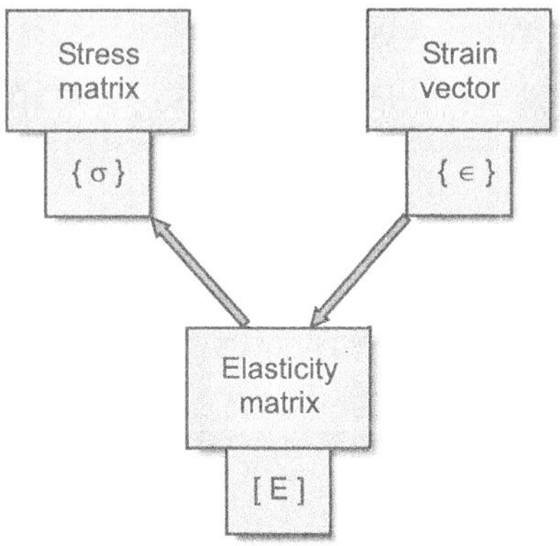

Fig. 4.2. Symbolic representation of stress-strain relationship.

For a linearly elastic solid the stress is related to strain by Hooke's Law, which states that stress is directly proportional to strain within the elastic limits of elongation. In the mathematical form.

$$\{\sigma\} = [E]\{\in\} \quad ...(4.2)$$

The strain and the deformations are related by the equation known as strain-displacements equation. Fig. 4.3 gives the symbolic representation of strain-displacement relation.

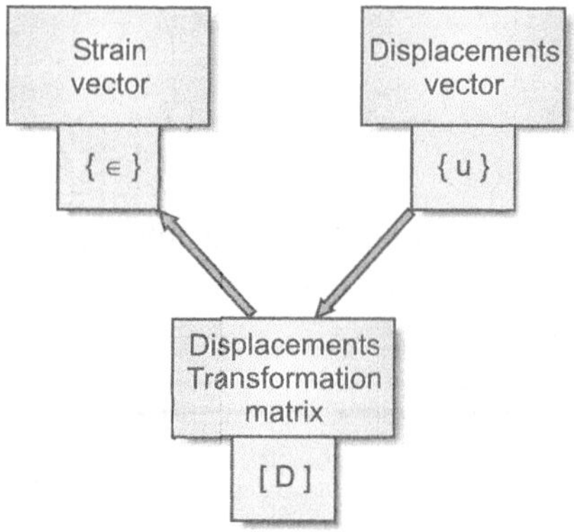

Fig. 4.3. Symbolic representation of strain-displacement equation.

The matrix equation for the strain-displacement relation is

$$\{\in\} = [D]\{u\} \quad ...(4.3)$$

The displacement transformation matrix $[D]$ is an operator matrix as is evident, for an axial bar, from the relation $\in = \frac{\partial u}{\partial x}$ where $D = \frac{\partial}{\partial x}$.

4.2 FORCE EQUILIBRIUM

According to Newton's Law of motion, for a system under equilibrium the sum of force components and moment components are zero.

i.e.,
$$\Sigma F_x = \Sigma F_y = \Sigma F_z = 0. \quad ...(4.4)$$

(Force components in x, y and z directions.)

and
$$\Sigma M_x = \Sigma M_y = \Sigma M_z = 0 \quad ...(4.5)$$

(Moment components in x, y and z-directions)

System considered can be a whole body or a part of it. Two types of forces can be identified to be internal resistive force and the external applied load on the system. The applied forces can be the surface force identified as force per unit area and the body force with force per unit volume.

4.3 STRESS COMPONENTS

The resistive force developed inside the body to provide an equilibrium to the external forces on the system, are identified by force per unit area, that is called the stress. The stress may be oriented in certain resultant direction on a surface cut internally in the body. The resolution of such stress, normal and parallel to the surfaces results in stress components. The normal components are termed as tensile or compressive stress and the parallel components of stress are termed the tangential or shear-stress components. If ΔF_n and ΔF_t are the normal and tangential force components respectively, acting on a elemental area ΔA then the stress components.

$$\sigma = \lim_{\Delta A \to 0} \frac{\Delta F_n}{\Delta A} \text{ and } \tau = \lim_{\Delta A \to 0} \frac{\Delta F_t}{\Delta A} \quad ...(4.6)$$

The complete state of stress, in a cartesian co-ordinate system, at a point can be depicted by representation on a cubical element. There are totally nine rectangular stress components that are categorized as three normal components in three cartesian directions and six shear-stress components on three planes. To identify these stress components in terms of direction and plane, double-subscript notations are used. The first subscript denotes the direction normal to the plane on which it acts and the second denotes the direction of the stress itself. The nine stress components are shown in the Fig. 4.4.

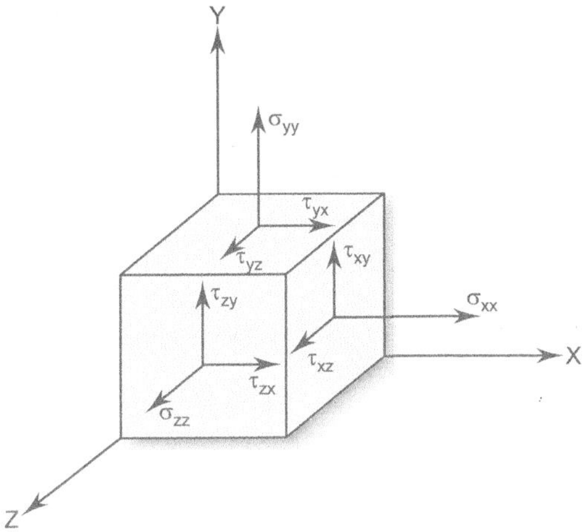

Fig. 4.4. Stress components.

The nine stress components can be represented in a square matrix called the stress tensor as

$$[\sigma_{ij}] = \begin{bmatrix} \sigma_{xx} & \tau_{xy} & \tau_{xz} \\ \tau_{yx} & \sigma_{yy} & \tau_{yz} \\ \tau_{zx} & \tau_{zy} & \sigma_{zz} \end{bmatrix} \quad ...(4.7)$$

4.4 STRAIN COMPONENTS

The displacement of a point in a solid body when resolved in the x, y and z directions in the cartesian co-ordinate system can be denoted by u, v and w respectively. If the dx, dy and dz are the original length of an differential element in the solid body then the rate of change of displacement with respect to respective original lengths the components of strain are obtained. The change in the linear dimension results in three strains $\in_x$, $\in_y$, and $\in_z$ which are called normal strains. The change in angular orientation of the point in the solid body results in three components γ_{xy}, γ_{yz} and γ_{zx} which are called the shear strains. Hence the components of strain can be written as,

normal strains,

$$\in_x = \frac{\partial u}{\partial x}, \in_y = \frac{\partial v}{\partial y}, \in_z = \frac{\partial w}{\partial z} \quad ...(4.8)$$

Shear strains

$$\gamma_{xy} = \frac{\partial u}{\partial y} + \frac{\partial v}{\partial x}$$

$$\gamma_{yz} = \frac{\partial v}{\partial z} + \frac{\partial w}{\partial y} \qquad ...(4.9)$$

and

$$\gamma_{zx} = \frac{\partial w}{\partial x} + \frac{\partial u}{\partial z}$$

The equation of compatibility are obtained by eliminating the displacement in the relations (4.9) and using equations (4.8)

$$\frac{\partial^2 \gamma_{xy}}{\partial x \partial y} = \frac{\partial^2}{\partial y^2}\left(\frac{\partial u}{\partial x}\right) + \frac{\partial^2}{\partial x^2}\left(\frac{\partial v}{\partial y}\right) = \frac{\partial^2 \in_x}{\partial y^2} + \frac{\partial^2 \in_y}{\partial x^2} \qquad ...(4.10)$$

$$\frac{\partial^2 \gamma_{yz}}{\partial y \partial z} = \frac{\partial^2}{\partial z^2}\left(\frac{\partial v}{\partial y}\right) + \frac{\partial^2}{\partial y^2}\left(\frac{\partial w}{\partial z}\right) = \frac{\partial^2 \in_y}{\partial z^2} + \frac{\partial^2 \in_z}{\partial y^2} \qquad ...(4.11)$$

$$\frac{\partial^2 \gamma_{zx}}{\partial z \partial x} = \frac{\partial^2}{\partial x^2}\left(\frac{\partial w}{\partial z}\right) + \frac{\partial^2}{\partial z^2}\left(\frac{\partial u}{\partial x}\right) = \frac{\partial^2 \in_z}{\partial x^2} + \frac{\partial^2 \in_x}{\partial z^2} \qquad ...(4.12)$$

The equations (4.10), (4.11) and (4.12) are called the compatibility equation of equilibrium.

4.5 STRESS-STRAIN RELATIONS

The equations derived in the previous sections are independent of the material properties of the solids of deformation. But Hooke related the stress and strain by the Hooke's law that states that the stress is directly proportional to strain within the elastic limit of elongation. If σ is the stress and $\in$ is the strain then by Hooke' Law

$$\sigma = E\in \qquad ...(4.13)$$

where E is the constant of proportionality which is called the Modulus of Elasticity or Young's Modulus, which is the technological property of the material. The equation (4.13) speaks about the uniaxial stress-strain relation. To relate the stress with the strains in other directions, y and z, the generalized Hooke's law is established as

$$\in_x = \frac{\sigma_x}{E},\ \in_y = \frac{\in_y}{\in_x}.\in_x = -\nu\frac{\sigma_x}{E} \text{ and } \in_z = \frac{\in_z}{\in_x}.\in_x = -\nu\frac{\sigma_x}{E} \qquad ...(4.14)$$

where ν is the Poisson's ratio defined by lateral strain to the axial strain. The negative sign is attributed to the fact that with the elongation in x direction, there is reduction in y and z directions. Now the triaxial stress-strain relations are given by

$$\in_x = \frac{1}{E}[\sigma_x - \nu(\sigma_y + \sigma_z)]$$

$$\in_y = \frac{1}{E}[\sigma_y - \nu(\sigma_z + \sigma_x)] \qquad ...(4.15)$$

$$\in_z = \frac{1}{E}[\sigma_z - \nu(\sigma_x + \sigma_y)]$$

By the existence of purely shearing stress, angular distortions are observed. Hence the shear strains are expressed symbolically as

$$\gamma_{xy} = \frac{\tau_{xy}}{G},\ \gamma_{yz} = \frac{\tau_{yz}}{G} \text{ and } \gamma_{zx} = \frac{\tau_{zx}}{G} \qquad ...(4.16)$$

where G is the shear modulus given by

$$G = \frac{E}{2(1+\nu)} \qquad ...(4.17)$$

The readers may refer books on strength of materials to know the derivation of equation (4.17).

From the equations (4.15), (4.16) and (4.17) the strain and stress are related by the matrix equation written as

$$\begin{bmatrix} \in_x \\ \in_y \\ \in_z \\ \gamma_{xy} \\ \gamma_{yz} \\ \gamma_{zx} \end{bmatrix} = \begin{bmatrix} \frac{1}{E} & -\frac{\nu}{E} & -\frac{\nu}{E} & 0 & 0 & 0 \\ -\frac{\nu}{E} & \frac{1}{E} & -\frac{\nu}{E} & 0 & 0 & 0 \\ -\frac{\nu}{E} & -\frac{\nu}{E} & \frac{1}{E} & 0 & 0 & 0 \\ 0 & 0 & 0 & \frac{1}{G} & 0 & 0 \\ 0 & 0 & 0 & 0 & \frac{1}{G} & 0 \\ 0 & 0 & 0 & 0 & 0 & \frac{1}{G} \end{bmatrix} \begin{bmatrix} \sigma_x \\ \sigma_y \\ \sigma_z \\ \tau_{xy} \\ \tau_{yz} \\ \tau_{zx} \end{bmatrix} \qquad ...(4.18)$$

or $$\{\in\} = [C]\{\sigma\} \qquad ...(4.19)$$

where $[C]$ is

The inverse relation relating stress with strain is given by

$$\{\sigma\} = [D]\{\in\} \qquad ...(4.20)$$

where $[D]$ is the elasticity matrix given by

$$[D] = \begin{bmatrix} \lambda+2G & \lambda & \lambda & 0 & 0 & 0 \\ \lambda & \lambda+2G & \lambda & 0 & 0 & 0 \\ \lambda & \lambda & \lambda+2G & 0 & 0 & 0 \\ 0 & 0 & 0 & G & 0 & 0 \\ 0 & 0 & 0 & 0 & G & 0 \\ 0 & 0 & 0 & 0 & 0 & G \end{bmatrix}$$

where $$\lambda = \frac{\nu E}{(1+\nu)(1-2\nu)} \qquad ...(4.21)$$

It may be noted that

$$[D] = [C^{-1}]$$

4.6 EQUILIBRIUM EQUATION

A differential element in three dimensions, dx, dy and dz, is considered that is dipited in Fig. 4.5. The stresses and the body forces act on the faces of the element as shown. The equations which establish equilibrium between the stresses and the forces is the equilibrium equation.

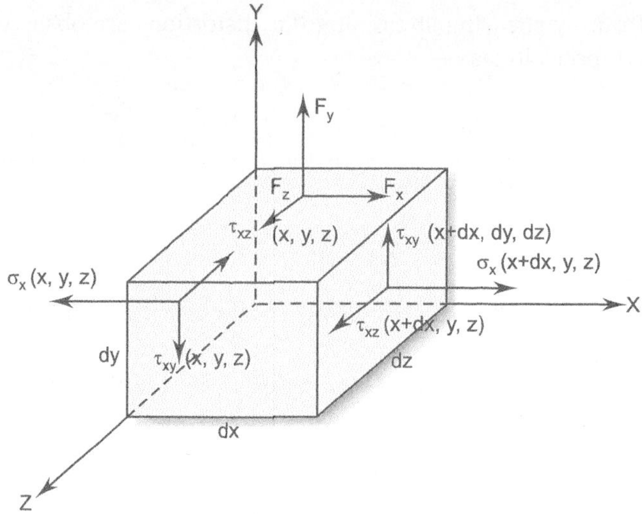

Fig. 4.5. Equilibrium state.

The Taylor's series expansion relating the stresses on one side of the element with that on the opposite face as

$$\begin{aligned}
\sigma_x\,(x + dx, y, z) &= \sigma_x\,(x, y, z) + \frac{\partial \sigma_x}{\partial x}\,.\,dx \\
\tau_{xy}\,(x + dx, y, z) &= \tau_{xy}\,(x, y, z) + \frac{\partial \tau_{xy}}{\partial x}\,.\,dx \\
\tau_{xz}\,(x + dx, y, z) &= \tau_{xz}\,(x, y, z) + \frac{\partial \tau_{xz}}{\partial x}\,.\,dx
\end{aligned} \qquad \text{...(4.22)}$$

The equation (4.22) neglects higher order terms in the Taylor series expansion. Similarly the equations can be written for the pair of surfaces in the y and z directions. From Newton's Law, for the element to be in equilibrium

$$\Sigma F_x = 0$$

Hence we arrive at

$$\left(\sigma_x + \frac{\partial \sigma_x}{\partial x} dx\right) dy\,dz - (\sigma_x\,.\,dx)\,dy\,dz + \left(\tau_{xy} + \frac{\partial \tau_{xy}}{\partial x}\right) dy\,dz - \tau_{xy}\,dx\,dy\,dz$$

$$+\left(\tau_{xz} + \frac{\partial \tau_{xz}}{\partial x}\right) dy\,dz - \tau_{xz}\,dx\,dy\,dz + f_x dx\,dy\,dz = 0 \qquad \text{...(4.23)}$$

Equation (4.23) simplifies to

$$\frac{\partial \sigma_x}{\partial x} + \frac{\partial \tau_{xy}}{\partial x} + \frac{\partial \tau_{yz}}{\partial x} + f_x = 0 \qquad \text{...(4.24)}$$

Similarly for the y and z directions

$$\frac{\partial \tau_{yx}}{\partial y} + \frac{\partial \sigma_y}{\partial y} + \frac{\partial \tau_{yz}}{\partial y} + f_y = 0 \qquad \text{...(4.25)}$$

and $$\frac{\partial \tau_{zx}}{\partial z} + \frac{\partial \sigma_z}{\partial z} + \frac{\partial \tau_{zy}}{\partial z} + f_z = 0 \qquad ...(4.26)$$

Similarly the moment $\Sigma M_x = 0$, from which we have

$$\tau_{xy} + \frac{\partial \tau_{xy}}{\partial x} \cdot \frac{dx}{2} - \tau_{yx} - \frac{\partial \tau_{yx}}{\partial y} \cdot \frac{dy}{2} = 0$$

In the limit $dx \rightarrow 0$ and $dy \rightarrow 0$, results in

$$\tau_{xy} = \tau_{yx}$$

Similarly $$\tau_{xz} = \tau_{zx} \text{ and } \tau_{yz} = \tau_{zx}. \qquad ...(4.27)$$

Hence the equation of equilibrium in the matrix form is

$$\begin{bmatrix} \frac{\partial}{\partial x} & 0 & 0 & \frac{\partial}{\partial y} & 0 & \frac{\partial}{\partial z} \\ 0 & \frac{\partial}{\partial y} & 0 & \frac{\partial}{\partial x} & \frac{\partial}{\partial z} & 0 \\ 0 & 0 & \frac{\partial}{\partial z} & 0 & \frac{\partial}{\partial y} & \frac{\partial}{\partial x} \end{bmatrix} \begin{bmatrix} \sigma_x \\ \sigma_y \\ \sigma_z \\ \tau_{xy} \\ \tau_{yz} \\ \tau_{zx} \end{bmatrix} = \begin{bmatrix} -f_x \\ -f_y \\ -f_z \end{bmatrix} \qquad ...(4.28)$$

The equilibrium equation in the compact form is

$$[E]\{\sigma\} = -\{f\}$$

4.7 PLANE STRESS AND PLANE STRAIN FORMULATION

The plane elasticity analysis is used more extensively in the finite element method. There are two types of plane elasticity problems. One is plane stress formulation and the other is plane strain formulation. Because of the assumptions made in these two cases to simplify the problem, the two formulations are considered approximate. The example for plane stress condition is thin flat elastic body. For the plain strain approximation a long cylindrical body is the good example.

4.7.1 Plane Stress Approximation

Consider a thin flat elastic plate as shown in Fig. 4.6 with z-axis normal to the plane of the body.

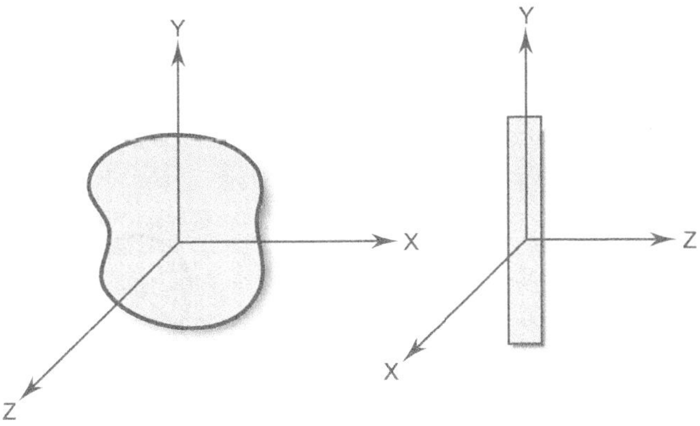

Fig. 4.6. Plane-stress example.

In the thin flat body the plane stress approximation is developed by assuming stresses in the direction normal to the mid-plane of the body to be equal to zero.

Hence

$$\sigma_z = \tau_{xz} = \tau_{yz} = 0 \qquad ...(4.29)$$

Then the stress-strain relations in (4.15) and (4.16) reduces to

$$\in_x = \frac{1}{E}(\sigma_x - \nu\sigma_y)$$

$$\in_y = \frac{1}{E}(\sigma_y - \nu\sigma_x) \qquad ...(4.30)$$

$$\in_z = -\frac{\nu}{E}(\sigma_x + \sigma_y)$$

and

$$\gamma_{xy} = \frac{\tau_{xy}}{G} \qquad ...(4.31)$$

By manipulation of equations (4.30) and (4.31), the relation between $\{\sigma\}$ and $\{\in\}$ may be expressed as

$$\{\sigma\} = [D]\{\in\} \qquad ...(4.32)$$

where

$$\{\sigma\} = \begin{bmatrix} \sigma_x \\ \sigma_y \\ \tau_{xy} \end{bmatrix} \text{ and } \{\in\} = \begin{bmatrix} \in_x \\ \in_y \\ \gamma_{xy} \end{bmatrix}$$

and

$$[D] = \begin{bmatrix} 1 & \nu & 0 \\ \nu & 1 & 0 \\ 0 & 0 & \frac{(1-\nu)}{2} \end{bmatrix}$$

The matrix $[D]$ is known as elasticity matrix, $\{\sigma\}$ is known as stress array and $\{\in\}$ is known as strain array.

4.7.2 Plane-Strain Approximation

This approximation is more useful with long cylindrical bodies as shown in Fig. 4.7 with z-axis along the axis of the cylinder. The assumption made in plane strain problems is to take the displacement in the direction of the cylinder axis to be zero and also the derivatives in the same direction to be zero.

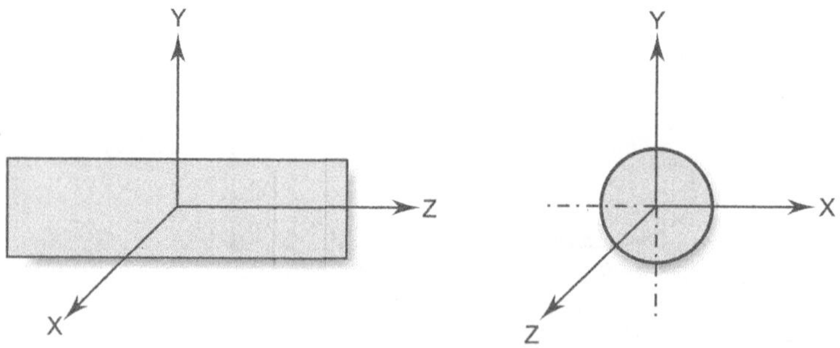

Fig. 4.7. Plane-strain problem.

The assumptions lead to the expressions

$$w = 0 \quad \text{and} \quad \frac{\partial(\omega)}{\partial z} = 0 \qquad \text{...(4.33)}$$

The expression (4.33) intend to mean

$$\epsilon_z = \gamma_{zx} = \gamma_{zy} = 0 \qquad \text{...(4.34)}$$

Hence, for the plane strain case, the stress-strain equations (4.15) and (4.16) reduce to

$$\epsilon_x = \frac{(1+\nu)}{E}[(1-\nu)\sigma_x - \nu\sigma_y] \qquad \text{...(4.35)}$$

$$\epsilon_y = \frac{(1+\nu)}{E}[(1-\nu)\sigma_y - \nu\sigma_x]$$

and

$$\gamma_{xy} = \frac{\tau_{xy}}{G} \qquad \text{...(4.36)}$$

In the plane-strain problems the stress array $\{\sigma\}$ and the strain array $\{\epsilon\}$ are related by the equation

$$\{\sigma\} = \{D_1\}\{\epsilon\}$$

and

$$\{\epsilon\} = [D_1^{-1}]\{\sigma\} \qquad \text{...(4.37)}$$

The elasticity matrix $[D_1]$ is given by

$$[D] = \frac{E}{(1+\nu)(1-2\nu)}\begin{bmatrix}(1-\nu) & \nu & 0 \\ \nu & (1-\nu) & 0 \\ 0 & 0 & \frac{(1-2\nu)}{2}\end{bmatrix}$$

Example 4.1. *The displacements and the constraints for a body shown in Fig. 4.8 are*

displacements – $u = \frac{1}{E}\left(x - \frac{a}{2}\right)y,\ \nu = \frac{1}{2E}(ax - x^2 - \nu y^2)$

$$w = -\frac{\nu yz}{E}$$

constraints – $u(0,0,0) = V(0, 0, 0) = v(a, 0, 0) = w\ (x, y\ 0) = 0.$

Find the stresses, strains and the applied load.

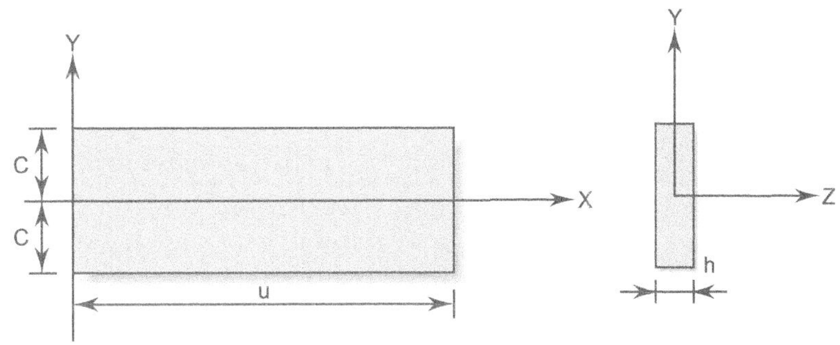

Fig. 4.8. Problem figure.

Solution. Using the strain-displacement relations

$$\epsilon_x = \frac{\partial u}{\partial x} = \frac{y}{E},\ \epsilon_y = \frac{\partial v}{\partial y} = -\frac{\nu y}{E} \text{ and } \epsilon_z = -\frac{\nu y}{E}$$

$$\gamma_{xy} = \frac{\partial u}{\partial y} + \frac{\partial v}{\partial x} = \frac{x}{E} - \frac{a}{2E} + \frac{a}{2E} - \frac{x}{E} = 0$$

$$\gamma_{yz} = \frac{\partial w}{\partial y} + \frac{\partial v}{\partial z} = -\frac{\nu z}{E}$$

$$\gamma_{zx} = \frac{\partial u}{\partial z} + \frac{\partial w}{\partial x} = 0$$

From the stress-strain relation of generalized Hooke's Law

$$\begin{bmatrix} \lambda+2G & \lambda & \lambda & 0 & 0 & 0 \\ \lambda & \lambda+2G & \lambda & 0 & 0 & 0 \\ \lambda & \lambda & \lambda+2G & 0 & 0 & 0 \\ 0 & 0 & 0 & G & 0 & 0 \\ 0 & 0 & 0 & 0 & G & 0 \\ 0 & 0 & 0 & 0 & 0 & G \end{bmatrix} \begin{bmatrix} \frac{\nu}{E} \\ -\frac{\nu y}{E} \\ -\frac{\nu y}{E} \\ 0 \\ -\frac{\nu z}{E} \\ 0 \end{bmatrix} = \begin{bmatrix} \sigma_x \\ \sigma_y \\ \sigma_z \\ \tau_{xy} \\ \tau_{yz} \\ \tau_{zx} \end{bmatrix}$$

This results in

$$\sigma_x = y, \quad \sigma_y = 0, \quad \sigma_z = 0$$

$$\tau_{xy} = 0, \quad \tau_{yz} = -\frac{vz}{2(1+v)}, \quad \tau_{zx} = 0$$

Now from equilibrium equations

$$\begin{bmatrix} \frac{\partial}{\partial x} & 0 & 0 & \frac{\partial}{\partial y} & 0 & \frac{\partial}{\partial z} \\ 0 & \frac{\partial}{\partial y} & 0 & \frac{\partial}{\partial x} & \frac{\partial}{\partial z} & 0 \\ 0 & 0 & \frac{\partial}{\partial z} & 0 & \frac{\partial}{\partial y} & \frac{\partial}{\partial x} \end{bmatrix} \begin{bmatrix} y \\ 0 \\ 0 \\ 0 \\ -\frac{\nu z}{2(1+\nu)} \\ 0 \end{bmatrix} = \begin{bmatrix} -f_x \\ -f_y \\ -f_z \end{bmatrix}$$

Hence $f_x = f_z = 0,\ f_y = \frac{\nu}{2(1+\nu)}$

For left and right ends, $x = 0$, $x = a$.

$$\sigma_x(0, y, z) = \sigma_x(a, y, z) = y$$

$$\tau_{xy} = \tau_{zx} = 0$$

On the top and bottom faces with $y = \pm c$

$$\sigma_y = 0 \quad \tau_{xy} = 0$$

$$\tau_{yz} = -\frac{\nu z}{2(1+\nu)}$$

On front and back faces with $z = \pm t/2$

$$\sigma_z = 0, \quad \tau_{zx} = 0, \quad \tau_{yz}(x, y, t/2) = \frac{-\nu t}{4(1+\nu)}, \quad \tau_{yz}(x, y, -t/2) = \frac{\nu t}{4(1+\nu)}$$

Example 4.2. *A thin rectangular plate of uniform thickness, t and length l, width 2a, has a uniform distributed load applied to two opposite edges as shown in Fig. 4.9. The boundary normal stresses have a value, S where S is a constant. Assuming no rigid body motion. Find the stresses, strains and deflection according to plane stress assumptions if no body forces are present.*

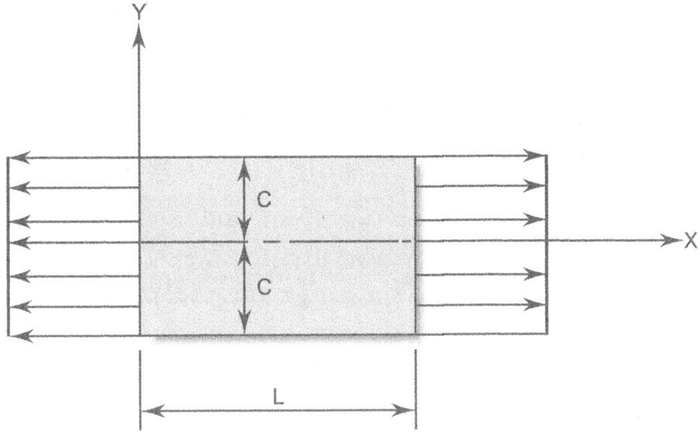

Fig. 4.9. Problem figure.

Solution. From the given conditions and plane stress approximations.

$$\sigma_x(x, y) = S \qquad \sigma_y(x, y) = 0 \qquad \tau_{xy}(x, y) = 0$$

From the force equilibrium, the equilibrium is satisfied, as, $[A]\{\sigma\} = -\{f\}$

$$\begin{bmatrix} \frac{\partial}{\partial x} & 0 & \frac{\partial}{\partial y} \\ 0 & \frac{\partial}{\partial y} & \frac{\partial}{\partial x} \end{bmatrix} \begin{bmatrix} S \\ 0 \\ 0 \end{bmatrix} = \begin{bmatrix} 0 \\ 0 \end{bmatrix}$$

The Hooke's Law for plane stress condition is

$$\begin{bmatrix} \in_x \\ \in_y \\ \gamma_{xy} \end{bmatrix} = \begin{bmatrix} \frac{1}{E} & -\frac{\nu}{E} & 0 \\ -\frac{\nu}{E} & \frac{1}{E} & 0 \\ 0 & 0 & \frac{1}{G} \end{bmatrix} \begin{bmatrix} S \\ 0 \\ 0 \end{bmatrix}$$

From the stress-strain relation, the strains are

$$\frac{\partial u}{\partial x} = \in_x(x, y) = \frac{S}{E} \qquad \frac{\partial v}{\partial y} = \in_y(x, y) = -\frac{\nu S}{E}$$

$$\frac{\partial u}{\partial y} + \frac{\partial v}{\partial x} = \gamma_{xy} = 0.$$

It may be noted that $\in_2 = -\frac{\nu S}{E}$

By integrating the strain relations

$$u(x, y) = \frac{S}{E}x + f(y)$$

$$v(x, y) = -\frac{\nu S}{E}y + g(x)$$

and
$$\frac{\partial f}{\partial y} = \frac{\partial g}{\partial x} = C_1$$
$$f(y) = C_1 y + C_2 \qquad g(x) = C_1 x + C_3$$

This implies
$$u(x, y) = \frac{S}{E}x + C_1 y + C_2$$
$$v(x, y) = -\frac{\nu S}{E}y + C_1 x + C_3$$

Since
$$u(0, 0) = 0, \quad v(0, 0) = 0 \quad v(L, 0) = 0$$
gives
$$C_1 = 0 \quad C_3 = 0 \quad \text{and} \quad C_2 = 0$$

Hence
$$u(x, y) = \frac{S}{E}x \qquad v(x, y) = -\frac{\nu S}{E}y$$

and
$$w(x, y, z) = -\frac{\nu S}{E}.z + h(x, y)$$

By boundary conditions $h(x, y) = 0$

A point at the right edge moves by a distance by substituting $x = L$
$$u(L, y) = \frac{S}{E}L.$$

Example 4.3. *A long rectangular prism with ends restrained is as shown in Fig. 4.10. The linearly varying traction forces applied to the opposite edges is as shown. If the body applied load is given by "by" where 'b' is the known constant. Assuming no body forces in the system and no rigid body displacements, find the stress, strain and deflection throughout the body.*

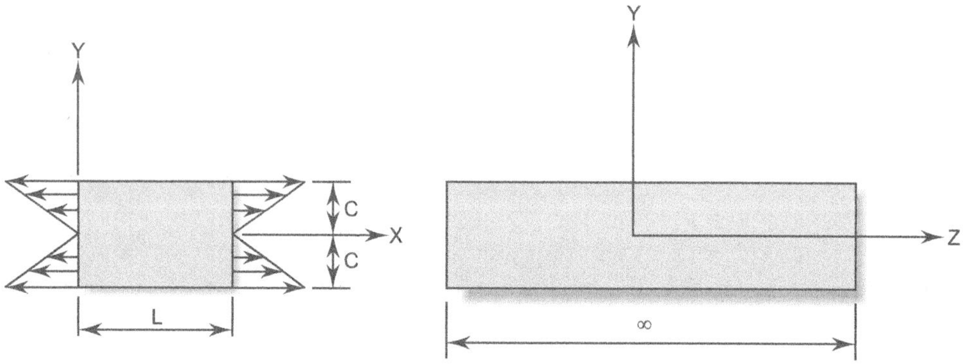

Fig. 4.10. Long rectangular prism.

Solution. The example is a plane-strain approximation problem first of all it is set to
$$w = \epsilon_z = \gamma_{xz} = \gamma_{yz} = f_z = 0$$

Knowing the stresses the strains can be obtained by stress-strain relation governed by Hooke's Law,
$$\begin{bmatrix} \epsilon_x \\ \epsilon_y \\ \gamma_{xy} \end{bmatrix} = \begin{bmatrix} \frac{1-\nu^2}{E} & -\frac{\nu(1+\nu)}{E} & 0 \\ -\frac{v(1+\nu)}{E} & \frac{1-\nu^2}{E} & 0 \\ 0 & 0 & \frac{1}{G} \end{bmatrix} \begin{bmatrix} by \\ 0 \\ 0 \end{bmatrix}$$

Hence the strains are

$$\frac{\partial u}{\partial x} = \epsilon_x = \frac{(1-\nu^2)}{E} by$$

$$\frac{\partial v}{\partial y} = \epsilon_y = -\frac{\nu(1+\nu)}{E} . by$$

$$\frac{\partial u}{\partial y} + \frac{\partial v}{\partial x} = 0 = \gamma_{xy}$$

From integration of strains

$$u(x, y) = \frac{(1-\nu^2)}{E} bxy + f(y)$$

$$v(x, y) = \frac{-(1+\nu)\,\nu b y^2}{2E} + g(x)$$

$$\frac{\partial u}{\partial y} + \frac{\partial v}{\partial x} = \frac{(1-\nu^2)}{E} bx + \frac{\partial f}{\partial y} + \frac{\partial g}{\partial x} = 0$$

It follows that

$$\frac{(1-\nu^2)}{E} bx + \frac{\partial g}{\partial x} = -\frac{\partial f}{\partial y} = C_1$$

$$f(y) = -C_1 y + C_2$$

$$g(x) = C_1 x - \frac{(1-\nu^2) b x^2}{2E} + C_3$$

For the boundary conditions $u(0, 0) = 0 \Rightarrow C_2 = 0$

$$u(0, 0) = 0 \quad \Rightarrow \quad C_3 = 0$$

$$v(L, 0) = 0 \quad \Rightarrow \quad C_1 = \frac{(1-\nu^2) bL}{2E}$$

Hence we have the deflections

$$u(x, y) = \frac{(1-\nu^2)}{E} bxy - \frac{(1-\nu^2)}{2E} bLy$$

$$v(x, y) = -\frac{(1+\nu) v\, b y^2}{2E} + \frac{(1-\nu^2)\, bLx}{2E} - \frac{(1-\nu^2) b x^2}{2E}$$

The stress in z-direction is given by

$$\sigma_z = \nu(\sigma_x + \sigma_y) = \nu by$$

The stress array is

$$[\sigma_x, \sigma_y, \sigma_z, \tau_{xy}, \tau_{yz}, \tau_{zx}] = [by, 0, \nu by, 0, 0, 0]$$

POINTS TO BE REMEMBER

- The balance between the applied loads and the internal resistive force developed results in the equilibrium equation.
- The stress is the resistive force developed per unit area.
- The strain is the result of deformation.

- Strain and deformation are related by strain displacement relation.
- Hooke's law establishes relation between stress and strain with the material property defined.
- Newton's law is used to define force equilibrium.
- The resolution of stress along the co-ordinate directions gives the stress components.
- There are nine stress components in the stress tensor.
- The relation between the shear strain and the normal strains is given by compatibility relation.
- The plane stress approximation assumes the stress in the direction normal to the mid-plane of the body to be equal to zero.
- In the plane-strain approximation, the displacement in the longitudinal direction is assumed to be zero.
- Generalized Hooke's law establishes the relation between components of strain and components of stress.
- The elasticity matrix relates stress matrix with strain matrix.

EXERCISE

1. State and explain the
   (*a*) Stress-strain relation
   (*b*) Strain-displacement equation
   (*c*) The equilibrium equation
   in the theory of elastic structures.
2. Explain the concept of force equilibrium.
3. What are the components of stresses in an elastic structure?
4. Define the components of strain.
5. Using the generalized Hooke's law establish the triaxial stress-strain relation.
6. Derive the equilibrium equation in cartesian co-ordinate system using a cubic differential element.
7. Discuss the plane-stress formulation with a suitable example.
8. Establish the plane-strain formulation with an example.
9. Derive elasticity matrix for plane-stress approximation.
10. Derive the elasticity matrix for plane strain approximation.
11. Fig. 1 shows the plane stress problem with the applied loads $\sigma_x(a, y) = A_y$ and $\sigma_y\,(x, b) = Bx$ and $\tau_{xy} = 0$, on all the surface. The geometric and the body forces are

$$u(0, 0) = v(0, 0) = 0$$

$$f_x(x, y) = f_y\,(x, y) = 0$$

Determine

(*i*) $\sigma_x(x, y)\ \tau_{yz}\,(x, y)$ and $\tau_{zx}\,(x, y)$

(*ii*) All the strain components in the plate.

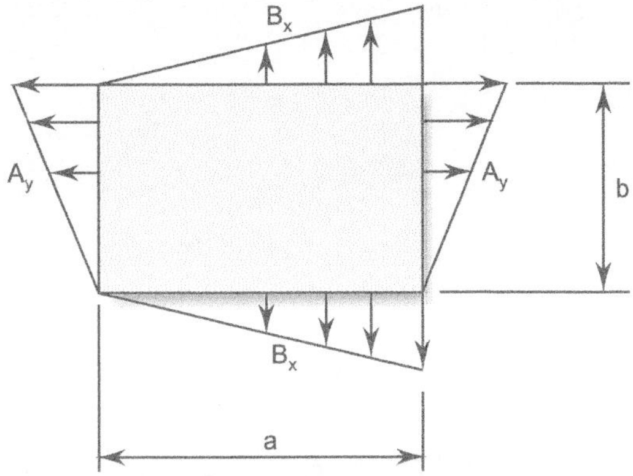

Fig. 1. Plane-stress problem.

12. A slender plate ($a >> 2c$) is as shown in Fig. 2. The applied loads at the two ends are given by

$$\sigma_x(0, y, z) = \sigma_x(a, y, z) = \frac{a}{2}\left(1+\frac{y}{x}\right)$$

Find the stresses in the plate, and sketch the loading on the two ends.

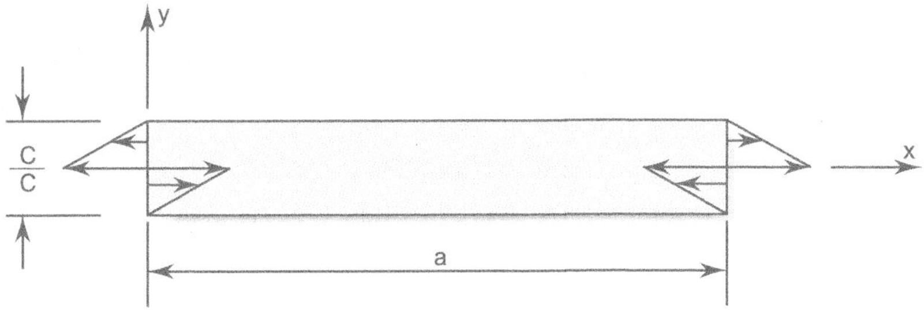

Fig. 2. Plane-strain problem.

Chapter 5

FEM FORMULATIONS OF BAR ELEMENT

वृश्चिकस्य विषं पुच्छम्।
मक्षिकस्य विषं पदम्॥
तक्षकस्य विषं दन्तां।
सर्वांगं दुर्जने विषमं॥

[Scorpio has poisson in tail. Housefly has poisson in its legs. The snake has poisson in its teeth. All parts of an evil man are poissionous.]

5.1 INTRODUCTION

Many problems in engineering and applied science are governed by differential or integral equations. The solution to these equations would provide exact closed-form solution to the particular problem being studied. However the complexity in the geometry, properties and in the boundary conditions that are seen in the most of the real world problems usually means that exact solution cannot be obtained or obtained in a reasonable amount of time. The current product design cycle times imply that the engineers must obtain design solutions in a short period of time. They have content to get approximate solutions that are easily obtained in a reasonable time frame and with reasonable effort. The FEM is one such approximate technique.

The FEM is a numerical procedure for obtaining approximate solutions to many of the problems encountered in engineering analysis. In FEM a complex region defining a continuum is discretized into simple geometric shapes called elements. The properties and the governing relationships are assumed over these elements and expressed mathematically in terms of unknown values at specific points in the elements called 'Nodes'. An assembly process is used to link, the individual elements to the given system. When the effects of loads and boundary conditions are considered a set of linear and non-linear algebraic equations is usually obtained. The solution to these equations gives the approximate behaviour of the continuum or the system.

The continuum has infinite number of degrees-of-freedom (DOF), write the discretized model has finite DOF. This is the origin of the name 'Finite Element Method (FEM). The number of equations to be solved in FEM is usually rather large for most of the real world applications of the FEM, and requires the computational power of digital computers. The FEM has little value if digital computers were not available.

Advances in and availability of computers and software has brought the FEM within reach of engineers working in small industries and students in academics.

Two features of finite element method are worth noting.

- The piecewise approximation of the physical field (continuum) on the finite elements provides good precision even with simple approximating functions. Simply increasing the elements can achieve increasing precision.
- The locality of the approximation leads to the sparse equation systems for a discretized problem. This helps to ease the problem solution having large number of nodal unknowns. It is not uncommon today to solve systems containing a million primary unknowns.

Hence it is understood that finite element method is a numerical method for solving problems of engineering and mathematical physics. FEM is specially useful in solving problems with complicated geometries, loading conditions and material properties where analytical solutions cannot be obtained. It is clear that FEM is a computer technique used to solve partial differential equations.

The chapter is dedicated to simple problems to predict deformation, stress fields within the solid bodies subjected to external forces. Complex structural systems are often too complicated to simply derive relationships between applied loads, deflection and internal stresses. Hence large structures are divided up into many individual finite element, that have much simpler structural form e.g., beams columns etc. The relation between load, displacements, stresses and strains in a finite element can be determined. Thus it is computationally possible for a complex structure to be modelled by assembly (aggregation) of many individual finite elements. The process of assembly must satisfy.

1. The equilibrium of the system
2. The continuity of the structure

5.1.1 Methods in Structural Analysis

The mechanism in mechanical engineering and the structures in civil engineering applications, are made of trusses and beam. The analysis of such structures were initially accomplished using the classical methods of mechanics. But later on when the elements of structures became large the matrix analysis methods were evolved. With the evolution of fast computers the matrix methods are transformed to computational methods like FEM.

The classical methods which help in bench marking used virtual work, approach, force method, slope-deflection procedure, and moment of area methods give exact solution to simple hypothetical problems. The matrix analysis of structures which led to FEM use stiffness by definition for simple problems. The direct stiffness approach which consists of formulation of stiffness for truss and beam element laid the foundation for FEM. Hence using the FEM and computers the approximate solution to complicated problems were made possible.

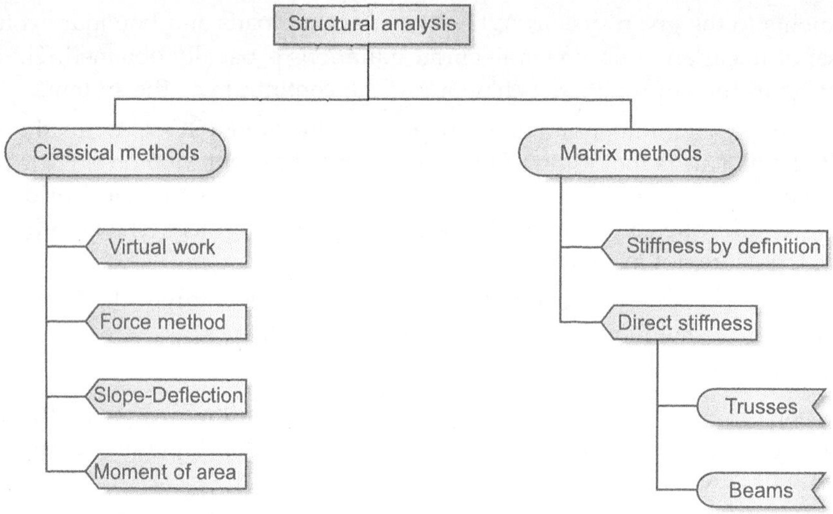

Fig. 5.1. Block diagram of structural analysis.

5.2 STIFFNESS BY DEFINITION

This method considers,

1. Stiffness from known displacements
2. Unknown displacements
3. Known loadings.

For each released degrees of freedom we get one equation that adds to the stiffness, displacement and loading matrices. The reactions (unknown forces) and the known displacements are discarded, in this method.

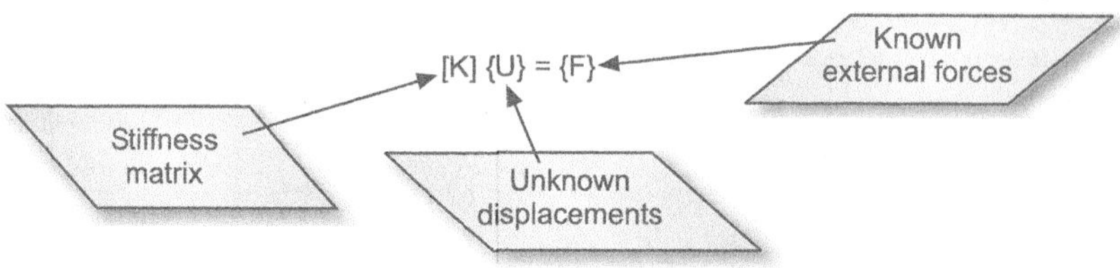

Fig. 5.1(*a*)

Direct stiffness Method

The basic procedure of direct stiffness method is listed as follows—

1. Calculate the stiffness matrix
   - Determine the local stiffness matrix
   - Transform into global co-ordinates
   - Assemble all the matrices to form Global stiffness matrix.
2. Solve for unknown displacements.
3. Use the computed unknown displacements to solve for unknown reactions.
4. Calculate the internal forces.

Illustration:

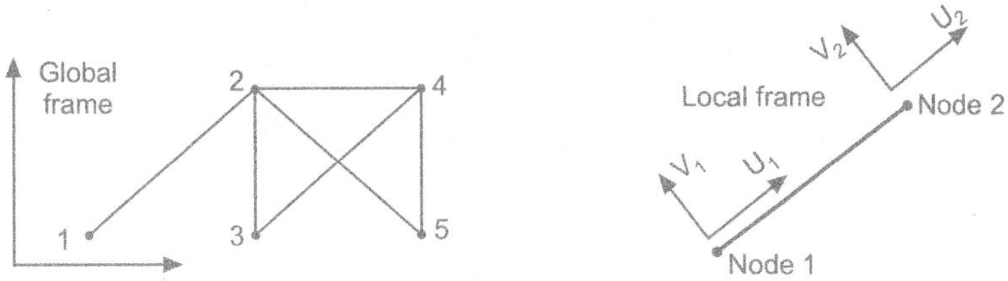

Fig. 5.1(b). Arbitrary Structure.

The procedure followed to obtain the global stiffness matrix is—Decompose the entire structure in Fig. 5.1*b* into set of finite elements (six elements in this case).

Build the stiffness matrix for each element. Transform these matrices to the global co-ordinate frame. Assemble the transformed matrices to form the global stiffness matrix.

Comparison between stiffness by definition and direct stiffness method.

	Stiffness by definition	**Direct stiffness method**
	DOF = 0	DOF = 2
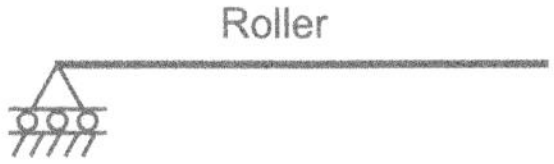	DOF = 1	DOF = 2
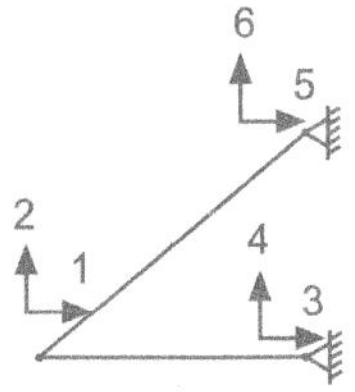	• Less number of equations to work with • For this truss system there are 2 degrees of freedom $u_1, u_2 \neq 0$ • Unknown reactions are not included. The force displacement relation is $[Kij]\ \{Uj\} = Fj$ $i = j = 2$ F_1 and F_2 are known U_1 and U_2 are determined	• More number of equation to work with Here in this there are six degrees of freedom $u_3, u_4, u_5, u_6 = 0$ • Unknown reactions to be solved are included. • The force displacement relation is $[K_{mn}]\ [U_n] = [F_n]$ $m = n = 6$ $U_3, U_4, U_5, U_6 = 0$ F_3 to F_6 are determine after obtaining U_1 and U_2.

5.3 STIFFNESS BY DEFINITION: SPRING ANALOGY

This is the direct method of structural analysis using trusses and beams. The stiffness of the axial bar element is compared with the arbitrary linear spring in the truss analysis of mechanical and civil structures. In the spring analogy the elements are modelled as linear spring with spring rate, *K*. Consider the axially, loaded bar and the linear spring as shown in Fig. 5.2.

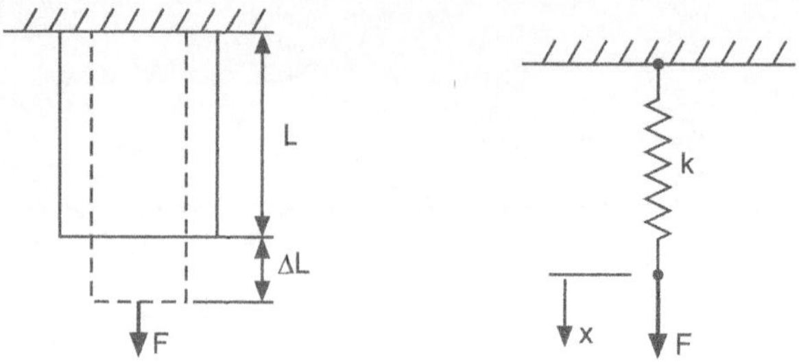

Fig 5.2. Bar and spring.

By the definition of stress,

$$\sigma = \frac{F}{A} \quad ...(5.1)$$

And the strain is defined by

$$\varepsilon = \frac{\Delta.L}{L} \quad ...(5.2)$$

In strength of materials the Hooke's Law gives the relation between stress and strain in the elastic limit. From equations (5.1) and (5.2)

$$F = \left(\frac{E.A}{L}\right)\Delta L \quad ...(5.3)$$

But the longitudinal force on a linear spring is expressed as

$$F = kx. \quad ...(5.4)$$

By comparison of the equations (5.3) and (5.4) the stiffness of an axial bar element is defined as

$$K = \frac{EA}{L} \quad ...(5.5)$$

The equation (5.5) shows that stiffness is determined by geometry and material property. The stiffness has the unit of force per length. The spring analogy for a axial element can be well related in the block diagram, Fig. 5.3.

Spring (Linear)

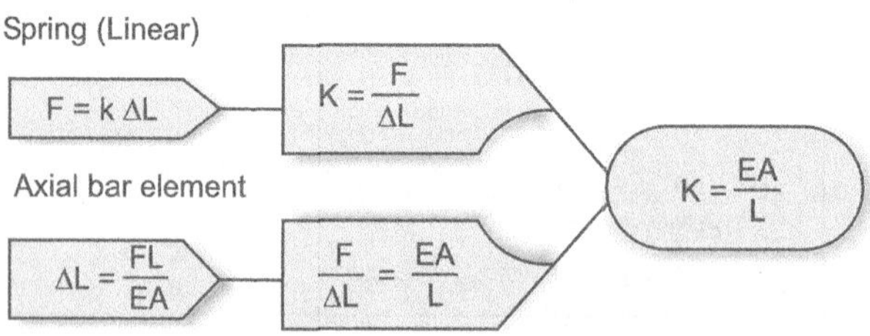

Fig. 5.3. Spring analogy.

The spring analogy relates the force in the spring to that of the resistive force in the axial bar element obtained from the basics of strength of material to obtain the expression for spring in terms of cross-sectional area and the length as the geometric property and Young's modulus as mechanical property of material.

5.3.1 Matrix Formulation

A single element spring has two nodes with one degree of freedom at each node and a nodal force on each node. This is shown as in Fig. 5.4*a*. In the Fig. 5.4*b* two linear springs of different stiffness are assembled and fixed at one end, to form a global system. To describe the behaviour of such a system in terms of a FEA model, the equations are represented in a form that consists of matrix of stiffness and vectors of displacement and loading.

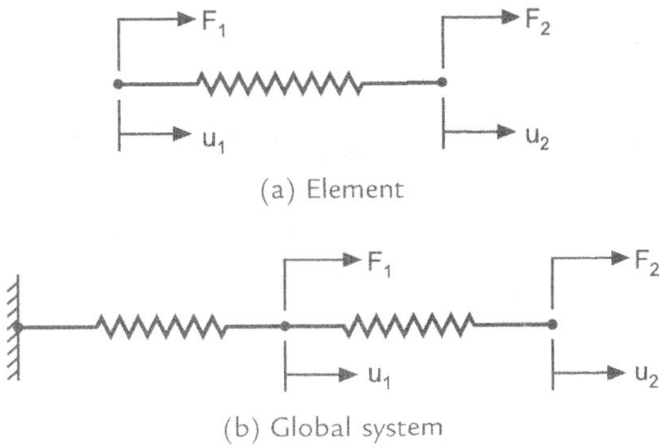

Fig. 5.4. Global system of linear springs.

By definition and force analysis of the global system of linear spring the following equations are derived.

$$F = kx \quad ...(5.6)$$

Here F represents the force, k the stiffness and x the displacement. Applying this to the global system represented in Fig. 5.4*b*.

$$F_1 - k_1 u_1 + k_2 (u_2 - u_1) = 0 \quad ...(5.7)$$

$$F_2 - k_2 (u_2 - u_1) = 0 \quad ...(5.8)$$

By manipulation of equations, (5.7) and (5.8) the obtained equations are expressed in the matrix form as

$$\begin{bmatrix} F_1 \\ F_2 \end{bmatrix} = \begin{bmatrix} k_1 + k_2 & -k_2 \\ -k_2 & k_2 \end{bmatrix} \begin{Bmatrix} u_1 \\ u_2 \end{Bmatrix} \quad ...(5.9)$$

In general the equation (5.9) is represented by

$$\{F\} = [K]\{U\} \quad ...(5.10)$$

where $\{F\}$ is column vector of forces, $[K]$ is the global stiffness matrix and $\{U\}$ the vector of nodal displacements. In the system of equations (5.10) if $[K]$ and $\{F\}$ are known, the unknown $\{U\}$ can be solved by numerical methods.

The FEA model of static analysis of global system of two linear springs connected in series obtained in equation (5.10) exhibits following behaviour to be observed before the analysis is done.

5.3.2 Analysis

The mesh is having

- two 1*D* elements (springs)
- two nodes in each element
- totally three nodes

The degree of Freedom (DOF)

- There is one DOF at each node
- There are totally three DOFs for the global system

The boundary conditions

- At the support, $U_0 = 0$ since the element is fixed
- Other conditions of displacement u_1 and u_2 are to be solved
- The solution of u_1 and u_2 requires the specification of external forces F_1 and F_2.

5.4 DISCRETIZATION—NODES AND ELEMENTS

Model the body by dividing it into an equivalent system of smaller bodies or units called "finite elements", interconnected at points common to two or more elements (nodes or nodal points) and/or boundary lines and/or surfaces.

A finite element mesh is defined by a set of nodes together with a set of finite elements as shown in Fig. 5.5.

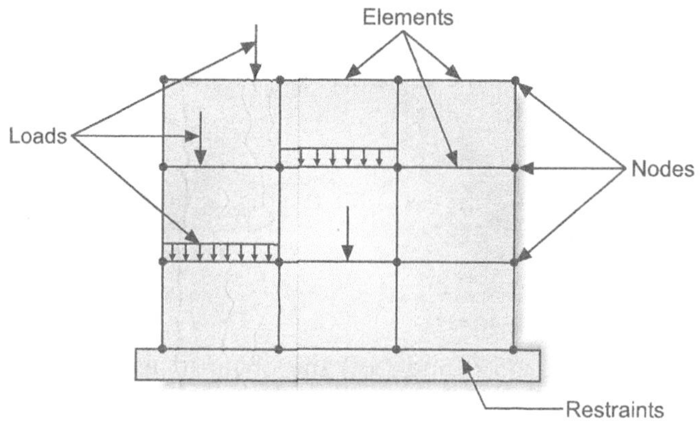

Fig. 5.5. Example of discretization.

5.4.1 Nodes

The nodes are set of discrete points within the solid body. Nodes have following properties:

- *A Node number*—Every node is assigned an integer number which is used to identify the node. Any convenient numbering scheme may be selected. The nodes need not be numbered in a sequence and numbers may be omitted. One could number a set of '*n*' nodes as 100, 200, 300...100*n* instead of 1, 2, 3, ...*n*.
- *Nodal Co-ordinates*—For three dimensional finite element analysis each node is assigned a set of nodal co-ordinates which specify the position of the node in the under formed body.
- *Nodal Variables*—The nodal variables are the unknowns at the start of the analysis and are computed by the finite element program. The unknowns can be displacement and forces as in structural analysis or temperature in heat transfer analysis or flow potential as in fluid flow analysis.

5.4.2 Elements

The elements are used portion the body into discrete regions. The body can be one dimensional like trusses, two dimensional like surfaces and three dimensional like solid surfaces. The elements are defined by the following properties:

- **An element number:** Every element is assigned an element number (integer) which is used to identify the element. Just as when numbering nodes any convenient scheme may be used to number the element.
- **A geometry:** There are many possible shapes for an element. A few of the more common element types are shown in the following Fig. 5.6. The node attached to the element are shown by dark dots. Element geometry is defined by node locations.

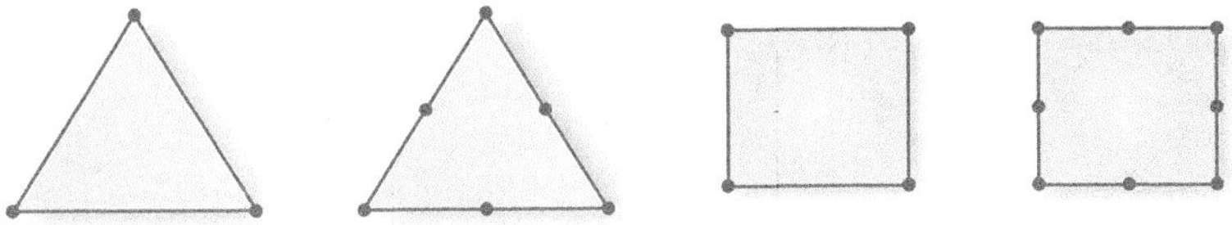

(a) Three noded triangle (b) Six noded triangle (c) Four noded quadrilateral (d) Eight noded quadrilateral

Fig. 5.6. 2D elements with nodes.

TABLE 5.1. Primitive Structural Elements

Physical structural components	Model Name	Finite element discretization
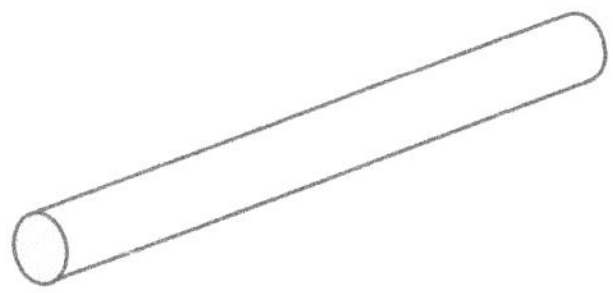	Bar	
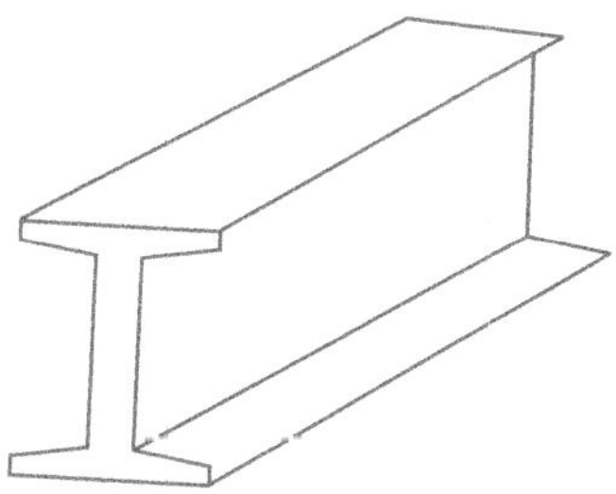	Beam	
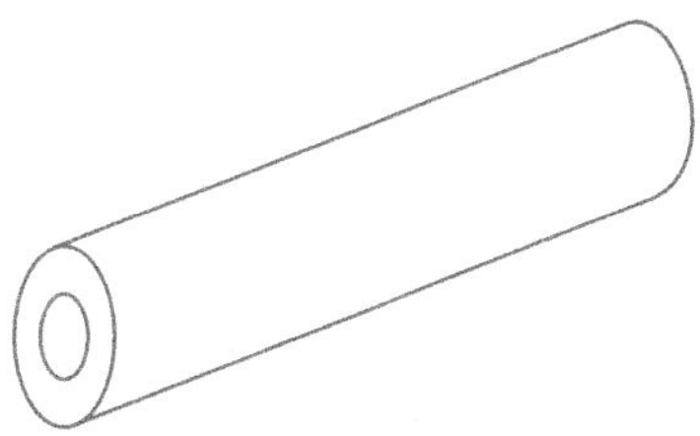	Tube/Pipe	

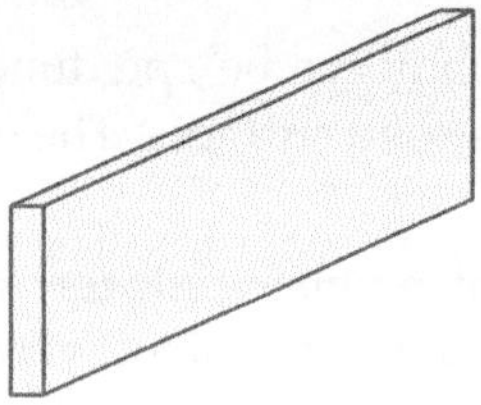	Spar (Web)	
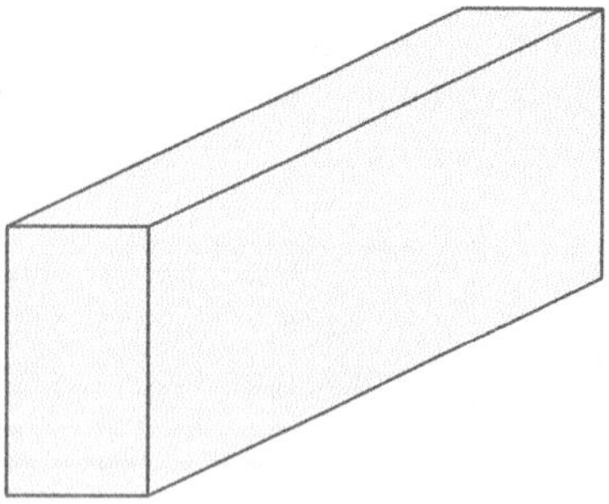	Shear panel (2D version of spar)	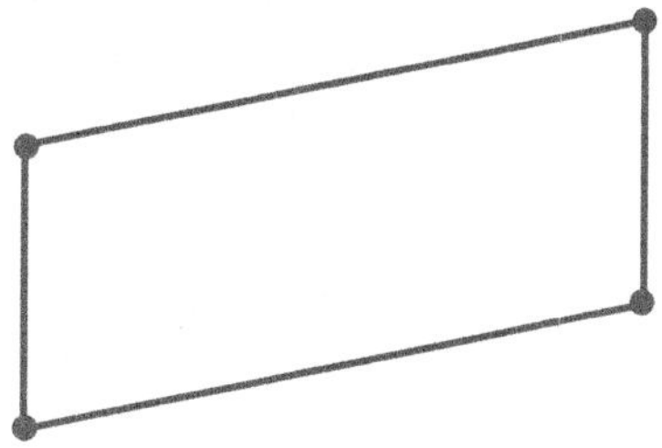

5.5 ONE DIMENSIONAL ELEMENT CONFIGURATIONS

The following one dimensional elements are recognized in the finite element analysis of structures. The definition of these elements reveal that

• Bars	transmit	Axial force
• Beams	transmit	bending force
• Shafts	transmit	torque
• Spars	transmit	shear force
• Beam-Columns	transmit	bending-cum-axial force

- Representation of Bars

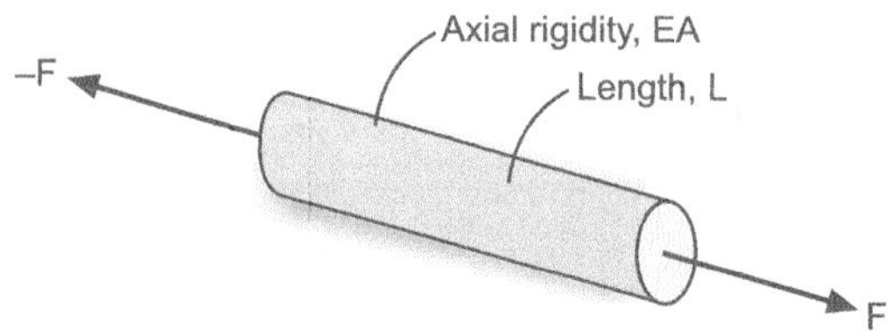

(a) Physical representation

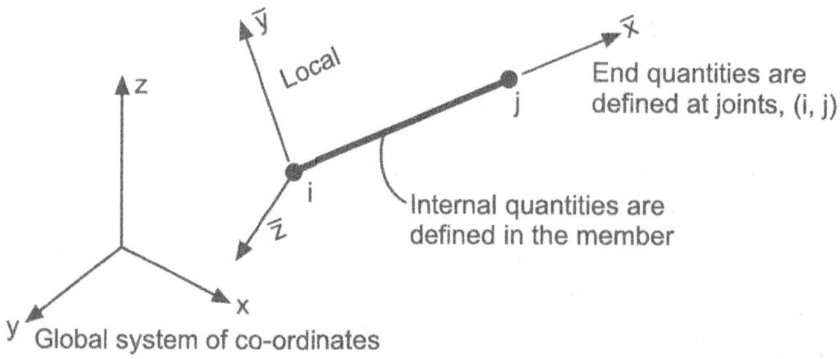

(b) Global representation

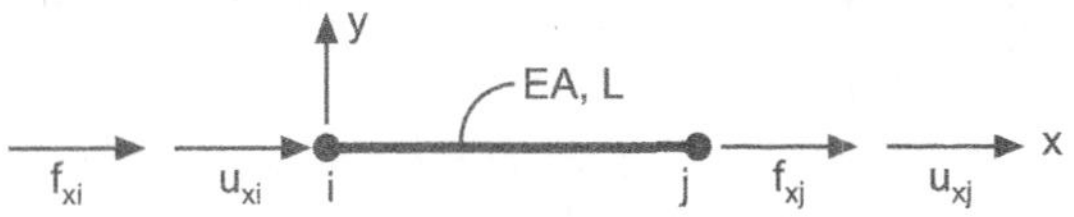

(c) Local representation with force and diplacements

Fig. 5.7. Representation for bars.

- Representation for Beams

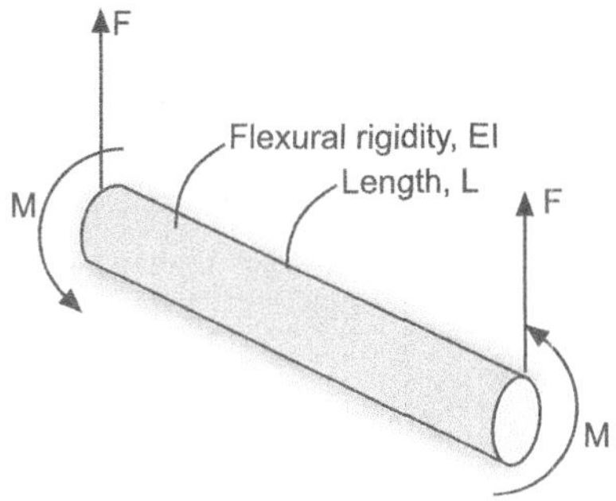

(a) Physical representation

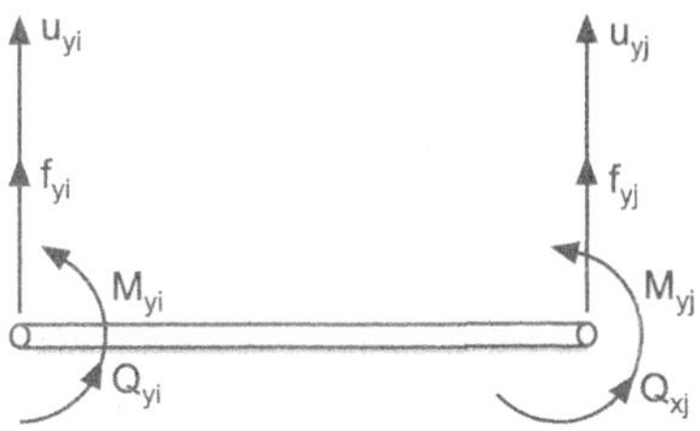

(b) Representation in local co-ordinates.

Fig. 5.8. Representation for beams.

- Representation for shafts

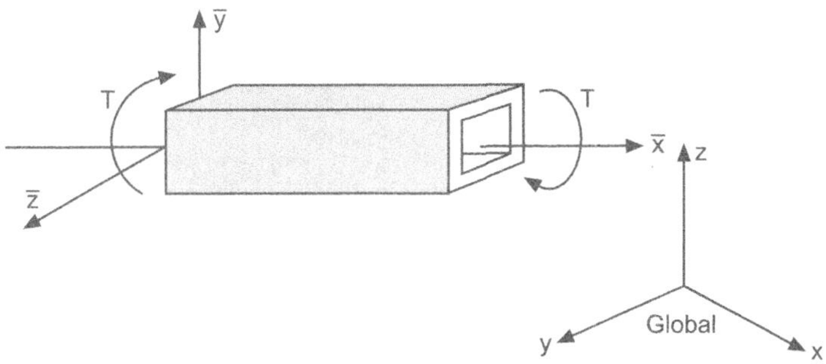

(a) Physical representation

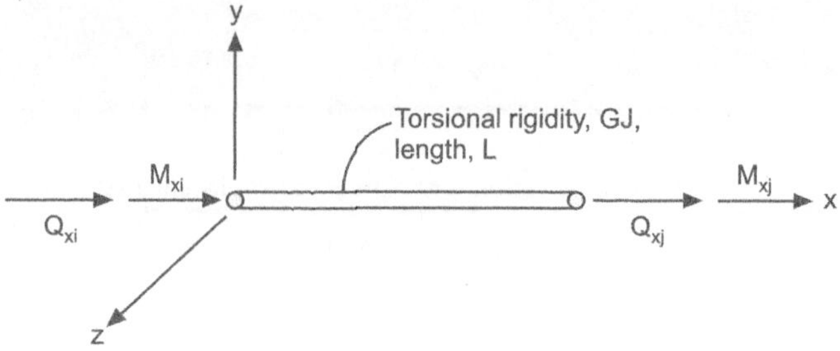

(b) Local representation in FEM.

Fig. 5.9. Representation for shaft.

- Representation for spar (shear web) element

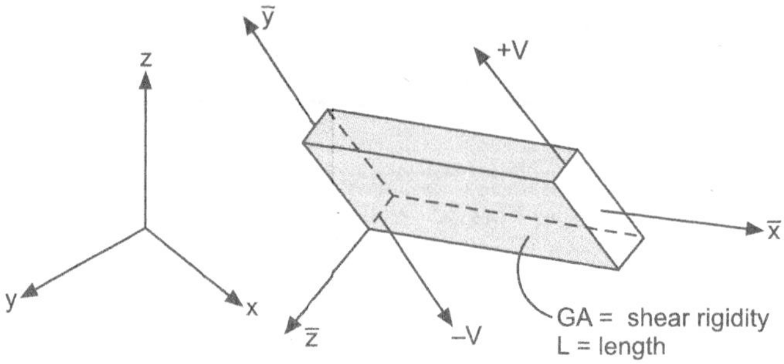

(a) Physical representation

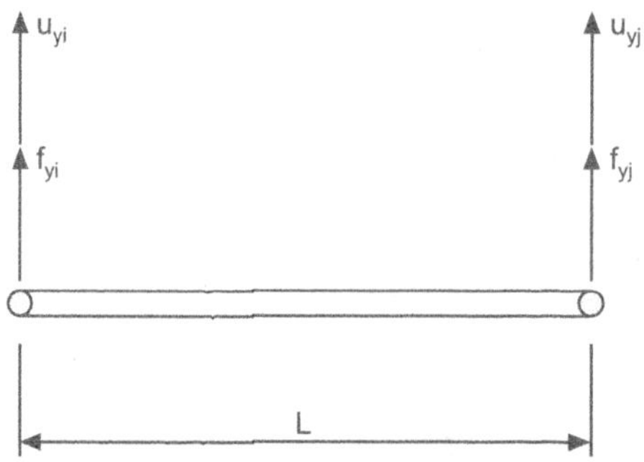

(b) FEM representation

Fig. 5.10. Spar element.

- Special Elements

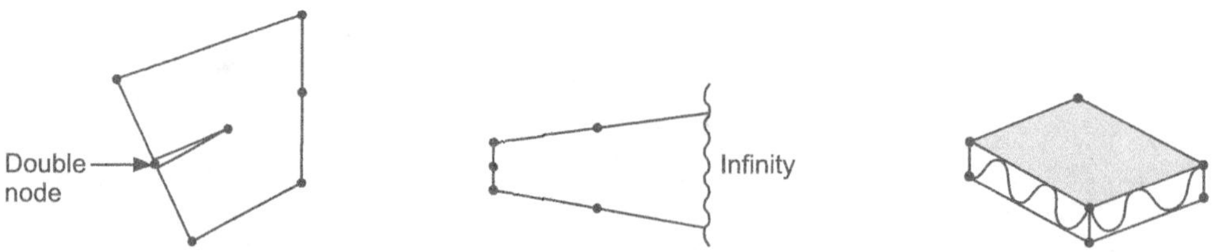

(a) Crack element (b) Infinite element (c) Honey comb element

Fig. 5.11. Special elements.

TABLE 5.2. Element Geometry in Discretization

Type of Element	Linear Element	Quadratic Element	Cubic Element
1D Element Bar			
2D Element Triangular	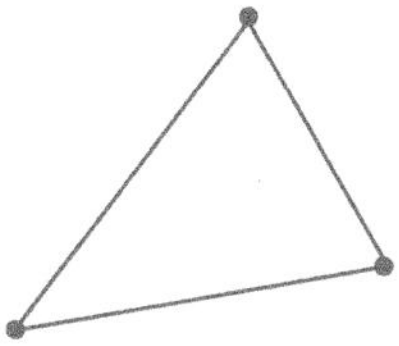	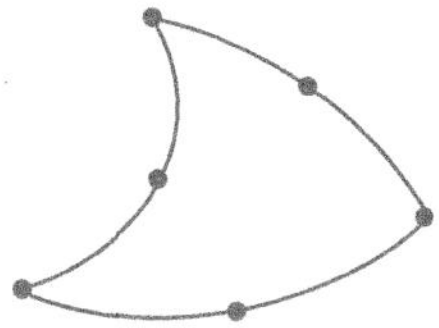	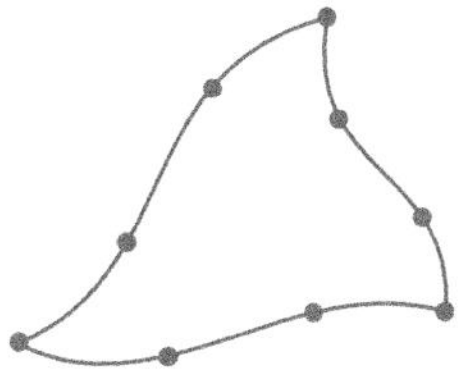
2D Element Quad	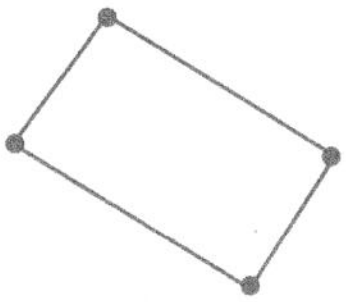	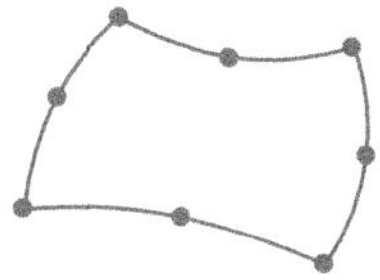	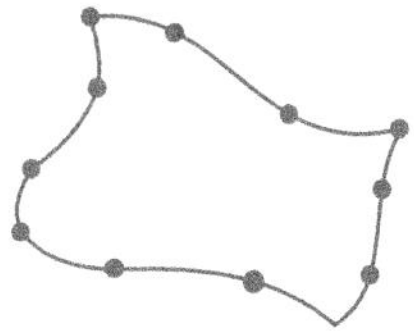
3D Tetra	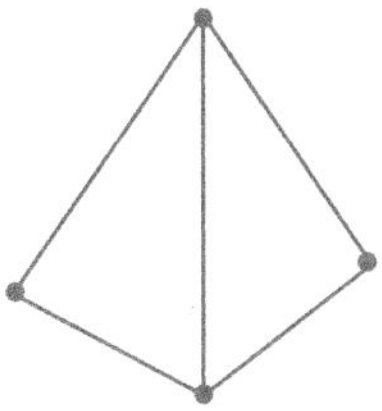	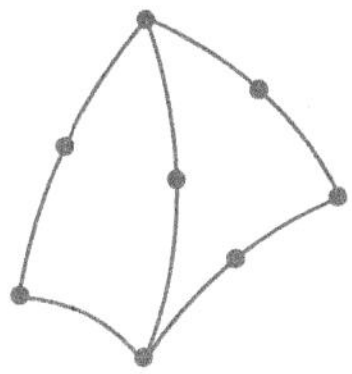	
3D HEX	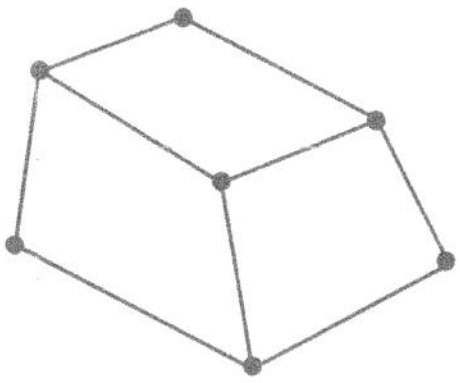	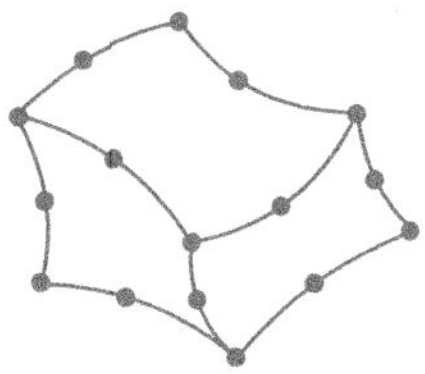	
3D Brick Element	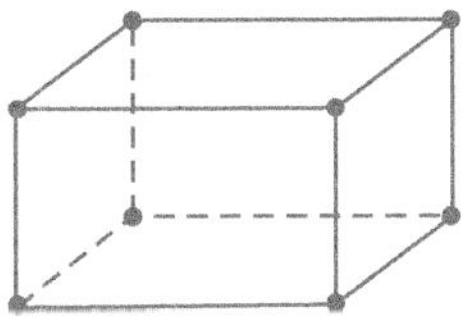	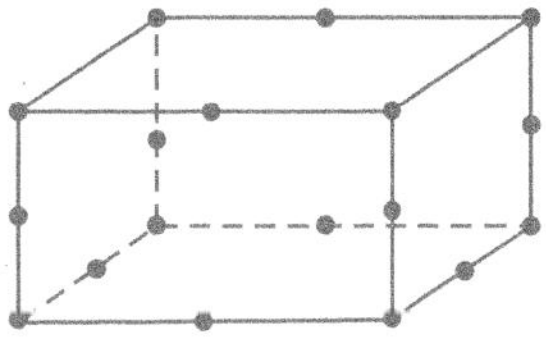	

Continuum Elements

Physical	Finite Element Idealization
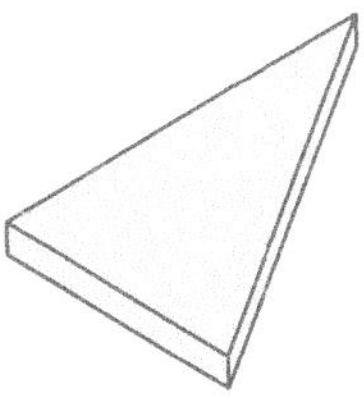	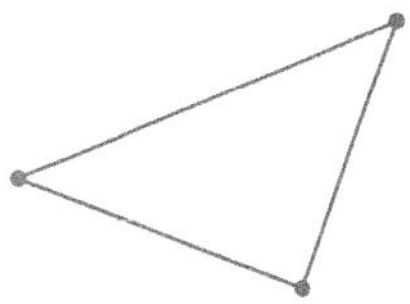

1. Plates	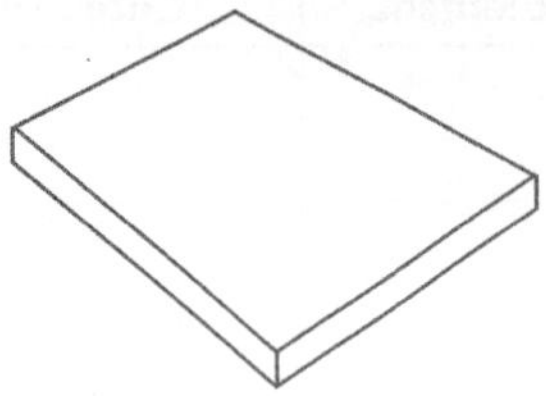	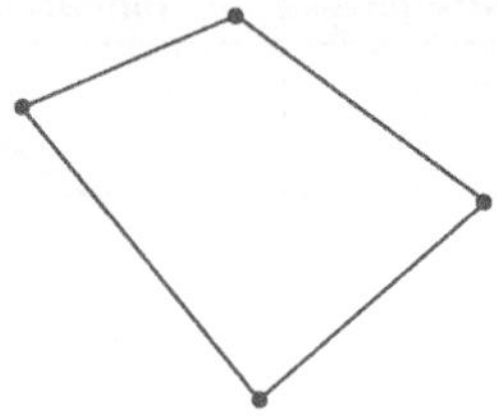
2. 3-D Solid	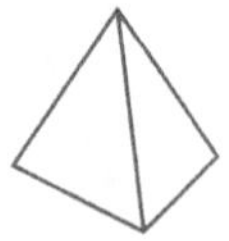 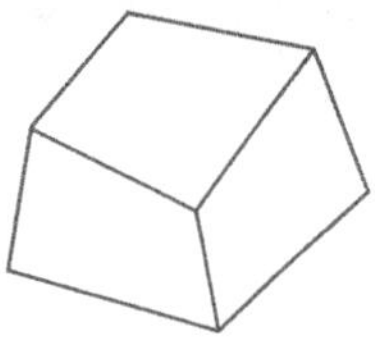	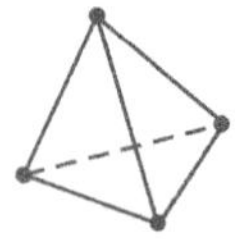 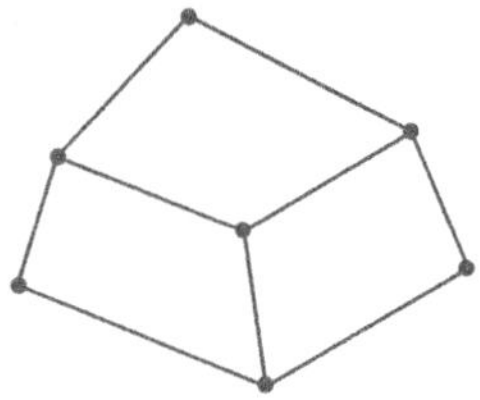

5.6 STIFFNESS BY DEFINITION: BAR ELEMENT

This method only uses small part of the information available to tackle the problem. It is attempted in this section to use this method to formulate the stiffness matrix of a bar element.

5.6.1 Stiffness Matrix of a Bar Element

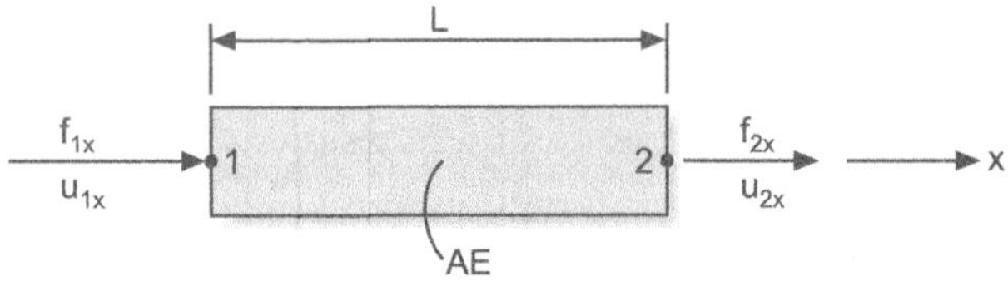

Fig. 5.12. Bar element.

Consider a bar element with the geometry and the property defined by the following representation.

Let L = Length of the bar

A = Cross-sectional area of the bar

E = Elastic (Young's) Modulus of bar

$u(x)$ = displacement of the bar as a function of local co-ordinate x of the bar

The strain in the bar at x distance is defined to be

$$\in(x) = \frac{du}{dx} \quad ...(5.11)$$

The stress in the bar from first principle (Hooke's Law)

$$\sigma(x) = E \in(x) \quad ...(5.12)$$

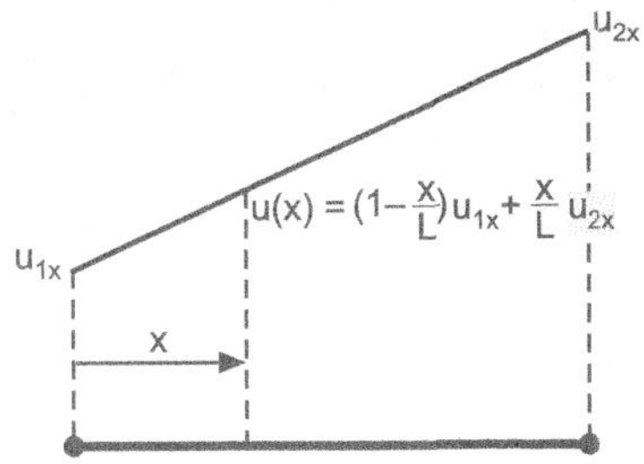

Fig. 5.13. Shape function of bar.

Assume that the displacement $u(x)$ is varying linearly along the bar as shown in Fig. 5.13. By applying analytical geometry with the similar triangles, the displacement function,

$$u(x) = \left(1 - \frac{x}{L}\right)u_{1x} \frac{x}{L}u_{2x} \quad ...(5.13)$$

The strain is constant along the bar

i.e.,

$$\in = \frac{du}{dx} = \frac{u_{2x} - u_{1x}}{L} \quad ...(5.14)$$

The stress is also constant along the bar

$$\sigma = E\in = \frac{E}{L}(u_{2x} - u_{1x}) \quad ...(5.15)$$

The tension along the length of the bar is also constant

$$F = EA\in = \frac{EA}{L}(u_{2x} - u_{1x}) \quad ...(5.16)$$

Thus the bar is acting like a linear spring with stiffness

$$K = \frac{EA}{L}$$

By the force analysis on the linear spring

$$f_{1x} - k\,(u_{1x} - u_{2x}) = 0$$

$$f_{2x} - k\,(u_{2x} - u_{1x}) = 0 \quad ...(5.17)$$

or

$$f_{1x} = ku_{1x} - ku_{2x}$$

$$f_{2x} = -ku_{1x} + ku_{2x} \quad ...(5.18)$$

In the matrix form

$$\underbrace{\begin{Bmatrix} f_{1x} \\ f_{2x} \end{Bmatrix}}_{F} = \underbrace{\begin{bmatrix} k & -k \\ -k & k \end{bmatrix}}_{[K]} \underbrace{\begin{Bmatrix} u_{1x} \\ u_{2x} \end{Bmatrix}}_{U}, \quad ...(5.19)$$

Hence equation of static equilibrium is

$$F = KU \quad ...(5.20)$$

We can identify following properties for the stiffness matrix $[K]$.

Property-1

A stiffness matrix accounts for the material and geometric properties of the structure.

Property-2

A stiffness matrix is a square (number of rows equal to number of column) and symmetric matrix, *i.e.*,

$$K_{ij} = K_{ji}$$

Property-3

The diagonal elements of the stiffness matrix are always positive.

Property-4

The stiffness matrix is independent of the load acting on the structure. Many loading cases may be tested without recalculating the stiffness matrix.

5.6.2 Stepped Bar Analysis

A stepped bar, as shown in Fig. 5.14, is fixed at both the ends. The step 1 has crossectional area A_1 and Young's modulus, E_1. The area of crossection and Modulus of elasticity for the step 2, are A_2 and E_2 respectively. The respective lengths of two steps are L_1 and L_2.

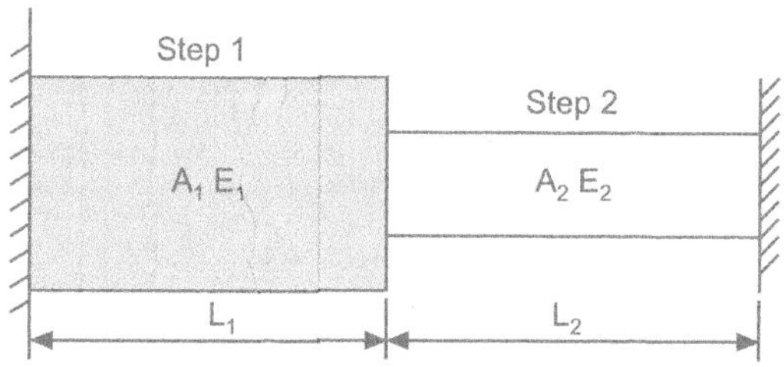

Fig. 5.14. Stepped bar.

The equivalent spring system for the stepped bar is shown in Fig. 5.15. The stepped bar is represented with two spring elements with stiffness k_1 and k_2. The equivalent system has three nodes and two elements. The stiffness of the elements,

$$k_1 = \frac{A_1E_1}{L_1} \qquad k_2 = \frac{A_2E_2}{L_2} \qquad \text{...(5.21)}$$

The system of equations with the global stiffness matrix formulated is written as

$$\begin{bmatrix} k_1 & -k_1 & 0 \\ -k_1 & k_1+k_2 & -k_2 \\ 0 & -k_2 & k_2 \end{bmatrix} \begin{Bmatrix} u_{1x} \\ u_{2x} \\ u_{3x} \end{Bmatrix} = \begin{bmatrix} F_{1x} \\ F_{2x} \\ F_{3x} \end{bmatrix} \qquad \text{...(5.22)}$$

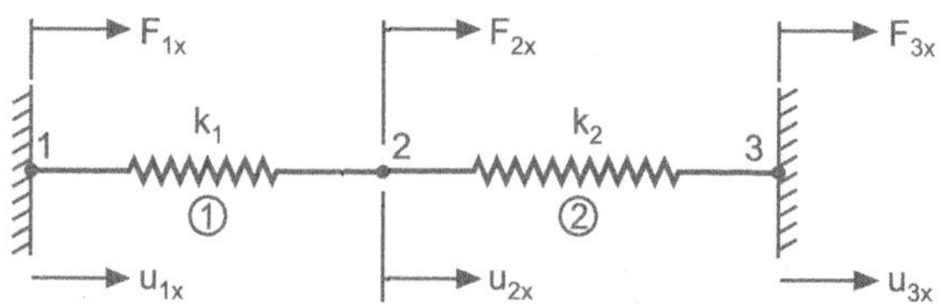

Fig. 5.15. Equivalent spring system.

Example 5.1. *A stepped bar with its geometry and material property is as shown by Fig. 5.16. Compute the displacement at node 2 and the stresses in each element.*

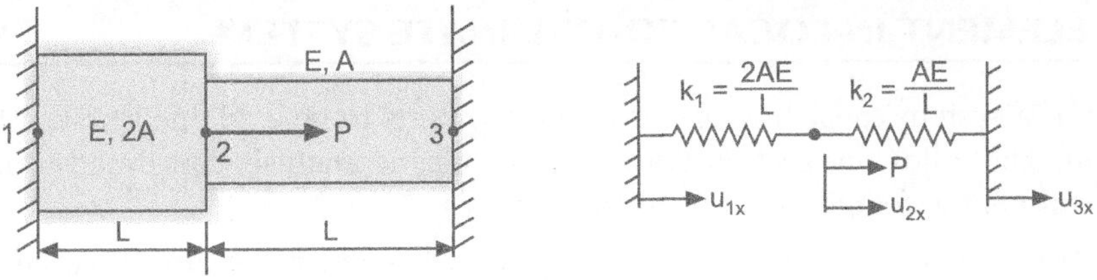

Fig. 5.16. Stepped bar and equivalent spring system.

Solution. The requirement of the problem is to determine the displacement at node 2, the reaction forces in the support. The strain and stress in each element by the method of stiffness by definition.

The global system of equations is

$$\begin{bmatrix} k_1 & -k_1 & 0 \\ -k_1 & k_1+k_2 & -k_2 \\ 0 & -k_2 & k_2 \end{bmatrix} \begin{Bmatrix} u_{1x} \\ u_{2x} \\ u_{3x} \end{Bmatrix} = \begin{Bmatrix} F_{1x} \\ F_{2x} \\ F_{3x} \end{Bmatrix} \qquad ...(5.23)$$

Since the two end supports are fixed

$$u_{1x} = u_{3x} = 0 \text{ and } F_{2x} = P.$$

Hence the reduced system of equations are

$$\begin{aligned} -k_1 u_{2x} &= F_{1x} \\ (k_1+k_2)u_{2x} &= P \\ -k_2 u_{2x} &= F_{3x} \end{aligned} \qquad ...(5.24)$$

The solution of these equations leads to

$$\begin{aligned} u_{2x} &= \frac{P}{k_1+k_2} = \frac{PL}{3AE} \\ F_{1x} &= -\left(\frac{2AE}{L}\right)\frac{PL}{3AE} = \frac{-2}{3}P \\ F_{3x} &= -\left(\frac{AE}{L}\right)\frac{PL}{3AE} = -\frac{1}{3}P \end{aligned} \qquad ...(5.25)$$

Strain in element number (1)

$$\in_1 = \frac{u_{2x}-u_{1x}}{L} = \frac{u_{2x}}{L} = \frac{P}{3AE} \qquad ...(5.26)$$

The stress in element (1)

$$\sigma_1 = E\in_1 = \frac{P}{3A} \qquad \text{(The element is in tension)}$$

Similarly in the element number (2)

$$\in_2 = \frac{u_{3x}-u_{2x}}{L} = \frac{-u_{2x}}{L} = -\frac{P}{3AE} \qquad ...(5.27)$$

The stress is estimated to be

$$\sigma_2 = E\in_2 = -\frac{P}{3A} \qquad \text{(The element is in compression)}$$

The elemental stresses even though are equal in magnitude the directions are opposite as element one is in tension and element (2) is in compression.

5.7 BAR ELEMENT IN LOCAL CO-ORDINATE SYSTEM

The behaviour of a system consisting of bar elements needs to be explained first in the local co-ordinate system. The following section addresses the elastic analysis concerned to bar element. The approach adopts the following steps in the analysis.

Task 1: Approximate, the displacement within each element. A linear variation is to be adapted here.

Task 2: Approximate the strain and stress developed in each element of the system.

Task 3: Derive the stiffness matrix of each element using principle of minimum potential energy.

- *Task 1:*

The simplest possible assumption about the displacement variation inside is to be linear. The assumption is mathematically represented by

$$u(x) = a_0 + a_1 x \qquad \text{...(5.28)}$$

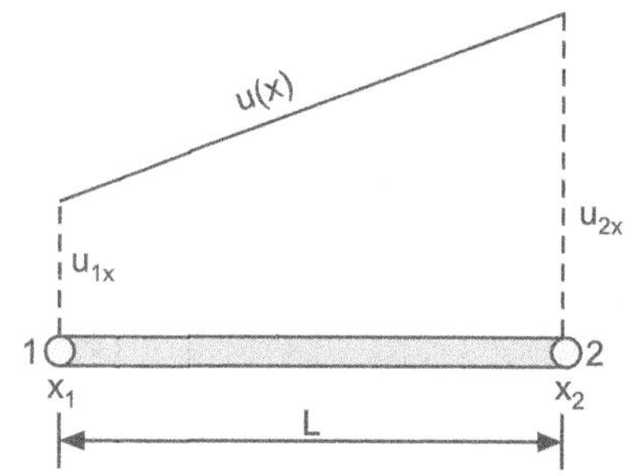

Fig. 5.17. Linear displacement variation.

In the expression (5.28) the constants a_0 and a_1 are to be determined to know the variation of displacement along the length of the bar element. It is clear from Fig. 5.17 that the displacements

$$u(x) = u_{1x} \text{ at } x = x_1$$

and

$$u(x) = u_{2x} \text{ at } x = x_2$$

Substituting this conditions in equation (5.28) the obtained system of equations are

$$a_0 + a_1 x_1 = u_{1x} \qquad \text{...(5.29)}$$

$$a_0 + a_1 x_2 = u_{2x}$$

When the equations are solved simultaneously, the constants get estimated to be

$$a_0 = \left(\frac{x_2}{x_2 - x_1}\right)u_{1x} - \left(\frac{x_1}{x_2 - x_1}\right)u_{2x}$$

$$a_1 = -\left(\frac{1}{x_2 - x_1}\right)u_{1x} + \frac{1}{(x_2 - x_1)}u_{2x} \qquad \text{...(5.30)}$$

Hence the displacement function is

$$u(x) = a_0 + a_{1x}$$

$$= \underbrace{\left(\frac{x_2 - x}{x_2 - x_1}\right)}_{N_1}u_{1x} + \underbrace{\left(\frac{x - x_1}{x_2 - x_1}\right)}_{N_2}u_{2x} \qquad \text{...(5.31)}$$

$$= N_1(x)\, u_{1x} + N_2(x)\, u_{2x} \qquad ...(5.32)$$

Where $N_1(x)$ and $N_2(x)$ are the shape functions. The displacement function in the matrix form

$$u(x) = [N]^T \{U\} \qquad ...(5.33)$$

Vector of nodal shape function

$$[N] = [N_1(x), N_2(x)] = \left[\frac{x_2 - x}{x_2 - x_1}, \frac{x - x_1}{x_2 - x_1}\right] \qquad ...(5.34)$$

Vector of nodal displacements

$$\{U\} = \begin{Bmatrix} u_{1x} \\ u_{2x} \end{Bmatrix}$$

- **Observations**

1. *Compatibility*

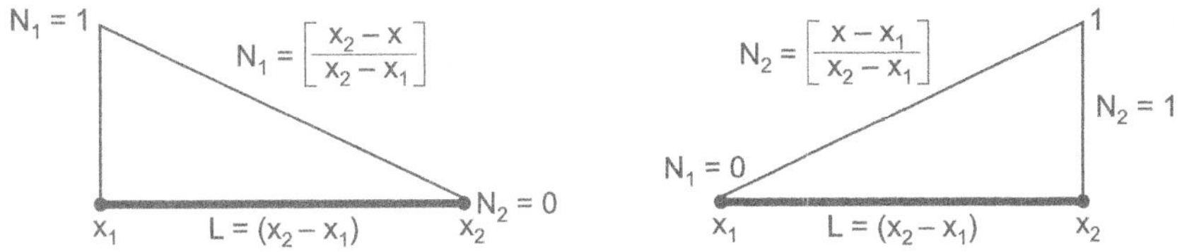

Fig. 5.18. Plot of the shape functions.

The displacement approximation for the function is continuous across the element boundaries. The displacement function is satisfied by each and every point along the length, $x_1 \le x \le x_2$.

2. *Completeness*

$$N_1(x) + N_2(x) = \Sigma N_i = 1 \text{ at all } x \text{ values of interval } (x_1, x_2) \qquad ...(5.35)$$

This condition represents the rigid body mode

Also

$$N_1(x).\, x_1 + N_2(x).\, x_2 = x \text{ at all values of } x \text{ in the interval } (x_1, x_2). \qquad ...(5.36)$$

This represents the constant strain state

Completeness = {Rigid body modes} + {Constant strain state}

3. *Convergence*

By satisfying compatibility condition and the completeness requirement the convergence is achieved. This ensures that solutions get better as more elements are introduced on in the limit approach the exact solution.

- *Task 2:*

This is concerned with the approximation of the strain and stress within each element. The equation (5.37) gives the expression for displacement in each element as

$$u(x) = [N]^T \{U\} \qquad ...(5.37)$$

By definition of strain in an axially loaded bar

$$\in = \frac{du}{dx} \qquad ...(5.38)$$

Hence from equation (5.37) and (5.38)

$$\in = \left[\frac{dN}{dx}\right]\{U\} = [B]\{U\} \qquad ...(5.39)$$

The matrix $[B]$ is known strain-displacement matrix

i.e.,
$$[B] = \left[\frac{dN}{dx}\right] \qquad ...(5.40)$$

For a linear element of an axial bar

$$[N] = \left[\frac{x_2 - x}{x_2 - x_1}, \frac{x - x_1}{x_2 - x_1}\right] \qquad ...(5.41)$$

Hence, from equation (5.40)

$$[B] = \left[\frac{-1}{(x_2 - x_1)}, \frac{1}{(x_2 - x_1)}\right] = \frac{1}{L}[-1, 1] \qquad ...(5.42)$$

The strain is expressed as

$$\in = \frac{1}{L}[-1, 1]\begin{Bmatrix} u_{1x} \\ u_{2x} \end{Bmatrix}$$

$$= \left[\frac{u_{2x} - u_{1x}}{L}\right] \qquad ...(5.43)$$

From this relation it is clear that the strain is constant within each element.

Displacement plot

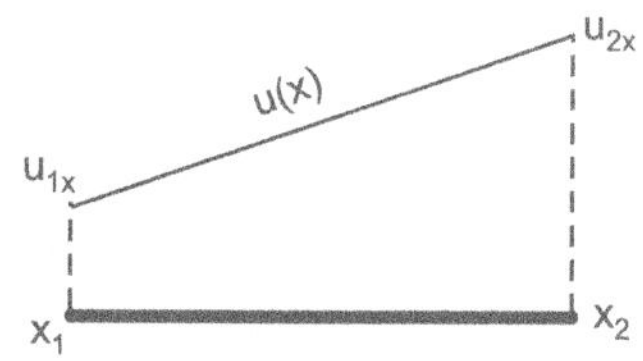

$$u(x) = a_0 + a_1 x$$

Strain plot

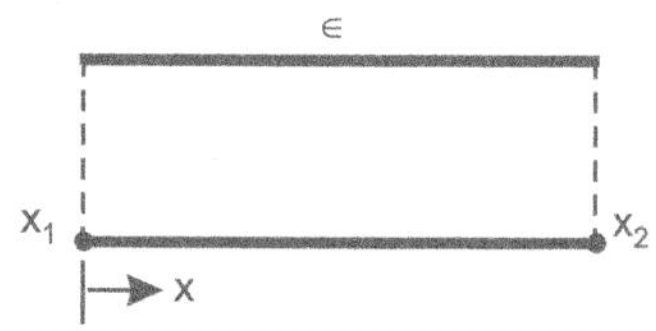

$$\in = \left[\frac{u_{2x} - u_{1x}}{L}\right] \qquad ...(5.44)$$

= Constant over the length

Recalling that the stress and strain are related by Hooke's Law of strength of material.

$$\sigma = E\in = E\frac{du}{dx} = E[B]\{U\} \qquad ...(5.45)$$

The expression (5.45) confirms that stress is also constant inside each element. This has the implication that the stress and strain are discontinuous accross element boundaries in general.

- **Observation**

1. Displacement approximation in terms of shape function,

$$u(x) = [N]\{U\} = \left[\frac{x_2 - x}{x_2 - x_1}, \frac{x - x_1}{x_2 - x_1}\right]\begin{Bmatrix} u_{1x} \\ u_{2x} \end{Bmatrix} \qquad ...(5.46)$$

2. Strain approximation in terms of strain-displacement matrix $[B]$

$$\in(x) = [B]\{U\} = \frac{1}{(x_2 - x_1)}[-1, 1]\begin{Bmatrix} u_{1x} \\ u_{2x} \end{Bmatrix} \qquad ...(5.47)$$

3. Stress approximation in terms of [B] and Young's modulus

$$\sigma = E[B]\{U\} = \frac{E}{(x_2 - x_1)}[-1, 1]\begin{Bmatrix} u_{1x} \\ u_{2x} \end{Bmatrix} \quad ...(5.48)$$

- *Task 3:*

Derivation of stiffness matrix of each element using principle of minimum potential energy.

The potential energy functional is the difference of strain energy and the work potential due to load.

$$\Pi_1 [U] = \frac{1}{2}\int_{x_1}^{x_2} \sigma \in Adx - \int_{x_1}^{x_2} budx \quad ...(5.49)$$

Let us plug in the informations obtained in the task 2.

$$u(x) = [N]\{U\},\ \in(x) = [B]\{U\} \text{ and } \sigma = E[B]\{U\}$$

Hence the equation (5.49) gets transformed to

$$\Pi_1 \{U\} = \frac{1}{2}U^T \underbrace{\left[\int_{x_1}^{x_2} B^T EB\, Adx\right]}_{\text{stiffness matrix}} U - U^T \underbrace{\left[\int_{x_1}^{x_2} N\, bdx\right]}_{\text{load vector}} \quad ...(5.50)$$

Thus for a 1D bar element

Stiffness matrix

$$[K] = \int_{x_1}^{x_2} [B]^T E[B]\, Adx \quad ...(5.51)$$

where the strain-displacement matrix

$$[B] = \frac{1}{(x_2 - x_1)}[-1, 1] \quad ...(5.52)$$

and

$$[B]^T E[B] = \frac{1}{(x_2 - x_1)}\begin{Bmatrix} -1 \\ 1 \end{Bmatrix} E \cdot \frac{1}{(x_2 - x_1)}[-1, 1]$$

$$= \frac{E}{(x_2 - x_1)^2}\begin{bmatrix} 1 & -1 \\ -1 & 1 \end{bmatrix}$$

$$\int_{x_1}^{x_2} (B^T EB) Adx = \frac{AE}{(x_2 - x_1)^2}\begin{bmatrix} 1 & -1 \\ -1 & 1 \end{bmatrix}\int_{x_1}^{x_2} dx$$

$$= \frac{AE}{(x_2 - x_1)^2}\begin{bmatrix} 1 & -1 \\ -1 & 1 \end{bmatrix}(x_2 - x_1)$$

Hence the stiffness matrix

$$[K] = \frac{AE}{(x_2 - x_1)}\begin{bmatrix} 1 & -1 \\ -1 & 1 \end{bmatrix} = \frac{AE}{L}\begin{bmatrix} 1 & -1 \\ -1 & 1 \end{bmatrix} \quad ...(5.53)$$

Element nodal force vector due to distributed body force is

$$[f_b] = \int_{x_1}^{x_2} [N]^T b dx \qquad ...(5.54)$$

The minimum potential energy

$$\pi_1(U) = \frac{1}{2} U^T [K] U - U^T f_b$$

$$\frac{\partial \pi_1}{\partial u_{x1}} = 0 \qquad \frac{\partial \pi_1}{\partial u_{x2}} = 0$$

Hence differentiating the equation for functional

$$\frac{\partial \pi}{\partial u_x} = [K]\, U - f_b = 0$$

Thus

$$[K]\,\{U\} = \{f_b\} \qquad ...(5.55)$$

- **Observations**
  1. Stiffness matrix is singular and therefore non-invertible.
  2. The stiffness matrix is symmetric *i.e.*, $[K] = [K]^T$
  3. Sum of the elements of any row or column of stiffness matrix zero.

5.8 ENERGY METHOD: STIFFNESS MATRIX

In static analysis of structures the relation is to be established between the force and the displacement. To determine the stress and strain in the element also it is required to determine the distribution of displacement in the finite element. This is accomplished through the assumption of an approximate polynomial for the displacement field. The displacement field is expressed in terms of the space variables and the generalized co-ordinates. By the application of boundary conditions it is possible to determine the generalized co-efficients in the polynomial expression assumed for the displacement function which approximates the distribution of displacement in the element. Hence it is made convenient to relate the degrees of freedom and the generalized constants to the displacement field. Then by the concept of virtual work the stiffness matrix and the load vectors are obtained. The virtual work, equates the energies due to internal work and the external work done on the system. The said procedure to obtain the shape function and the stiffness matrices are described in the mathematical form as stated in the following way.

The trial function for the displacement field is given as

$$u(x) = a_0 + a_1 x + a_2 x^2 + \ldots a_n x^n \qquad ...(5.56)$$

The same may be written in the matrix form as

$$u(x) = [1 \; x \; x^2 \; \ldots \; x^n] \begin{Bmatrix} a_0 \\ a_1 \\ a_2 \\ \vdots \\ a_n \end{Bmatrix} \qquad ...(5.57)$$

$$u(x) = [P]\{A\} \qquad ...(5.58)$$

where $[P]$ is the cartesian parameter matrix and $\{A\}$ is the generalized co-ordinate matrix.

Let the available boundary conditions be

$$\begin{aligned} u(x_0) &= u_0 = a_0 + a_1 x_0 + a_2 x_0^2 ... a_n x_0^n \\ u(x_1) &= u_1 = a_0 + a_1 x_1 + a_2 x_1^2 + \dots a_n x_1^n \qquad ...(5.59) \end{aligned}$$

so on $\quad u(x_n) = u_n = a_0 + a_1 x_n + a_2 x_n^2 + \dots a_n x_n^n$

These expressions can be written in the matrix form

$$\{U\} = \begin{Bmatrix} u_0 \\ u_1 \\ u_2 \\ \vdots \\ u_n \end{Bmatrix} = \begin{bmatrix} 1 & x_0 & x_0^2 & \cdots & x_0^n \\ 1 & x_1 & x_1^2 & \cdots & x_1^n \\ \vdots & \vdots & \vdots & \cdots & \vdots \\ 1 & x_n & x_n^2 & \cdots & x_n^n \end{bmatrix} \begin{Bmatrix} a_0 \\ a_1 \\ a_2 \\ \vdots \\ a_n \end{Bmatrix} \qquad ...(5.60)$$

$$\{U\} = [C]\{A\} = \text{nodal displacement parameters}$$

where $[C]$ is the connectivity matrix. By inverting the matrix $[C]$ the generalized co-ordinates are obtained as

$$\{A\} = [C]^{-1}\{U\} \qquad ...(5.61)$$

Substituting the same in equation (5.58) we arrive at

$$\begin{aligned} u(x) &= [P][C]^{-1}\{U\} \\ &= [N]\{U\} \qquad ...(5.62) \end{aligned}$$

where $[N] = [P][C]^{-1}$ is the rectangular matrix of shape functions and $[C]^{-1}$ is the displacement transformation matrix. The number of generalized co-ordinates is equal to total number of degrees of freedom in the element, which is equal to number of nodes multiplied by degrees of freedom at each node. Hence the connectivity and displacement transformation matrices are square matrices. The number of shape functions is equal to number of nodes in the element. To determine shape function $[P]$ and $[C]$ are needed.

The expression for strain energy in structural mechanics is one of the basis methods to arrive at the relation for stiffness matrix infinite element method. The force-displacement relation in theory of static analysis has the mathematical form

$$\{F\} = [K]\{U\} \qquad ...(5.63)$$

where $\quad \{F\} = \begin{Bmatrix} F_1 \\ F_2 \\ F_3 \\ \vdots \\ F_n \end{Bmatrix}$ and $\{U\} = \begin{Bmatrix} u_1 \\ u_2 \\ u_3 \\ \vdots \\ u_n \end{Bmatrix}$

$u_1, u_2, u_3 ... u_n$ are the degrees of freedom for an arbitrary element. $F_1, F_2, F_3 \dots F_n$ are the corresponding nodal forces, $[K]$ is the stiffness matrix that relates the nodal forces with nodal displacements. To derive the relation for stiffness matrix we equate the internally developed strain energy with the external work done on the system at equilibrium.

The average work done by the force vector to displace the nodes through displacement vector is given by

$$W = \frac{1}{2}F_1u_1 + \frac{1}{2}F_2u_2 + \dots \frac{1}{2}F_nu_n$$

In the matrix form

$$W = \frac{1}{2}[F]^T\{u\} \qquad ...(5.64)$$

Substituting equation (5.63) in equation (5.64)

$$W = \frac{1}{2}\{[K]\{u\}\}^T\{u\}$$

$$= \frac{1}{2}\{u\}^T[K]\{u\} \qquad ...(5.65)$$

The expression for total strain energy is :

$$U = \int_V \frac{1}{2}\{\sigma\}\{\in\}dV \qquad ...(5.66)$$

For simplicity we consider 1D axial bar for which the stress strain relation is obtained from Hooke's Law in mechanics of materials.

$$[\sigma] = [E]\{\in\} = [D]\{\in\} \qquad ...(5.67)$$

where $[D]$ is the elasticity matrix.

Hence we can write

$$U = \int_V \frac{1}{2}\{\in\}^T[D]\{\in\}dV \qquad ...(5.68)$$

But for a bar element

$$\in = \frac{du}{dx}(x) = [L][N]\{u\} \qquad ...(5.69)$$

Because $\qquad u(x) = [N]\{u\}$

Representing $[L]$ $[N]$ = $[B]$ equation (5.69) is rewritten as ($[B]$ is the strain-displacement matrix)

$$\{\in\} = [B]\{U\} \qquad ...(5.70)$$

Hence total strain energy takes the form

$$U = \frac{1}{2}\int_V [B]^T\{u\}^T[D][B]\{u\}dV \qquad ...(5.71)$$

But from the principle of virtual work

$$W = U$$

From equation (5.65) and equation (5.71)

$$\frac{1}{2}\{U\}^T[K]\{u\} = \frac{1}{2}\{u\}^T\left[\int_V [B]^T[D][B]dV\right]\{u\}$$

Hence

$$\text{Stiffness matrix} = [K] = \int_V [B]^T[D][B]dV \qquad ...(5.72)$$

Example 5.2. *If for a bar element the shape functions are given as* $N_1 = 1 - \frac{x}{L}$ *and* $N_2 = \frac{x}{L}$ *determine the stiffness matrix given the ratio* $\frac{AE}{L} = 200$ *units.*

Solution. For an axial bar with two nodes the given shape functions are

$$N_1 = 1-\frac{x}{L}$$

$$N_2 = \frac{x}{L}$$

But the stiffness matrix is expressed as

$$[K] = \int_V [B]^T [B]\, dv$$

In this expression $[D] = E$ and $Adx = dv$ for an axial bar of uniform diameter.

The same can be rewritten as

$$[K] = AE\int_0^L [B]^T [B]\, dv$$

But $$[B] = [L]\,[N] \quad \text{and} \quad [L] = \frac{d}{dx}$$

$$= \frac{d}{dx}\left[1-\frac{x}{L}\;\frac{x}{L}\right] = \left[-\frac{1}{L},\frac{1}{L}\right]$$

Hence $$[K] = AE\int_0^L \begin{bmatrix}-1/L\\ 1/L\end{bmatrix}[-1/L \;\; 1/L]\,dx$$

$$= \frac{AE}{L^2}\int_0^L \begin{bmatrix}1 & -1\\ -1 & 1\end{bmatrix} dx$$

$$\text{stiffness matrix} = \frac{AE}{L}\begin{bmatrix}1 & -1\\ -1 & 1\end{bmatrix} = 200\begin{bmatrix}1 & -1\\ -1 & 1\end{bmatrix}$$

5.9 BAR ELEMENT-REVISITED

Consider a bar element of length L with one end in origin and longitudinal axis along the x-direction. The co-ordinates of node 1 is (0,0) and node 2 is $(L,0)$. Let the displacement at a distance x from node 1 be $U(x)$. Let us conjecture a displacement function to represent $u(x)$. The configuration of the bar element is as shown in Fig. 5.19.

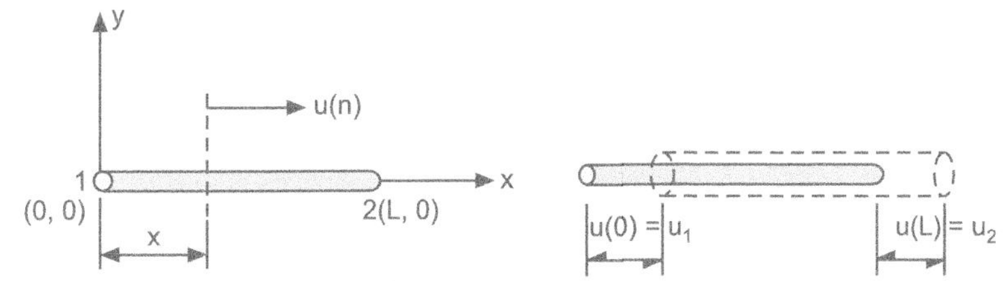

Fig. 5.19. Bar element deformation.

The linear displacement function assumed for the deformation variation is

$$u(x) = a_1 + a_2x$$

$$= [1, x]\begin{bmatrix} a_1 \\ a_2 \end{bmatrix}$$

$$= [P]\,[A] \qquad ...(5.73)$$

where $[P]$ = parameter matrix and $[A]$ is the matrix of generalized co-efficients.

Incorporating the boundary conditions $u(x = 0) = u_1$ and $u(x = L) = u_2$ the system of equations obtained are

$$u_1 = a_1$$

$$u_2 = a_1 + a_2 L$$

To express these in the matrix form

$$\begin{bmatrix} u_1 \\ u_2 \end{bmatrix} = \begin{bmatrix} 1 & 0 \\ 1 & L \end{bmatrix}\begin{bmatrix} a_1 \\ a_2 \end{bmatrix} \qquad ...(5.74)$$

$$\{U\} = [C]\,[A]$$

where $\{U\}$ is nodal displacement vector and [C] is the connectivity matrix. The generalized co-efficients, from equation (5.73)

$$[A] = [C]^{-1}\,\{U\}$$

Substitution of [A] in equation (5.73)

$$u(x) = [P]\,[C]^{-1}\,\{U\}$$

$$= [1, x]\begin{bmatrix} 1 & 0 \\ 1 & L \end{bmatrix}^{-1}\begin{Bmatrix} u_1 \\ u_2 \end{Bmatrix}$$

$$= [1, x]\begin{bmatrix} 1 & 0 \\ -\frac{1}{L} & \frac{1}{L} \end{bmatrix}\begin{Bmatrix} u_1 \\ u_2 \end{Bmatrix}$$

$$u(x) = \left[\left(1 - \frac{x}{L}\right), \frac{x}{L}\right]\begin{Bmatrix} u_1 \\ u_2 \end{Bmatrix}$$

$$= [N]\begin{Bmatrix} u_1 \\ u_2 \end{Bmatrix} \qquad ...(5.74a)$$

where $[N]$ is the shape function matrix. The shape functions are given by

$$N_1(x) = 1 - \frac{x}{L} \quad \text{and} \quad N_2(x) = \frac{x}{L}$$

It may be noted that $N_1 + N_2 = 1$

The strain-displacement relationship by using the mechanics theory of materials gives the axial strain.

$$\epsilon(x) = \frac{du}{dx} = \frac{d}{dx}([N]\{U\})$$

$$= [B]\,\{U\} \qquad ...(5.75)$$

$[B]$ is the strain displacement matrix

$$\epsilon(x) = \left[-\frac{1}{L}, \frac{1}{L}\right]\{U\} \qquad ...(5.76)$$

Hooke's law from theory of elasticity yields the stress displacement relation to be

$$\sigma(x) = E\in(x) = E[B]\{U\} \quad ...(5.77)$$

$\sigma(x)$ is the axial stress and E is the elastic modulus in the equation (5.77)

Using the principle of virtual work

$$\text{Inter work} = \text{Stress} \times \text{strain} \times \text{volume}$$

$$W_I = \iiint \sigma(x) \in(x)dx\, dy\, dz$$

But $\iint dy\, dz$ = Bar cross-sectional area = A

$$W_I = \int \sigma(x) \in(x) . A . dx \quad ...(5.78)$$

Using equations (5.77) and (5.78)

$$W_I = A\left[\int B^T EBdx\right]U^T U$$

External work done by the forces is given by

$$W_E = \text{force} \times \text{displacement}$$

$$= \begin{bmatrix} f_1 \\ f_2 \end{bmatrix} [u_1, u_2]$$

$$= [U]^T [F] \quad ...(5.79)$$

Equating the internal work (virtual) and the external work form equations (5.78) and (5.79)

$$A\left[\int [B]^T E[B]\right]U^T U\, dx = U^T F$$

Hence

$$[F] = [K]\{U\} \quad ...(5.80)$$

where

$$[K] = \text{stiffness matrix} = A\int [B]^T E[B]\, dx$$

- Derivation of stiffness matrix.

  The resultant stiffness matrix is obtained in the following way L

$$[K] = AE\int_0^L \begin{bmatrix} -1/L \\ 1/L \end{bmatrix}\left[-\frac{1}{L}, \frac{1}{L}\right] dx \quad ...(5.81)$$

$$= AE\int_0^L \begin{bmatrix} \frac{1}{L^2} & -1/L^2 \\ -1/L^2 & 1/L^2 \end{bmatrix} dx$$

$$= \frac{AE}{L}\begin{bmatrix} 1 & -1 \\ -1 & 1 \end{bmatrix} \quad ...(5.82)$$

5.10 TONTI DIAGRAM

Tonti diagram shown in Fig. 5.20 relates

1. Change in length with nodal displacement vector
2. Nodal force with the change in length.
3. Unknown forces with nodal force vector.

4. Force vector with nodal displacement through the stiffness matrix.

The preceding points represent the kinematic condition constitutive relation, equilibrium condition and stiffness relation respectively.

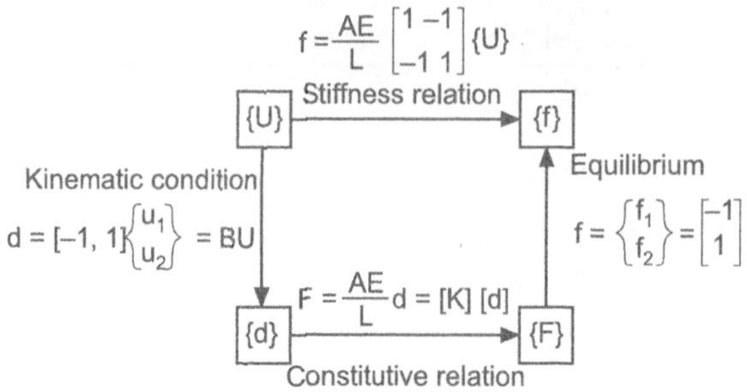

Fig. 5.20. Tonti diagram for bar element.

Bar element: Relation between natural and global variables.

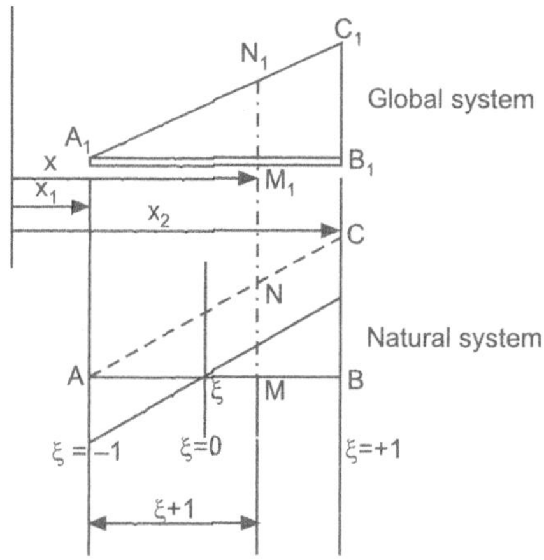

Fig. 5.21. Natural system.

From triangles A_1, B_1, C_1 and A_1, M_1, N_1 and comparing with triangles ABC and AMN (They are similar triangles)

$$\frac{A_1M_1}{A_1B_1} = \frac{AM}{AB}$$

$$\frac{(x - x_1)}{(x_2 - x_1)} = \frac{(\xi + 1)}{+1-(-1)}$$

Hence the relation between ξ and x is

$$\xi = \frac{2(x - x_1)}{(x_2 - x_1)} - 1 \qquad \ldots(5.83)$$

5.11 1-D BAR ELEMENT IN NATURAL CO-ORDINATES

An one-dimensional bar element with two nodes and one degree of freedom at each node is a linear element. For the finite element analysis the same element is considered here, with natural co-ordinate system. The formulation and derivation of shape functions, the stiffness matrix and consistent load vectors for different loading conditions have been accomplished in natural co-ordinate system.

The natural co-ordinates are the dimensionless system of variables that are defined and related to Cartesian system by

$$\xi = \frac{2(x - x_1)}{(x_2 - x_1)} - 1$$

where $(x_1, 0)$ and $(x_1, 0)$ are Cartesian local co-ordinates of the node 1 and node 2 respectively as shown in Fig. 5.22. To define the co-ordinates of node 1 and node 2 in natural co-ordinate,

when $x = x_1$, at node 1; $\xi = -1$

when $x = x_2$, at node 2; $\xi = +1$

The system is shown in Fig. 5.22.

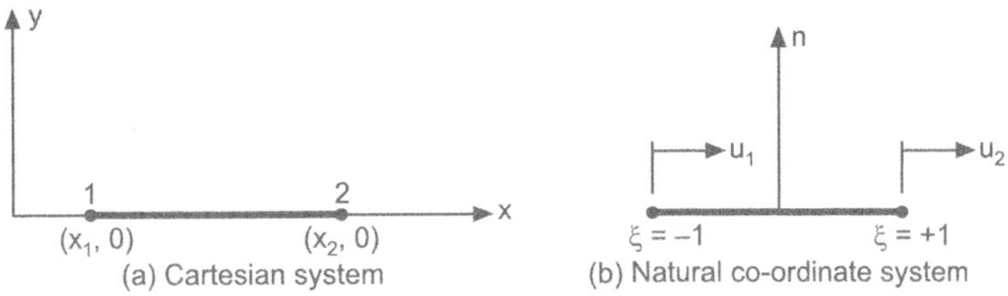

Fig. 5.22. Natural co-ordinates.

To obtain the differential of the natural co-ordinate ξ the expression (5.83) is differentiated as

$$d\xi = \frac{2}{(x_2 - x_1)} dx \qquad \text{...(5.84)}$$

or $\frac{d\xi}{dx} = \frac{2}{L}$, where L is the length of the element.

The admissible function approximated over the parent element system using linear polynomial expressed in terms of the natural co-ordinates ξ is

$$u(\xi) = a_1 + a_2\xi \qquad \text{...(5.85)}$$

This can be written in matrix form as

$$[1, \xi] \begin{Bmatrix} a_1 \\ a_2 \end{Bmatrix} = [P][A] \qquad \text{...(5.86)}$$

where $[P]$ is the parameter matrix and $[A]$ is the matrix of generalized co-ordinates. The generalized co-efficients a_1 and a_2 are to be determined to obtain the polynomial admissible function of displacement with the use of boundary condition

$$u = u_1 \text{ at } \xi = -1$$

and

$$u = u_2 \text{ at } \xi = +1$$

the system of equations arrived at is

$$\begin{aligned} a_1 - a_2 &= u_1 \\ a_1 + a_2 &= u_2 \end{aligned} \qquad \text{...(5.87)}$$

Expressing in the matrix form

$$\begin{bmatrix} +1 & -1 \\ +1 & +1 \end{bmatrix} \begin{Bmatrix} a_1 \\ a_2 \end{Bmatrix} = \begin{Bmatrix} u_1 \\ u_2 \end{Bmatrix}$$

$$\begin{bmatrix} +1 & -1 \\ +1 & +1 \end{bmatrix} = [C] \text{ is called the connectivity matrix.}$$

This leads to the formation of $[A] = [C]^{-1}\{U\}$

where $$[C]^{-1} = \frac{1}{2}\begin{bmatrix} +1 & +2 \\ -1 & 1 \end{bmatrix}$$

Substitution of this in equation (5.86) leads to

$$u(\xi) = [P]\,[A] = [P][C]^{-1}\{U\}$$

$$= \frac{1}{2}[1, \xi]\begin{bmatrix} +1 & +1 \\ -1 & +1 \end{bmatrix}\begin{Bmatrix} u_1 \\ u_2 \end{Bmatrix}$$

$$= \left(\frac{1-\xi}{2}\right)u_1 + \left(\frac{1+\xi}{2}\right)u_2 \quad \text{...(5.88)}$$

Hence it can be confirmed that

$$u(\xi) = N_1u_1 + N_2u_2$$

where $N = \left(\frac{1-\xi}{2}\right)$ and $N_2 = \left(\frac{1+\xi}{2}\right)$ are the shape functions. In the matrix representation of equation (5.88).

$$u(\xi) = [N_1, N_2]\begin{Bmatrix} u_1 \\ u_2 \end{Bmatrix} = [N]\{U\} \quad \text{...(5.89)}$$

From the fundamentals of strength of materials, the strain displacement relation is defined to be

$$\in(x) = \frac{du}{dx} = \frac{\partial}{\partial x}\{u(x)\} = [L]\,u(x)$$

But $$u(x) = [N]U \text{ which leads to}$$

$$\in(x) = [L]\,[N]\{U\} = [B]\{U\}$$

where $[B]$ is called the strain-displacement matrix.

$$[B] = \left[\frac{\partial N_1}{\partial x}, \frac{\partial N_2}{\partial x}\right] = \left[\frac{\partial N_1}{\partial \xi}\cdot\frac{\partial \xi}{\partial x}, \frac{\partial N_2}{\partial \xi}\cdot\frac{\partial \xi}{\partial x}\right] \quad \text{...(5.90)}$$

$$= \left[-\frac{1}{2}\cdot\frac{2}{L}, \frac{1}{2}\cdot\frac{2}{L}\right]$$

$$= \frac{1}{L}[-1, +1] \quad \text{...(5.91)}$$

From the theory of elasticity the establish stress-strain relation is given by

$$\sigma(x) = E[\in(x)] = E\,[B]\,\{U\}$$

$$= \frac{E}{L}[-1, +1]\begin{Bmatrix} u_1 \\ u_2 \end{Bmatrix} \quad \text{...(5.92)}$$

The strain energy,

$$U = \frac{1}{2}\int_V \sigma(x)\in(x)\,dV$$

$$= \frac{1}{2}\int_V E[B]^T\{U\}^T[B]\{U\}dV$$

But for an element of uniform cross-section, $A\,dv = Adx$. Hence

$$U = \frac{1}{2}U^T\left(\int_0^L [B]^T AE[B]dx\right)U$$

$$= \frac{1}{2}U^T[K]_e U \qquad \ldots(5.93)$$

where $[K]_e$ is the elemental stiffness matrix.

$$[K]_e = \int_0^L AE\frac{1}{L}\begin{bmatrix}-1\\+1\end{bmatrix}\frac{1}{L}[-1,+1]\,dx$$

$$= \frac{AE}{L^2}\int_0^L \begin{bmatrix}+1 & -1\\-1 & +1\end{bmatrix}dx$$

$$= \frac{AE}{L}\begin{bmatrix}1 & -1\\-1 & 1\end{bmatrix} \qquad \ldots(5.94)$$

This is the elemental stiffness matrix which depends on the geometry (A = area, L = length) and the mechanical property of material (E = Young's modulus) of the element.

5.11.1 Tapering Bar

For a uniformly varying bar with area A_1, and A_2 at the two ends and having length L the variable area at any distance x.

$$A(x) = N_1A_1 + N_2A_2$$

$$= \left(\frac{1-\xi}{2}\right)A_1 + \left(\frac{1+\xi}{2}\right)A_2$$

Further

$$A(x)dx = \left[\frac{1-\xi}{2}, \frac{1+\xi}{2}\right]\begin{bmatrix}A_1\\A_2\end{bmatrix}dx$$

But this is shown that $dx = \frac{L}{2}d\xi$.

The elemental stiffness matrix for the tapering bar is given as

$$[K]_e = \int_0^L [B]^T E[B]\,A(x)\,dx$$

$$= \frac{E}{L^2}\begin{bmatrix}-1\\1\end{bmatrix}[-1,1]\int_{-1}^{+1}\left[\left(\frac{1-\xi}{2}\right),\left(\frac{1+\xi}{2}\right)\right]\begin{Bmatrix}A_1\\A_2\end{Bmatrix}\frac{L}{2}d\xi$$

$$= \frac{E}{L}\left(\frac{A_1+A_2}{2}\right)\begin{bmatrix}1 & -1\\-1 & 1\end{bmatrix} \qquad \ldots(5.95)$$

Again it is relevant from the relation (5.95) that the elemental stiffness matrix is the function of geometry and mechanical property of the material.

5.11.2 The Force Vector

The force acting on an element can be of the following type:

1. Body force due to self-weight
2. Body force due to centrifugal loading
3. The fraction force acting on the surface of the member.
4. The concentrated (point) force acting on the element (nodes of the element).
5. Force due to temperature variation.

For these type of loads the elemental force vectors are derived as follows.

The body forces are developed by the self-weight of the body or the centrifugal force developed in the rotating body. The elemental force vector generally represented by and related by:

$$[f^b] = \int_V [N]^T [b]\, dV \qquad \text{...(5.96)}$$

$[a]$ Is the body force acting on the volume? When constant area of cross-section is considered for the element

$$dV = A dx$$

In natural co-ordinates

$$dV = A \cdot \frac{1}{2} d\xi$$

To estimate the force vector with the substitution $[N]^T$ in natural co-ordinates

$$[f^b] = \int_{-1}^{+1} \begin{bmatrix} (1-\xi)/2 \\ (1+\xi)/2 \end{bmatrix} [b]\, A \cdot \frac{L}{2}\, d\xi \qquad (5.96a)$$

Integrating the definite integral with respect to ξ

$$[f^b] = \frac{AL}{2}[b]\begin{bmatrix} 1 \\ 1 \end{bmatrix} \qquad \text{...(5.97)}$$

- When the body force is due to the self-weight

$[b] = \rho g$ where ρ is the mass density of the material in kg/mm^3.

Hence

$$[f^b] = \frac{A\rho g L}{2}\begin{bmatrix} 1 \\ 1 \end{bmatrix} \qquad \text{...(5.98)}$$

- When the body force is due to the centrifugal force

$$[b] = \rho w^2 r$$

where w = angular velocity in rad/sec.

r = is the average radius $= \dfrac{(r_1 + r_2)}{2}$

$$[f^b] = \frac{A\rho w^2 r L}{2}\begin{bmatrix} 1 \\ 1 \end{bmatrix} \qquad (5.99)$$

Force vector due to surface force:

Let the surface force, p, due to traction forces and uniformly distributed loads act on the bar element in the axial direction. The force vector, then is defined with

$$[f^b] = \int_S [N]^T [p] ds$$

For an element with unit width $ds = dx$. The dx in the natural co-ordinates in $\frac{L}{2}d\xi$. The substitution in equation (5.96b) leads to

$$[f^b] = \int_{-1}^{+1} \begin{bmatrix} (1-\xi)/2 \\ (1+\xi)/2 \end{bmatrix} \frac{pL}{2} dz$$

$$= \frac{pL}{2} \begin{bmatrix} 1 \\ 1 \end{bmatrix} \qquad \ldots(5.100)$$

Force vector: concentrated load

The concentrated load acts at various points in the element let the concentrated (point) loads be denoted by $[P_i]$. The effect of these loads on the elements is given by the force vector as:

$$[f^b] = \sum_{i=1}^{n} [N_i]^T [P_i] = \begin{bmatrix} (1-\xi)/2 \\ (1+\xi)/2 \end{bmatrix} [P_i] \qquad \ldots(5.101)$$

If a force P_1 acts at a distance 'd' from the origin $\xi = 0$

$$[f^b] = \frac{P_1}{2} \begin{bmatrix} (1-d) \\ (1+d) \end{bmatrix}$$

with $d = -1$, $[f^b] = \frac{P_1}{2} \begin{bmatrix} 2 \\ 0 \end{bmatrix}$

with $d = 0$, $[f^b] = \frac{P_1}{2} \begin{bmatrix} 1 \\ 1 \end{bmatrix}$

with $d = +1$, $[f^b] = \frac{P_1}{2} \begin{bmatrix} 0 \\ 2 \end{bmatrix}$

with $d = +3/4$, $[f^b] = \frac{P_1}{8} \begin{bmatrix} 1 \\ 7 \end{bmatrix}$

The force vector here is independent of the geometry but depends on the location of the point force.

Force due to temperature change

The temperature variation results in expansion or contraction of the system made of elements. This leads to the development of initial strain in the member. The initial strain is given as:

$$\epsilon^{in} = \propto \Delta T$$

where α is the thermal expansion co-efficient, 1°C

ΔT = is the temperature change at that node.

But $\Delta T = \Delta N_i T_i$

For a bar with uniform cross-section, A

$$[f^b] = \int_V [B]^T [D] \propto \sum [N_i] T_i (Adx) \qquad \ldots(5.102)$$

In the natural co-ordinate system

$$[f^b] = \int_{-1}^{+1} \frac{1}{L}\begin{bmatrix}-1\\1\end{bmatrix} E \propto \left[\frac{(1-\xi)}{2}, \frac{(1+\xi)}{2}\right]\begin{bmatrix}\Delta T_1\\ \Delta T_2\end{bmatrix} A \frac{L}{2} d\xi$$

$$= EA \propto \frac{(\Delta T_1 + \Delta T_2)}{2}\begin{bmatrix}-1\\+1\end{bmatrix}$$

$$= \frac{EA \propto \Delta T}{2}\begin{bmatrix}-1\\+1\end{bmatrix} \qquad \ldots(5.103)$$

Here ΔT is the change in temperature at the two nodes of the axial bar element. E is modulus of elasticity. It may be noted that the force vector is dependent on the thermal property apart from geometry and the mechanical property of material.

5.12 LOCAL TO GLOBAL TRANSFORMATIONS

Let the two dimensional local co-ordinate system be represented at the nodes as $(\bar{x}, \bar{y})$. The corresponding global system be (x, y) as shown in Fig. 5.23.

Let the displacements in local co-ordinate system at node i be $\bar{u}_{ix}$ and $\bar{u}_{iy}$. The displacement of node I in the global system be u_{ix} and u_{iy}. The corresponding forces be $\bar{f}_{ix}$ and $\bar{f}_{iy}$ in local and f_{ix} and f_{iy} in global system.

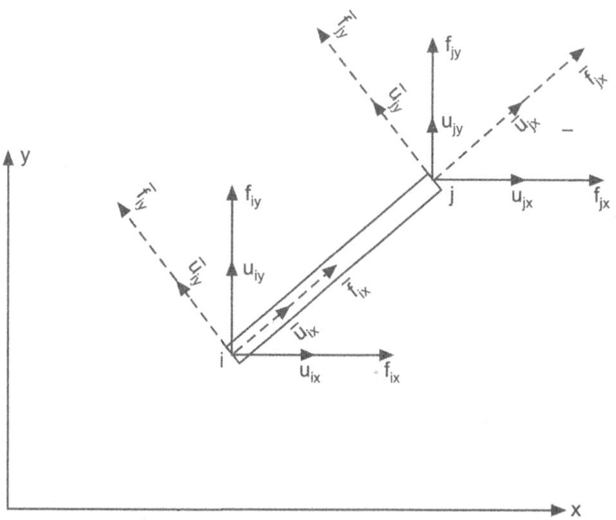

Fig. 5.23. Local and global co-ordinate system.

Consider the transformation between the global system and local system of co-ordinate frames. Consider a vector drawn from the origin to node i as u. The components of u are (u_x, u_y) and $(\bar{u}_x, \bar{u}_y)$ in the global and local frame respectively. From Fig. 5.24 it is obtained as

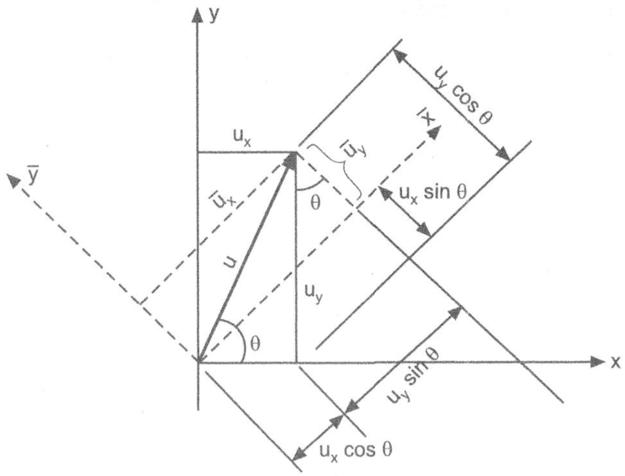

Fig. 5.24. Transformations.

$$\bar{u}_x = u_x \cos\theta + u_y \sin\theta$$
$$\bar{u}_y = -u_x \sin\theta + u_y \cos\theta \qquad ...(5.104)$$

The set of equations can be written in the matrix form as

$$\begin{Bmatrix} \bar{u}_x \\ \bar{u}_y \end{Bmatrix} = \begin{bmatrix} \cos\theta & \sin\theta \\ -\sin\theta & \cos\theta \end{bmatrix} \begin{Bmatrix} u_x \\ u_y \end{Bmatrix}$$

Since it is evident that $\sin^2\theta + \cos^2\theta = 1$ the system is orthogonal. Hence it can written that

$$\begin{Bmatrix} u_x \\ u_y \end{Bmatrix} = \begin{bmatrix} \cos\theta & -\sin\theta \\ \sin\theta & \cos\theta \end{bmatrix} \begin{Bmatrix} \bar{u}_x \\ \bar{u}_y \end{Bmatrix} = [T^*]\{\bar{U}\}$$

Assuming $l = \cos\theta$, $m = \sin\theta$

$$[T^*] = \begin{bmatrix} l & -m \\ m & l \end{bmatrix} \qquad ...(5.105)$$

At node i, consider the resolution of the displacement and force vectors.

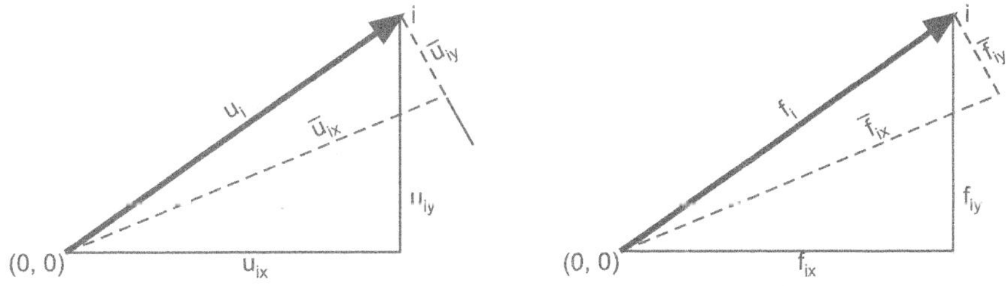

Fig. 5.25. Force vector and displacement vector.

Similarly when the force and displacement vectors are resolved

(u_{jx}, u_{jy}) = Displacement in global co-ordinates

$(\bar{u}_{jx}, \bar{u}_{jy})$ = Displacement in local co-ordinates

(f_{jx}, f_{jy}) = force vector in global co-ordinates

$(\bar{f}_{jx}, \bar{f}_{jy})$ = forces in local co-ordinates.

For an bar element with nodes i and j

Global

$$[u] = \begin{Bmatrix} u_{ix} \\ u_{iy} \\ u_{jx} \\ u_{jy} \end{Bmatrix}$$

Local

$$\{\bar{u}\} = \begin{Bmatrix} \bar{u}_{ix} \\ \bar{u}_{iy} \\ \bar{u}_{jx} \\ \bar{u}_{jy} \end{Bmatrix}$$

The force vectors

Global

$$\{f\} = \begin{Bmatrix} f_{ix} \\ f_{iy} \\ f_{jx} \\ f_{jy} \end{Bmatrix}$$

Local

$$\{\bar{f}\} = \begin{Bmatrix} \bar{f}_{ix} \\ \bar{f}_{iy} \\ \bar{f}_{jx} \\ \bar{f}_{jy} \end{Bmatrix}$$

By transformations

$$\{u\} = \begin{Bmatrix} u_{ix} \\ u_{iy} \\ u_{jx} \\ u_{jy} \end{Bmatrix} = \begin{bmatrix} l & -m & 0 & 0 \\ m & l & 0 & 0 \\ 0 & 0 & l & -m \\ 0 & 0 & m & l \end{bmatrix} \begin{Bmatrix} \bar{u}_{ix} \\ \bar{u}_{iy} \\ \bar{u}_{jx} \\ \bar{u}_{jy} \end{Bmatrix} \quad ...(5.106)$$

Hence

$$\{u\} = \begin{bmatrix} T^* & 0 \\ 0 & T^* \end{bmatrix} \{\bar{u}\} = [T]\{\bar{u}\}$$

Similarly for the force transformations

$$\{f\} = \begin{bmatrix} T^* & 0 \\ 0 & T^* \end{bmatrix} \{\bar{f}\} = [T]\{\bar{f}\} \quad ...(5.107)$$

5.13 GLOBAL STIFFNESS MATRIX

Let $[K]$ and $[\bar{K}]$ be the global and local stiffness matrices respectively. We know that, by transformation,

$$\{f\} = [T]\,\{\bar{f}\}$$

$$\{u\} = [T]\,\{\bar{u}\} \quad ...(5.108)$$

From the force displacement relation for an element in the local system

$$\{\bar{f}\} = [\bar{K}]\{\bar{u}\} \quad ...(5.109)$$

From equations (5.108) and (5.109)

$$[T]^{-1}\{f\} = [\bar{K}][T]^{-1}\{u\}$$

By simplification

$$\{f\} = [T]\,[\bar{K}][T]^{-1}\{u\} \quad ...(5.110)$$

We know that for an orthogonal system with $l^2 + m^2 = 1$

$$[T]^{-1} = [T]^T$$

Hence $\{f\} = [T][\bar{K}][T]^T\{u\}$

This leads to $[f] = [K]\{u\}$ ...(5.111)

where $[K]$ is the global elemental stiffness matrix.

The force displacement relation in local co-ordinate system for a bar element in two dimensional frame.

$$\{\bar{f}\} = \begin{bmatrix} k & 0 & -k & 0 \\ 0 & 0 & 0 & 0 \\ -k & 0 & k & 0 \\ 0 & 0 & 0 & 0 \end{bmatrix} \begin{Bmatrix} \bar{u}_{1x} \\ \bar{u}_{1y} \\ \bar{u}_{2x} \\ \bar{u}_{2y} \end{Bmatrix}$$

The axial stiffness for a bar, $k = \frac{AE}{L}$.

The elemental stiffness matrix in a global system

$$[k] = \begin{bmatrix} l & -m & 0 & 0 \\ m & l & 0 & 0 \\ 0 & 0 & l & -m \\ 0 & 0 & m & l \end{bmatrix} \begin{bmatrix} k & 0 & -k & 0 \\ 0 & 0 & 0 & 0 \\ -k & 0 & k & 0 \\ 0 & 0 & 0 & 0 \end{bmatrix} \begin{bmatrix} l & m & 0 & 0 \\ -m & l & 0 & 0 \\ 0 & 0 & l & m \\ 0 & 0 & -m & l \end{bmatrix}$$

Hence the stiffness matrix in global system

$$[k] = \frac{AE}{L} \begin{bmatrix} l^2 & lm & -l^2 & -lm \\ lm & m^2 & -lm & -m^2 \\ -l^2 & -lm & l^2 & lm \\ -lm & -m^2 & lm & m^2 \end{bmatrix} = \begin{Bmatrix} \text{Stiffness matrix} \\ \text{for truss element} \end{Bmatrix} \quad ...(5.112)$$

Direction co-sines

From the trigonometric relation

$$l = \cos\theta = \left(\frac{x_2 - x_1}{L}\right)$$

$$M = \sin\theta = \left(\frac{y_2 - y_1}{L}\right)$$

where L is the length of the element (x_1, y_1) are the co-ordinates of node 1 and (x_2, y_2) are the co-ordinates of node 2 in the global co-ordinate system. The angle θ is assumed positive in the anti-clockwise direction measured from x-axis.

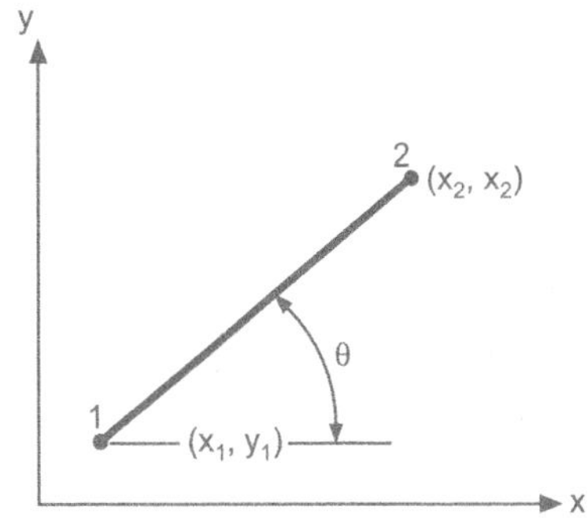

Fig. 5.26. Element co-ordinates.

Hence the elemental stiffness matrix for truss element depends on the element geometry (*i.e.*, length and area of cross-section) the mechanical property of material (*i.e.*, Young's modulus) and also the orientation of the element in the global frame.

5.13.1 Elemental Strain

To obtain the elemental strain in the global co-ordinate system the start point is the definition of strain. For an axial member subjected to axial loading , the strain,

$$\in(x) = \frac{\bar{u}_{2x} - \bar{u}_{1x}}{L}$$

$$= \frac{1}{L}[-1 \ 0 \ 1 \ 0] \begin{Bmatrix} \bar{u}_{1x} \\ \bar{u}_{1y} \\ \bar{u}_{2x} \\ \bar{u}_{2y} \end{Bmatrix}$$

$$= \frac{1}{L}[-1\ 0\ 1\ 0]\{\bar{u}\}$$

But by transformation of local system into global system

$$\{\bar{u}\} = [T]^{-1}\{u\} = [T]^T\{u\}$$

Hence

$$\bar{\epsilon} = \frac{1}{L}[-1,0,1,0]\begin{bmatrix} l & m & 0 & 0 \\ -m & l & 0 & 0 \\ 0 & 0 & l & m \\ 0 & 0 & -m & l \end{bmatrix}\{u\}$$

$$\bar{\epsilon} = \frac{1}{L}[-l\ -m\ l\ m]\{u\}$$

or

$$\bar{\epsilon} = \frac{1}{L}[-l\ -m\ l\ m]\begin{Bmatrix} u_{1x} \\ u_{2y} \\ u_{2x} \\ u_{2y} \end{Bmatrix} \quad ...(5.113)$$

5.13.2 Elemental Stress

The element stress distribution governed by the fundamentals of elasticity theory states that

$$\sigma = E\bar{\epsilon} = \frac{E}{L}(\bar{u}_{2x} - \bar{u}_{1x})$$

By the substitution of the elemental strain from equation (5.113)

$$\sigma = \frac{E}{L}[-l\ -m\ l\ m]\{u\} \quad ...(5.114)$$

Elemental Tension:

Further the tensile force in the element is estimated using the cross-sectional area of the element. For an element of uniform cross section, A.

$$\{t\} = \sigma.A = \frac{EA}{L}[-l\ -m\ l\ m]\{u\} \quad ...(5.115)$$

Example 5.3. *A bar element is inclined at 30° to x-axis. The properties for the element are E = 30 × 10*^{6} *N/mm*^{2}*, A = 2 mm*^{2}*, L = 60 mm. Compute the elemental stiffness matrix in global co-ordinates.*

$$l = \cos 30 = \frac{\sqrt{3}}{2} \qquad m = \sin 30 = \frac{1}{2}$$

Solution. The elemental stiffness matrix in global co-ordinates

$$[k] = \frac{(30\times 10^6)(2)}{60}\begin{bmatrix} 3/4 & \sqrt{3}/4 & -3/4 & -\sqrt{3}/4 \\ \sqrt{3}/4 & 1/4 & -\sqrt{3}/4 & -1/4 \\ -3/4 & -\sqrt{3}/4 & 3/4 & +\sqrt{3}/4 \\ -\sqrt{3}/4 & -1/4 & +\sqrt{3}/4 & 1/4 \end{bmatrix}$$

Example 5.4. *A linear elastic bar is having tapering section given by* $A(x) = \left(1+\frac{x}{40}\right)^2$ *cm*^{2}*. The configuration of the bar is as shown in Fig. 5.27. Compare the exact solution with FEM solution considering E to be the Young's modulus.*

Solution. The governing equilibrium equation is

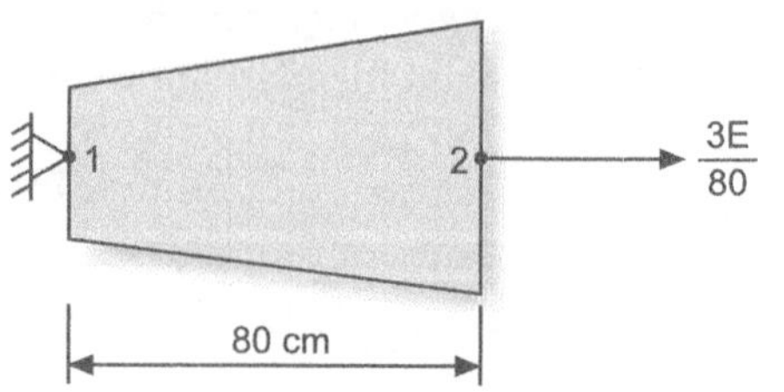

Fig. 5.27. Tapering bar.

$$E\frac{d}{dx}\left(A(x)\frac{dy}{dx}\right) = 0$$

and

$$EA\frac{dy}{dx} = P = \frac{3E}{80} \qquad \text{...}(a)$$

The analytical solution to equation (*a*) results in

$$u^{\text{exact}} = 3/2\left[1-\frac{1}{\left(1+\frac{x}{40}\right)}\right]_{x=80} = 1\text{ cm}$$

For the two noded bar element the FEM approximation

$$u^{\text{FEM}} = N_1(x)\, u_{1x} + N_2(x)\, u_{2x}$$

$$N_1(x) = \left(1-\frac{x}{80}\right) \qquad N_2(x) = \frac{x}{80}$$

at $x = 0,\ u_{1x} = 0$

Hence

$$u^{\text{FEM}} = \frac{x}{80}\, u_{2x}$$

But

$$K = \frac{E}{L^2}\int_0^L A(x)\,dx \begin{bmatrix} 1 & -1 \\ -1 & 1 \end{bmatrix}$$

$$= \frac{E}{(80)^2}\int_0^{80}\left[\left(1+\frac{x}{40}\right)^2 dx\right]\begin{bmatrix} 1 & -1 \\ -1 & 1 \end{bmatrix}$$

$$= \frac{13E}{240}\begin{bmatrix} 1 & -1 \\ -1 & 1 \end{bmatrix}$$

From force displacement relation

$$[K]\,\{U\} = \{F\}$$

$$\frac{13E}{240}\begin{bmatrix} 1 & -1 \\ -1 & 1 \end{bmatrix}\begin{Bmatrix} u_{1x} \\ u_{2x} \end{Bmatrix} = \begin{Bmatrix} f_{1x} \\ P \end{Bmatrix}$$

But at the fixed end node (1) $u_{1x} = 0$

$$u_{2x} = \frac{240}{13E}\left(\frac{3E}{80}\right) = \frac{9}{13}\text{ cm}$$

u_{2x}^{FEM} is less than the exact solution $u^{\text{exact}} = 1\text{cm}$

$$f_{1x} = -\frac{13E}{240}\cdot u_{2x}$$

$$= -\frac{13E}{240}\cdot\frac{240}{13E}P = -P$$

Observations:

1. Elemental equilibrium
2. Nodal equilibrium

Example 5.5. *A stepped bar as shown in Fig. 5.28 is fixed at one end and a tensile point load acts at the other end. Both the steps have same material and length. But the area of first step is two times the second step. Determine the unknown displacements.*

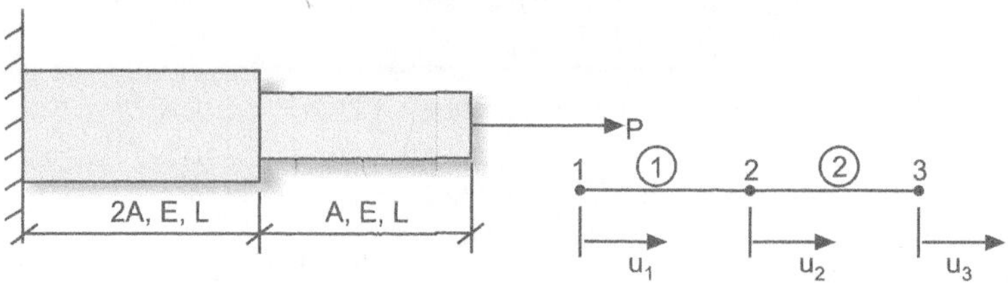

Fig. 5.28. Stepped bar.

Solution. The system of equation for element –1

$$\frac{AE}{L}\begin{bmatrix} 2 & -2 & 0 \\ -2 & 2 & 0 \\ 0 & 0 & 0 \end{bmatrix}\begin{Bmatrix} u_1 \\ u_2 \\ u_3 \end{Bmatrix} = \begin{Bmatrix} P_1^{(1)} \\ P_2^{(1)} \\ 0 \end{Bmatrix}$$

The system of equations for element –2

$$\frac{AE}{L}\begin{bmatrix} 0 & 0 & 0 \\ 0 & 1 & -1 \\ 0 & -1 & 1 \end{bmatrix}\begin{Bmatrix} u_1 \\ u_2 \\ u_3 \end{Bmatrix} = \begin{Bmatrix} 0 \\ P_1^{(2)} \\ P_2^{(2)} \end{Bmatrix}$$

Assembled global system of equations

$$\frac{AE}{L}\begin{bmatrix} 2 & -2 & 0 \\ -2 & 3 & -1 \\ 0 & -1 & -1 \end{bmatrix}\begin{Bmatrix} u_1 \\ u_2 \\ u_3 \end{Bmatrix} = \begin{Bmatrix} P_1^{(1)} \\ P_2^{(1)} + P_1^{(2)} \\ P_2^{(2)} \end{Bmatrix}$$

The boundary conditions are given at the support as:

$$u_1 = 0 \qquad P_2^{(2)} = P \qquad P_2^{(1)} + P_1^{(2)} = 0$$

By the method of elimination the retained system of equations are

$$\frac{AE}{L}\begin{bmatrix} 3 & -1 \\ -1 & 1 \end{bmatrix}\begin{Bmatrix} u_2 \\ u_3 \end{Bmatrix} = \begin{Bmatrix} 0 \\ P \end{Bmatrix}$$

Solving the simultaneous equation by Gauss elimination method,

$$\frac{AE}{L}\begin{bmatrix} 3 & -1 \\ 0 & 2 \end{bmatrix}\begin{Bmatrix} u_2 \\ u_3 \end{Bmatrix} = \begin{Bmatrix} 0 \\ 3P \end{Bmatrix}$$

From last row

$$u_3 = \frac{3PL}{2AE}$$

Substituting this in row 1

$$u_2 = \left(\frac{3P}{2}\right)\cdot\frac{L}{3AE} = \frac{PL}{2AE}$$

Strain in element –1

$$\in^{(1)} = \frac{u_2 - u_1}{L} = \frac{(PL/2AE) - 0}{L} = \frac{P}{2AE}$$

Stress in element – 1

$$\sigma^{(1)} = E \in^{(1)} = \frac{P}{2A}$$

Strain in element – 2

$$\in^{(2)} = \frac{u_3 - u_2}{L} = \frac{3/2\left(\frac{PL}{AE}\right) - \frac{1}{2}\left(\frac{PL}{AE}\right)}{L} = \frac{P}{AE}$$

Stress in element – 2

$$\sigma^{(2)} = E \in^{(2)} = \frac{P}{A}$$

Example 5.6. *Consider a bar of uniform cross-sectional area of 20 mm^2. For the bar shown in Fig. 5.29 has the Young's modulus of 2 × 10^5 N/mm^2. If the displacements at the two ends are 0.5 mm and 0.6 mm respectively. Determine*

1. *Displacement at M at a distance x = 60 mm*
2. *Strain and stress*
3. *Element stiffness matrix*
4. *Strain energy in the element.*

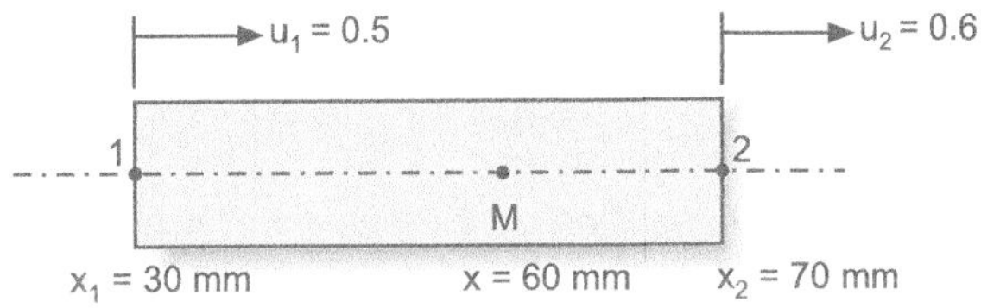

Fig. 5.29

Solution.

$$E = 2 \times 10^5 \text{ N/mm}^2$$
$$u_1 = 0.5 \text{ mm}$$
$$u_2 = 0.6 \text{ mm}$$
$$A = 20 \text{ mm}^2$$

The natural co-ordinate at $M(x = 60 \text{ mm})$

$$\xi = \frac{2(x - x_1)}{(x_2 - x_1)} - 1 = \frac{2(60 - 30)}{(70 - 30)} - 1 = 0.5$$

The shape function in natural co-ordinates

$$N_1(\xi) = \left(\frac{1-\xi}{2}\right) \quad \text{and} \quad N_2(\xi) = \left(\frac{1+\xi}{2}\right)$$

$$= \frac{1-0.5}{2} = 0.25 \qquad = \frac{1+0.5}{2} = 0.75$$

1. Let $u(\xi)$ be the displacement at M

$$u(\xi) = N_1 u_1 + N_2 u_2$$
$$u = 0.25\,(0.5) + 0.75\,(0.6) = 0.575$$

The displacement at M is $= 0.575$ mm.

2. Stress and strain

$$\in = \frac{(u_2 - u_1)}{L} = \frac{0.6 - 0.5}{40} = 0.0025$$

$$\sigma = E\in = 2 \times 10^5 \times 25 \times 10^{-4} = 500 \text{ N/mm}^2$$

3. Elemental stiffness matrix

$$[K]^e = \frac{AE}{L}\begin{bmatrix} 1 & -1 \\ -1 & 1 \end{bmatrix} = \frac{20 \times 2 \times 10^5}{40}\begin{bmatrix} 1 & -1 \\ -1 & 1 \end{bmatrix}$$

$$= 10^5\begin{bmatrix} 1 & -1 \\ -1 & 1 \end{bmatrix} \text{N/mm}$$

4. Strain energy in the element

$$U = \frac{1}{2}\{U\}^T[K]^e\{u\}$$

$$= \frac{1}{2} \times 10^5 [0.5 \;\; 0.6]\begin{bmatrix} 1 & -1 \\ -1 & 1 \end{bmatrix}\begin{bmatrix} 0.5 \\ 0.6 \end{bmatrix}$$

$$= 0.5 \times 10^5 [-0.1 \;\; +0.1]\begin{bmatrix} 0.5 \\ 0.6 \end{bmatrix}$$

$$U = 500 \text{ N–mm}$$

Example 5.7. *Tapering bar.*

A tapering hanging bar with given dimensions is as shown in Fig. 5.30. If the data for the material are $E = 2 \times 10^5$ N/mm^2, density $= 8 \times 10^{-6}$ kg/mm^3 and the thickness of the plate, t = 25 mm determine the following considering two elements for the tapering hanging plate.

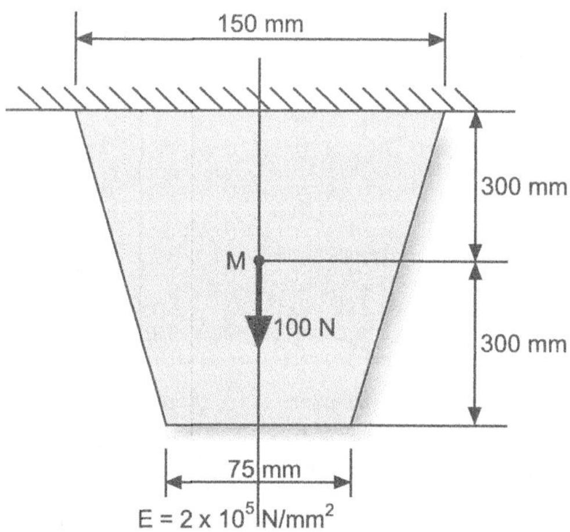

Fig 5.30. Tapering bar.

1. *The expression for the element stiffness matrix and the force vectors.*
2. *Assembled stiffness matrix and the load vector.*
3. *The unknown displacements.*
4. *Stress in each element.*
5. *Reaction at the support.*

Solution. The tapering bar with two elements is analogous to a stepped bar with two steps and three nodes. Let A_1, A_2 and A_3 be the areas at base, at point M and bottom end.

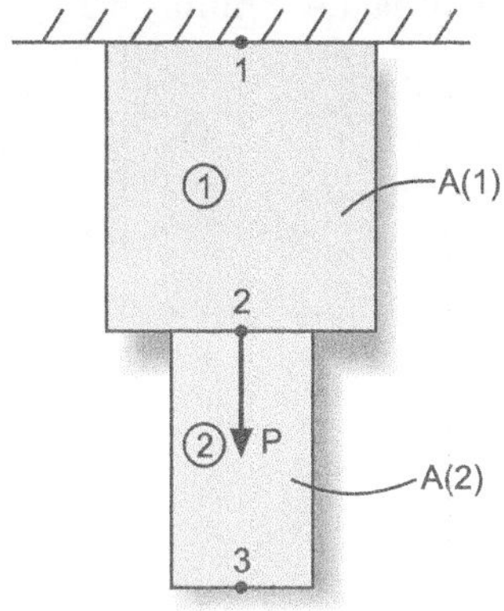

Fig. 5.31. Stepped approximation.

Now $$A_2 = \frac{A_1 + A_3}{2} = \frac{25(150+75)}{2} = 2812.5 \text{ mm}^2$$

The average area of step 1

$$A^{(1)} = \frac{A_1 + A_2}{2} = 3281.25 \text{ mm}^2$$

The average area of step 2

$$A^{(2)} = \frac{A_2 + A_3}{2} = 2343.75 \text{ mm}^2$$

Node-element connectivity matrix

Element No.	*Node i*	*Node j*
1	1	2
2	2	3

Element stiffness matrix for element number 1.

$$[K^{(1)}] = \frac{A^{(1)}E}{L}\begin{bmatrix} 1 & -1 \\ -1 & 1 \end{bmatrix}$$

$$= 21875 \times 10^2 \begin{bmatrix} 1 & -1 \\ -1 & 1 \end{bmatrix}$$

For element 2

$$[K^{(2)}] = \frac{A^{(2)}E}{L}\begin{bmatrix} 1 & -1 \\ -1 & 1 \end{bmatrix}$$

$$= 15625 \times 10^2 \begin{bmatrix} 1 & -1 \\ -1 & 1 \end{bmatrix}$$

Elemental load vector : Body weight

$$f_b = \int_{x_1}^{x_2} N^T b \, dx$$

$$= \int_0^{300} \begin{bmatrix} N_1 \\ N_2 \end{bmatrix} \rho (A^{(1)}) g \, dx$$

$$N_1 = 1 - \frac{x}{L} = 1 - x/300$$

$$N_2 = \frac{x}{L} = x/300$$

Hence

$$f_b = \rho A^{(1)} g \begin{bmatrix} 150 \\ 150 \end{bmatrix} = \begin{bmatrix} 38.62 \\ 38.62 \end{bmatrix} N$$

For element number 2

$$N_2 = \frac{x_3 - x}{(x_3 - x_2)} = \frac{600 - x}{(600 - 300)} = 2 - x/300$$

$$N_3 = \frac{x - x_2}{(x_3 - x_2)} = \frac{x - 300}{300} = \frac{x}{300} - 1$$

$$[f_b^{(2)}] = \rho g A^{(2)} \int_{300}^{600} \begin{bmatrix} 2 - x/300 \\ x/300 - 1 \end{bmatrix} dx$$

$$= 8 \times 10^{-6} (2343.75)\, 9.81 \begin{bmatrix} 150 \\ 150 \end{bmatrix} = \begin{bmatrix} 27.6 \\ 27.6 \end{bmatrix} N$$

Assembled global stiffness matrix

$$[K] = \begin{bmatrix} k_1 & -k_1 & 0 \\ -k_1 & k_1 + k_2 & -k_2 \\ 0 & -k_2 & k_2 \end{bmatrix} = \begin{bmatrix} 21875 & -21875 & 0 \\ -21875 & 37500 & -15625 \\ 0 & -15625 & 15625 \end{bmatrix} \times 10^2$$

Force vector due to body weight

$$f_b = f_b^{(1)} + f_b^{(2)} = \begin{Bmatrix} 38.62 \\ 38.62 + 27.6 \\ 27.6 \end{Bmatrix}$$

The force vector due to concentrated loads

$$[f_c] = \begin{Bmatrix} 0 \\ 100 \\ 0 \end{Bmatrix}$$

Resultant force vector

$$[f] = \begin{Bmatrix} 38.62 \\ 166.2 \\ 27.6 \end{Bmatrix}$$

The system of equation for static analysis

$$[K]\{U\} = \{F\}$$

$$10^2 \begin{bmatrix} 21875 & -21875 & 0 \\ -21875 & 37500 & -15625 \\ 0 & -15625 & 15625 \end{bmatrix} \begin{Bmatrix} u_1 \\ u_2 \\ u_3 \end{Bmatrix} = \begin{Bmatrix} 38.62 + R_1 \\ 166.2 \\ 27.2 \end{Bmatrix}$$

From the boundary condition at the support $u_1 = 0$

$$10^2 \begin{bmatrix} 37500 & -15625 \\ -15625 & 15625 \end{bmatrix} \begin{Bmatrix} u_2 \\ u_3 \end{Bmatrix} = \begin{Bmatrix} 166.2 \\ 27.2 \end{Bmatrix}$$

Using Gauss Elimination method

$$10^2 \begin{bmatrix} 37500 & -15625 \\ 0 & 21875 \end{bmatrix} \begin{Bmatrix} u_2 \\ u_3 \end{Bmatrix} = \begin{Bmatrix} 166.2 \\ 231.48 \end{Bmatrix}$$

$$u_3 = 0.106 \times 10^{-3} \text{ mm}$$
$$u_2 = 0.0885 \times 10^{-3} \text{ mm}$$
$$-(21875 \times 10^2)\, 0.0885 \times 10^{-3} = 38.62 + R_1$$
$$-193.56 - 38.62 = R_1$$

Hence the reaction is, $R_1 = 232.18$ N

Stress in element (1) $\quad \sigma^{(1)} = EB^{(1)} \{U^{(1)}\}$

$$\sigma^{(1)} = \frac{E}{(x_2 - x_1)}[-1 \quad 1]\begin{Bmatrix} u_1 \\ u_2 \end{Bmatrix}$$
$$= \frac{E}{L} u_2 = \frac{2 \times 10^5}{300}(0.0885 \times 10^{-3})$$
$$= 0.059 \text{ N/mm}^2$$

Stress in element (2)

$$\sigma^{(2)} = \frac{E}{(x_3 - x_2)}[-1 \quad 1]\begin{Bmatrix} u_2 \\ u_3 \end{Bmatrix}$$
$$= \frac{2 \times 10^5}{300}(0.106 - 0.0885) \times 10^{-3}$$
$$= 0.0116 \text{ N/mm}^2$$

Reaction at the support may be verified as

$$\text{R} = P + \int_0^{60} \rho\, A_{av} dx$$
$$= 100 + 8 \times 10^{-6} \frac{(150 + 75)}{2} \times 25 \times 600$$
$$= 235 \text{ N.}$$

5.13.3 Problems on 2D-Truss Element

The following brief procedure may be adapted to solve the problems related to 2D Trusses.

Step 1: Write down the node-element connectivity table linking local and global nodal co-ordinate and also form the table of direction cosines.

Step 2: Write down the stiffness matrix of each element in the global co-ordinate system with global numbering.

Step 3: Assemble element stiffness matrices to form the global stiffness matrix for the entire structure using the node-element connectivity table.

Step 4: Incorporate appropriate boundary conditions.

Step 5: Solve resulting set of reduced equations for the unknown displacements.

Step 6: Compute the unknown nodal forces.

Example 5.8. *A two dimensional truss structure with two elements is as given in Fig. 5.32. If the length, cross-sectional area and the material of the two members are same solve for displacement and stresses in each bar.*

Solution. Node-Element connectivity table

Element#	*First Node*	*Second Node*
1	1	2
2	2	3

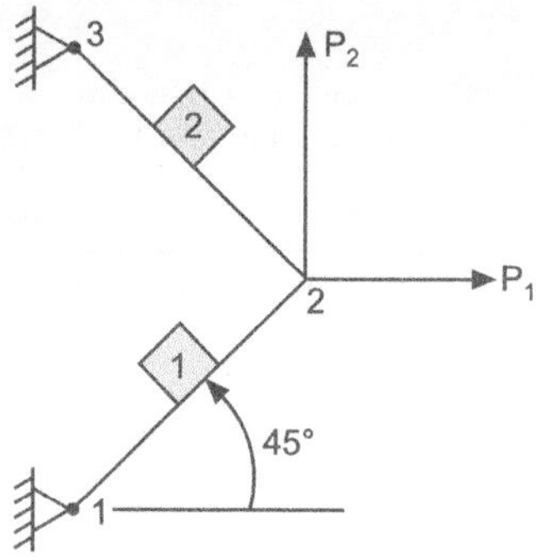

Fig 5.32. Truss member.

Table of nodal co-ordinates

Node	*x*	*y*
1	0	0
2	L cos 45°	L sin 45°
3	0	$2L$ sin 45°

Direction cosines

Element #	*length*	$l = \dfrac{x_2 - x_1}{L}$	$m = \dfrac{y_2 - y_1}{L}$
1	L	cos 45°	sin 45°
2	L	– cos 45°	sin 45°

Stiffness matrix for element number (1)

$$[K^{(1)}] = \frac{EA}{L}\begin{bmatrix} l^2 & lm & -l^2 & -lm \\ lm & m^2 & -lm & -m^2 \\ -l^2 & -lm & l^2 & lm \\ -lm & -m^2 & lm & m^2 \end{bmatrix}$$

$$= \frac{EA}{2L}\begin{array}{c} \begin{matrix} u_{1x} & u_{1y} & u_{2x} & u_{2y} \end{matrix} \\ \left[\begin{matrix} 1 & 1 & -1 & -1 \\ 1 & 1 & -1 & -1 \\ -1 & -1 & 1 & 1 \\ -1 & -1 & 1 & 1 \end{matrix}\right] \end{array} \begin{matrix} u_{1x} \\ u_{1y} \\ u_{2x} \\ u_{2y} \end{matrix}$$

Element stiffness matrix for element (2)

$$[K^{(2)}] = \frac{EA}{2L}\begin{array}{c} \begin{matrix} u_{2x} & u_{2y} & u_{3x} & u_{3y} \end{matrix} \\ \left[\begin{matrix} 1 & -1 & -1 & 1 \\ -1 & 1 & 1 & -1 \\ -1 & 1 & 1 & -1 \\ 1 & 1 & -1 & 1 \end{matrix}\right] \end{array} \begin{matrix} u_{2x} \\ u_{2y} \\ u_{3x} \\ u_{3y} \end{matrix}$$

Assembled global stiffness matrix

$$[K] = \frac{AE}{2L}\begin{bmatrix} 1 & 1 & -1 & -1 & 0 & 0 \\ 1 & 1 & -1 & -1 & 0 & 0 \\ -1 & -1 & 2 & 0 & -1 & 1 \\ -1 & -1 & 0 & 2 & 1 & -1 \\ 0 & 0 & -1 & 1 & 1 & -1 \\ 0 & 0 & 1 & -1 & -1 & 1 \end{bmatrix}$$

The final governing equation for static analysis

$$[K]\{U\} = \{F\}$$

The displacement boundary condition

$$\{U\} = [0 \quad 0 \quad u_{2x} \quad u_{2y} \quad 0 \quad 0]^T$$
$$\{F\} = [0 \quad 0 \quad P_1 \quad P_2 \quad 0 \quad 0]^T$$

Reduced set of equations

$$\frac{EA}{2L}\begin{bmatrix} 2 & 0 \\ 0 & 2 \end{bmatrix}\begin{Bmatrix} u_{2x} \\ u_{2y} \end{Bmatrix} = \begin{Bmatrix} P_1 \\ P_2 \end{Bmatrix}$$

Solving for unknown displacements u_{2x} and u_{2y}

$$\begin{Bmatrix} u_{2x} \\ u_{2y} \end{Bmatrix} = \begin{Bmatrix} P_1 L/AE \\ P_2 L/AE \end{Bmatrix}$$

The stresses in the two elements

For element (1)

$$\sigma^{(1)} = \frac{E}{L}[-l \quad -m \quad l \quad m]\begin{Bmatrix} u_{1x} \\ u_{1y} \\ u_{2x} \\ u_{2y} \end{Bmatrix}$$

$$= \frac{E}{\sqrt{2}L}[-1 \quad -1 \quad 1 \quad 1][0 \quad 0 \quad u_{2x} \quad u_{2y}]^T$$

$$= \frac{E}{\sqrt{2}L}[u_{2x} + u_{2y}] = \frac{P_1 + P_2}{A\sqrt{2}}$$

For element (2)

$$\sigma^{(2)} = \frac{E}{\sqrt{2}L}[1 \quad -1 \quad -1 \quad 1]\begin{Bmatrix} u_{2x} \\ u_{2y} \\ u_{3x} = 0 \\ u_{3y} = 0 \end{Bmatrix}$$

$$= \frac{E}{\sqrt{2}L}(u_{2x} - u_{2y}) = \frac{P_1 - P_2}{A\sqrt{2}}$$

Example 5.9. *For the two dimensional truss shown in Fig. 5.33 the following data are available: length L = 1m, Young's modules, E = 10 GPa.*

Area of cross section for element 1, 2 is $6 \times 10^{-4} m^2$ and for element 3, is $6\sqrt{2} \times 10^{-4} m^2$. The load P = 1000 KN. Determine unknown displacement and reaction forces.

Solution. Node-element connectivity table

Element #	Nodes	
1	1	2
2	2	3
3	1	3

Nodal co-ordinates

Nodes	x	y
1	0	0
2	0	L
3	L	L

Direction cosine table

Element	*Length*	*l*	*m*	l^2	*lm*	m^2
1	L	0	1	0	0	1
2	L	1	0	1	0	0
3	$\sqrt{2}L$	$1/\sqrt{2}$	$1/\sqrt{2}$	1/2	1/2	1/2

Element stiffness matrix in global co-ordinates

$$[k^e] = \frac{EA}{L}\begin{bmatrix} l^2 & lm & -l^2 & -lm \\ lm & m^2 & -lm & -m^2 \\ -l^2 & -lm & l^2 & lm \\ -l^2 & -m^2 & lm & m^2 \end{bmatrix}$$

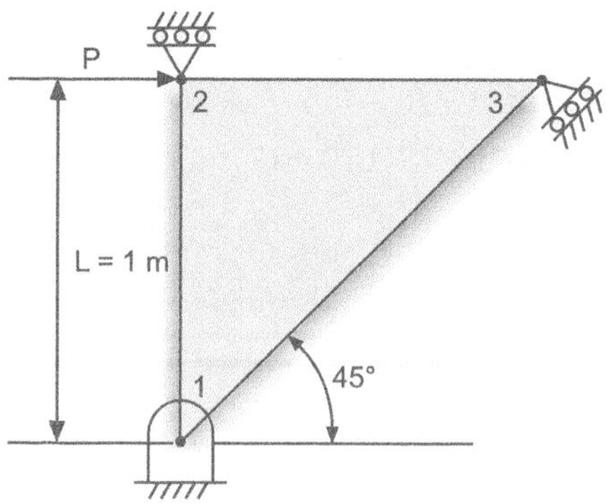

Fig. 5.33. Two dimensional truss example.

Substituting the values from direction cosine table

The stiffness matrices

For element (1)

$$[k^{(1)}] = \frac{AE}{L}\begin{bmatrix} 0 & 0 & 0 & 0 \\ 0 & 1 & 0 & -1 \\ 0 & 0 & 0 & 0 \\ 0 & -1 & 0 & 1 \end{bmatrix} \quad \begin{matrix} A = 6 \times 10^{-4} \\ E = 210 \times 10^9 \\ L = 1m \end{matrix}$$

For element (2)

$$[k^{(2)}] = \frac{210 \times 10^9 \times 6 \times 10^{-4}}{(1)}\begin{bmatrix} 1 & 0 & -1 & 0 \\ 0 & 0 & 0 & 0 \\ -1 & 0 & 1 & 0 \\ 0 & 0 & 0 & 0 \end{bmatrix}$$

For element (3)

$$[k^{(3)}] = \frac{210 \times 10^9 \times 6\sqrt{2} \times 10^{-4}}{(1)\sqrt{2}}\begin{bmatrix} 0.5 & 0.5 & -0.5 & -0.5 \\ 0.5 & 0.5 & -0.5 & -0.5 \\ -0.5 & -0.5 & 0.5 & 0.5 \\ -0.5 & -0.5 & 0.5 & 0.5 \end{bmatrix}$$

Assembled global stiffness matrix

$$[k] = 210 \times 10^9 \times 6 \times 10^{-4}\begin{bmatrix} 0.5 & 0.5 & 0 & 0 & -0.5 & -0.5 \\ 0.5 & 1.5 & 0 & -1 & -0.5 & -0.5 \\ 0 & 0 & 1 & 0 & -1 & 0 \\ 0 & -1 & 0 & 1 & 0 & 0 \\ -0.5 & -0.5 & -1 & 0 & 1.5 & 0.5 \\ -0.5 & -0.5 & 0 & 0 & 0.5 & 0.5 \end{bmatrix}$$

The final set of equations for the static analysis of the two dimensional truss system is

$$[K]\{U\} = \{F\}$$

Incorporate the boundary conditions

$$\{U\} = [0 \;\; 0 \;\; u_{2x} \;\; 0 \;\; u_{3x} \;\; u_{3y}]^T$$

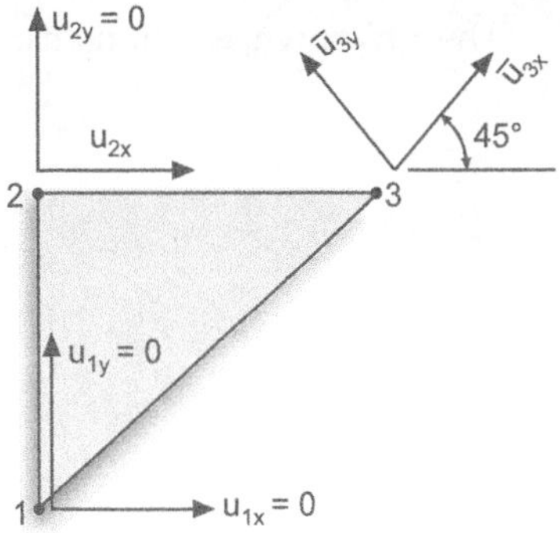

Fig. 5.34. Analysis.

It is given in the local co-ordinate system that $\bar{u}_{3y} = 0$. This has to be converted into boundary condition with respect to global system.

$$\begin{Bmatrix} \bar{u}_{3x} \\ \bar{u}_{3y} \end{Bmatrix} = \begin{bmatrix} l & m \\ -l & m \end{bmatrix} \begin{Bmatrix} u_{3x} \\ u_{3y} \end{Bmatrix} \qquad l = m = 1/\sqrt{2}$$

$$\bar{u}_{3y} = \frac{1}{\sqrt{2}}(u_{3y} - u_{3x}) = 0$$

Hence $\qquad u_{3y} = u_{3x}$

Force boundary condition

$$[F] = [F_{1x} \;\; F_{1y} \;\; P \;\; F_{2y} \;\; F_{3x} \;\; F_{3y}]^T$$

$$\bar{F}_{3x} = 0$$

This is in the local co-ordinate system.

By transformation
$$\begin{Bmatrix} \bar{F}_{3x} \\ \bar{F}_{3y} \end{Bmatrix} = \begin{bmatrix} l & +m \\ -l & +m \end{bmatrix} \begin{Bmatrix} F_{3x} \\ F_{3y} \end{Bmatrix}$$

$$= \begin{bmatrix} 1/\sqrt{2} & 1/\sqrt{2} \\ -1/\sqrt{2} & 1/\sqrt{2} \end{bmatrix} \begin{Bmatrix} F_{3x} \\ F_{3y} \end{Bmatrix}$$

$$\frac{1}{\sqrt{2}}[F_{3x} + F_{3y}] = \bar{F}_{3x} = 0$$

Hence $\qquad F_{3x} = -F_{3y}$

The reduced set of equations after applying the boundary conditions

$$1260 \times 10^5 \begin{bmatrix} 1 & -1 & 0 \\ -1 & 1.5 & 0.5 \\ 0 & 0.5 & 0.5 \end{bmatrix} \begin{Bmatrix} u_{2x} \\ u_{3x} \\ u_{3x} \end{Bmatrix} = \begin{Bmatrix} P \\ F_{3x} \\ -F_{3x} \end{Bmatrix}$$

On further manipulation

$$1260 \times 10^5 \, [u_{2x} - u_{3x}] = P \qquad \text{...}(a)$$

$$1260 \times 10^5 \, [-u_{2x} + 1.5u_{3x} + 0.5\, u_{3x}] = F_{3x} \qquad \text{...}(b)$$

$$1260 \times 10^5 \, [0.5u_{3x} + 0.5u_{3x}] = -F_{3x} \qquad \text{...}(c)$$

Adding equation (*b*) and (*c*)

$$1260 \times 10^5 \, (-u_{2x} + 3u_{3x}) = 0$$

$$u_{2x} = 3u_{3x}$$

From equation (*a*)

$$1260 \times 10^5 \, (3u_{3x} - u_{3x}) = P$$

$$u_{3x} = \frac{10^6}{2520 \times 10^5} = 0.003968 \text{ m.}$$

$$u_{2x} = 3u_{3x} = 0.0119 \text{ m.}$$

The unknown reactions can be obtained as follows

$$\begin{Bmatrix} F_{1x} \\ F_{1y} \\ F_{2y} \\ F_{3x} \\ F_{3y} \end{Bmatrix} = \begin{bmatrix} 0 & -0.5 & -0.5 \\ 0 & -0.5 & -0.5 \\ 0 & 0 & 0 \\ -1 & 1.5 & 0.5 \\ 0 & 0.5 & 0.5 \end{bmatrix} \begin{Bmatrix} u_{2x} \\ u_{3x} \\ u_{3x} \end{Bmatrix}$$

$$= [-500 \quad -500 \quad 0 \quad -500 \quad 500]^T \ N$$

Example 5.10. *For the truss system shown in Fig. 5.35 the two elements have area of cross-section, A = 10 cm² and the Young's modulus, E = 2 × 10⁷N/cm². Calculate the unknown displacements and unknown reactions.*

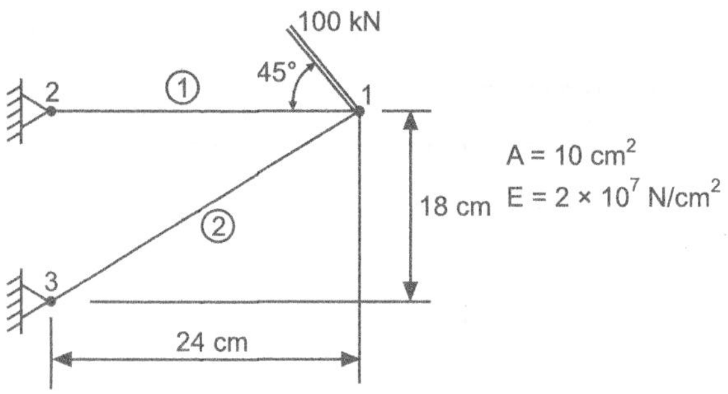

Fig 5.35. Truss example.

Solution.

Node – Element connectivity

Element	Nodes	
1	2	1
2	3	1

Nodal co-ordinates

Nodes	x	y
1	24	18
2	0	18
3	0	0

The elemental stiffness matrix in global co-ordinates is written as

$$[K^e] = \frac{A_e E_e}{L_e} \begin{bmatrix} l^2 & lm & -l^2 & -lm \\ lm & m^2 & -lm & -m^2 \\ -l^2 & -lm & l^2 & lm \\ -lm & -m^2 & lm & m^2 \end{bmatrix}$$

Direction cosines table

Element #	Le	l	m	l^2	lm	m^2
1	24	1	0	1	0	0
2	30	0.8	0.6	0.64	0.48	0.36

The element stiffness matrices

$$[K^{(1)}] = \frac{2\times10^8}{24}\begin{bmatrix}1 & 0 & -1 & 0\\ 0 & 0 & 0 & 0\\ -1 & 0 & 1 & 0\\ 0 & 0 & 0 & 0\end{bmatrix}$$

$$[K^{(2)}] = \frac{2\times10^8}{30}\begin{bmatrix}0.64 & 0.48 & -0.64 & -0.48\\ 0.48 & 0.36 & -0.48 & -0.36\\ -0.64 & -0.48 & 0.64 & 0.48\\ -0.48 & 0.36 & 0.48 & 0.36\end{bmatrix}$$

Assembled global stiffness matrix

$$[K] = \frac{2\times10^8}{6}\begin{bmatrix}0.25 & 0 & -0.25 & 0 & 0 & 0\\ 0 & 0 & 0 & 0 & 0 & 0\\ -0.25 & 0 & 0.378 & 0.096 & -0.128 & -0.096\\ 0 & 0 & 0.096 & 0.072 & -0.096 & -0.072\\ 0 & 0 & -0.128 & -0.096 & 0.128 & 0.096\\ 0 & 0 & -0.096 & -0.072 & 0.096 & 0.072\end{bmatrix}$$

The displacement and the force boundary conditions are

$$\{U\} = \{0 \quad 0 \quad U_{1x} \quad U_{1y} \quad 0 \quad 0\}^T$$

$$\{F\} = [F_{2x} \quad F_{2y} \quad 707.1 \quad -707.1 \quad F_{3x} \quad F_{3y}]^T$$

From the system of equations

$$[K]\{U\} = \{F\}$$

$$\frac{10^8}{3}\times(0.378\,u_{1x} + 0.096u_{1y}) = 707.1$$

$$\frac{10^8}{3}(0.096\,u_{1x} + 0.072\,u_{1y}) = -707.1$$

Solving these system of equations by Gauss-elimination

$$U_{1y} = -\,0.445 \times 10^{-3} \text{ mm}$$

$$U_{1x} = 0.479 \times 10^{-4} \text{ mm}$$

Substituting the values of these unknown displacement and the force boundary conditions.

$$\begin{Bmatrix}F_{2x}\\ F_{2y}\\ F_{3x}\\ F_{3y}\end{Bmatrix} = \begin{bmatrix}-0.25 & 0\\ 0 & 0\\ -0.128 & -0.096\\ -0.096 & -0.072\end{bmatrix}\begin{Bmatrix}u_{1x}\\ u_{1y}\end{Bmatrix}\times\frac{10^8}{3}$$

$$= \begin{Bmatrix}-399.16\\ 0\\ 1745.38\\ 1309.04\end{Bmatrix}.$$

POINTS TO BE REMEMBER

- **Discretization:** Dividing a continuum into an equivalent system of smaller bodies or units called finite elements interconnected at points common to two or more elements or nodes and/or boundary lines and or surfaces.
- **Nodes:** The nodes are set of discrete points with in the solid body.
- **Properties of node:** The node number, the nodal co-ordinates and nodal variables are the properties of nodes.
- **Elements:** The discrete regions of the one, two or three dimensional body.
- **Properties of elements:** The element number, element geometry, elemental variables are the properties of elements.
- One dimensional linear bar element has two nodes and one degree of freedom at each node.
- Two and three dimensional bar elements are called truss elements.
- It is advantageous to use natural co-ordinate system because it makes the numerical integration easy.
- Shape function establishes the relation between displacement function and the nodal displacement.
- Stiffness matrix establishes relation between force vector and the displacement vector.
- Area and length are the geometric properties of stiffness matrix.

EXERCISE

1. Explain the method of stiffness by definition.
2. Describe the procedure of direct stiffness method.
3. Compare the features of stiffness by definition and direct stiffness method.
4. Using spring analogy obtain the stiffness of a bar element.
5. Obtain the stiffness matrix of two springs arranged in series fixed at one end and forced at the other end.
6. What is discretization and what is its advantage and disadvantage?
7. What is a node and what are its properties?
8. What is an element and what are its properties?
9. Give the physical and FE representation of 1D structural elements.
10. Give physical and global representation of bar and beam elements.
11. Give the linear, quadratic and cubic representation of each of 1D bar and a 2D element.
12. Formulate the stiffness matrix of a linear bar element by the method of stiffness by definition.
13. State four properties of stiffness matrix.
14. Derive the shape functions of a linear bar element in local Cartesian co-ordinate system.
15. Discuss the convergence, completeness and compatibility of a 1D linear bar element.

16. Obtain the strain-displacement matrix and strain relation for a 1D linear bar element.
17. Derive the stiffness matrix of 1D linear bar element using principle of minimum potential energy.
18. Explain the approach of energy method to obtain the shape function.
19. Obtain the integral expression of stiffness matrix using the principle of virtual work.
20. Obtain the shape functions of 1D linear bar element in Cartesian co-ordinate system using displacement approach.
21. Derive the stiffness matrix of a bar element with two node by the principle of virtual work.
22. Explain the procedure of FEM for a bar element using Tonti diagram.
23. Establish the relation between natural and global Cartesian variables considering a linear bar element.
24. Obtain the shape function for a 1D linear bar element in natural co-ordinate system.
25. Given the shape functions in natural co-ordinates for the 1D linear bar element obtain the strain-displacement matrix and stiffness matrix.
26. Obtain the expression for force vector for body force due to self weight.
27. Obtain the expression for force vector for body force due to centrifugal force.
28. Starting with the shape functions derive the expression for the stiffness matrix for the tapering bar-with the given area of cross-section at the ends.
29. Obtain the force vector due to temperature variation in a bar element with two nodes.
30. Obtain the matrix for transformation of local to global system for a *2D* bar element.
31. Obtain the transformation matrix that converts the elemental local stiffness matrix to global systiffness matrix in global co-ordinate system.
32. Obtain in the matrix form the expressions for elemental strain and stress distribution in a 1D linear bar element.
33. What is meant by shape function and what are its properties?
34. A stepped bar fixed at one end has two steps. The first step has the area, young's modulus and length to be $2A$, E and $2L$. The second step has corresponding properties to be A, $2E$ and L. If a force P acts at the interface in the tensile manner, obtain the nodal displacement elemental stress and strain.
35. List down the steps to be followed in the analysis of a *2D* truss problem.

Chapter 6

FINITE ELEMENT FORMULATION OF BEAM ELEMENT

अनुष्ठानेन रहिता।
पाठ मात्रेण केवलम्॥
रंजयत्येव या लोकान्।
किं तया शुक विद्यया॥ (दर्पदलनम्)॥

[There is no use of parrot talk-like education that is preached to look glamours but not implemented in practice]

6.1 INTRODUCTION

The finite element procedure applied to beam element is presented in this chapter. A two noded beam element with two degrees of freedom is considered for the analysis. To establish the force-displacement and moment-rotation relationship the basics from the elementary mechanics of materials has been used. The flexure equation, the deflection relations and the shear force and bending moment formulations have been used in the analysis. The shape functions that represent the deformation pattern are derived by displacement transformation method. The derivation of shape functions in local Cartesian as well as natural co-ordinate system have been demonstrated. The easier method to derive shape function using the Hermitian polynomial and the respective boundary condition is also presented. The consistent load vectors are obtained for different load cases established in the mechanics. The different loads like body force, centrifugal force and distributed and point load cases are highlighted.

The stiffness matrix using the principles of mechanics and the approaches of finite element method have demonstrated. The two dimensional Euler-Bernoulli beam element in global co-ordinate system with transformation from local system is highlighted. The problems solved give the feel of importance of the numerical implementation given at the end of this chapter.

6.2 BEAM THEORY

The detailed study on analysis of bending of beams is available in theory of elastic structures. The Euler-Bernoulli's beam is considered here for brief study. Consider a beam acted upon by the distributed load that develops bending moment and the shear forces as the internal effect. The sense of these internal loads is defined according to specific sign convention.

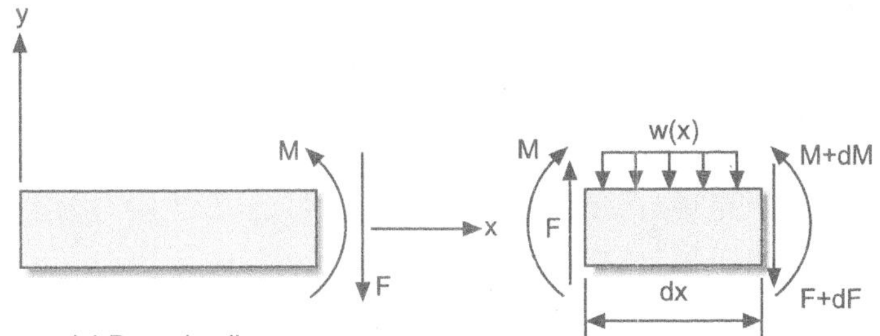

Fig. 6.1. Beam analysis.

Consider a beam acted upon by bending moment M, and the shear force, F. Let $w(x)$ be the distributed force acting on the member. Consider a differential element dx for the development of flexure equation. The assumptions of bending theory hold good in this analysis. Based on the assumption that the plane sections remain plane in the process of bending. The flexure equation form used here is

$$M = \frac{EI}{R} \quad ...(6.1)$$

where $1/R$ is the curvature. EI is the flexural rigidity that depends both on mechanical property and the cross-section of the beam.

From the mechanics of materials, the shear force has the definition given by

$$F = \frac{dM}{dx}. \quad ...(6.2)$$

Further the distributed load developed internally is defined as

$$w(x) = -\frac{dF}{dx}. \quad ...(6.3)$$

From the deflection mechanics for beams, the governing kinematic equation establishing relation between displacement and motion irrespective of casual external forces.

$$\frac{1}{R} = \frac{M}{EI} \approx \frac{d^2V(x)}{dx^2} \quad ...(6.4)$$

This leads to

$$M = EI\frac{d^2}{dx^2}V(x). \quad ...(6.5)$$

The shear force is obtained by differentiating M from equation (6.5) as:

$$F = EI\frac{d^3V}{dx^3}. \quad \text{...(6.6)}$$

The distributed force (load) developed

$$w(x) = -EI\frac{d^4V}{dx^3}. \quad \text{...(6.7)}$$

6.3 FINITE ELEMENT METHOD: BEAMS

The Euler-Bernoulli's beam element with two nodes is as shown in Fig. 6.2. The element equations are formed by approximation to the beam mechanics. All types of structural problems made of beams should be approximated with the use of beam elements. The Euler-Bernoulli's beam considered here has two degrees of freedom at each node: one is the transverse deflection called the translatory DOF and the other is rotation called rotary DOF. These motions result from the shear forces and bending moments respectively.

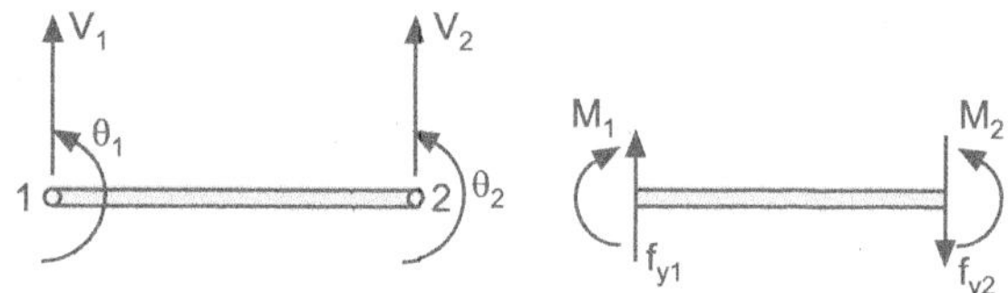

(a) Notations of DOF (b) Forces & moments

Fig. 6.2. Configuration of Euler-Bernoulli's beam.

The forces and moments are related to displacements from the following relations:

$$F_{1y} = +F = EI\left.\frac{d^3V}{dx^3}\right|_{x=0} \quad \text{...(6.8)}$$

$$M_1 = -M = -EI\left.\frac{d^2V}{dx^2}\right|_{x=0} \quad \text{...(6.9)}$$

$$F_{2y} = -F = -EI\left.\frac{d^3V}{dx^3}\right|_{x=L} \quad \text{...(6.10)}$$

$$M_2 = +M = EI\left.\frac{d^2V}{dx^2}\right|_{x=L} \quad \text{...(6.11)}$$

When we try to establish relation between force and moments with the displacement (*i.e.*, deflections and rotations) through equations (6.8) to (6.11) the relation for stiffness matrix is arrived at.

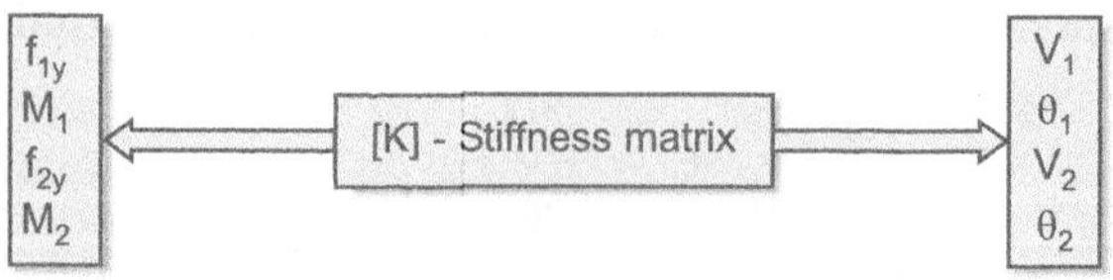

The displacement function used to approximate the deformation in the element should satisfy certain conformation requirements.

- **Compatibility/Continuity**

  A continuous displacement and rotation should exist within a single element (beam) and also across element boundaries. This implies a smooth (continuous) first derivative of the displacement function $V(x)$ should exist.

- **Completeness**

  There should be the need of a constant shear force existing within the element. This is possible if the third derivative of $V(x)$ can be obtained to be existing in the domain x.

6.4 BEAM ELEMENT: SHAPE FUNCTION

The requirement of compatibility/continuity and completeness favourably leads us to use a cubic function to approximate the variation of $V(x)$ in the element.

Hence $$V(x) = a_0 + a_1x + a_2x^2 + a_3x^3 \quad ...(6.12)$$

The displacement function in equation (6.12) as assumed must interpolate the nodal generalized displacements to ensure compatibility and completeness so that the element works out be conforming. The equation (6.12) can be written in the matrix form

$$v(x) = [1, x, x^2, x^3] \begin{Bmatrix} a_0 \\ a_1 \\ a_2 \\ a_3 \end{Bmatrix} \quad ...(6.13)$$

$$\{v\} = [P]\{A\} \quad ...(6.14)$$

where $\{V\}$ is the vector of deformations at node 1 and node 2 as shown Euler-Bernoulli's beam, Fig. 6.2. This vector is given by

$$\{V\} = [V_1, \theta_1, V_2, \theta_2]^T \quad ...(6.15)$$

The displacement boundary conditions (also known as Dirichle boundary conditions)

$$V(x) = V_1 \quad \text{at } x = 0 \quad \text{hence } V_1 = 90$$

$$V(x) = V_2 = a_0 + a_1L + a_2L^2 + a_3L^3 \quad \text{at } x = L$$

By the force boundary condition (*i.e.*, the Neumann boundary condition)

$$\frac{dV(x)}{dx} = \tan\theta_1 \quad \text{at } x = 0$$

$$\frac{dV(x)}{dx} = \tan\theta_2 \quad \text{at } x = L.$$

But for small displacements $\tan\theta_1 = \theta_1$ and $\tan\theta_2 = \theta_2$, thus

$$\theta_1 = a_1 \quad \theta_2 = a_1 + 2a_2L + 3a_3L^2$$

The set of equations arrived at are

$$V_1 = a_0$$

$$\theta_1 = a_1$$

$$V_2 = a_0 + a_1L + a_2L^2 + a_3L^3$$

$$\theta_2 = a_1 + 2a_2L + 3a_3L^2$$

The equation in the matrix form

$$\begin{Bmatrix} V_1 \\ \theta_1 \\ V_2 \\ \theta_2 \end{Bmatrix} = \begin{bmatrix} 1 & 0 & 0 & 0 \\ 0 & 1 & 0 & 0 \\ 1 & L & L^2 & L^3 \\ 0 & 1 & 2L & 3L^2 \end{bmatrix} \begin{Bmatrix} a_0 \\ a_1 \\ a_2 \\ a_3 \end{Bmatrix} \qquad ...(6.16)$$

This is equivalent to

$$\{U\} = [C]\{A\} \qquad ...(6.17)$$

Hence $\{A\} = [C]^{-1}\{U\}$

where $$[C]^{-1} = \begin{bmatrix} 1 & 0 & -3/L^2 & 2/L^3 \\ 0 & 1 & -2/L & 1/L^2 \\ 0 & 0 & 3/L^2 & -2/L^3 \\ 0 & 0 & -1/L & 1/L^2 \end{bmatrix}^T$$

But $$V(x) = [P]\{A\}$$
$$= [P][C]^{-1}\{U\}$$

$$= [1, x, x^2, x^3]^T \begin{bmatrix} 1 & 0 & -3/L^2 & 2/L^3 \\ 0 & 1 & -2/L & 1/L^2 \\ 0 & 0 & 3/L^2 & -2/L^3 \\ 0 & 0 & -1/L & 1/L^2 \end{bmatrix}^T \{U\}$$

$$= \frac{1}{L^3}\begin{bmatrix} L^3 - 3x^2L + 2x^3 \\ xL^3 - 2x^2L^2 + x^3L \\ 3x^2L - 2x^3 \\ -x^2L^2 + x^3L \end{bmatrix}^T \begin{bmatrix} V_1 \\ \theta_1 \\ V_2 \\ \theta_2 \end{bmatrix} \qquad ...(6.18)$$

This equation can be written as

$$V(x) = \Sigma N_i U_i = [N_1\ N_2\ N_3\ N_4]\{U\} \qquad ...(6.19)$$

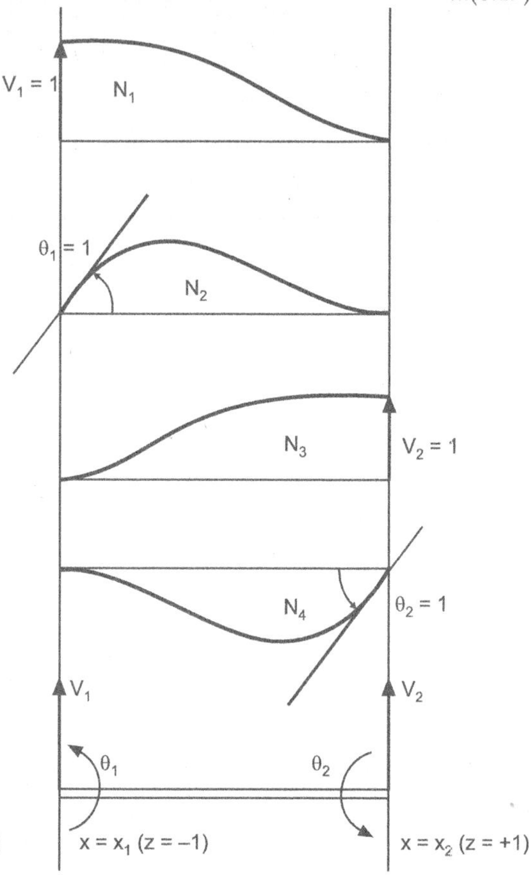

Fig. 6.3. Plot of the shape functions.

where N_i are the shape functions given by

$$N_1(x) = \frac{1}{L^3}(L^3 - 3x^2L + 2x^3)$$

$$N_2(x) = 1/L^3\,(xL^3 - 2x^2L^2 + x^3L) \qquad ...(6.20)$$

$$N_3(x) = (3x^2L - 2x^3)/L^3$$

$$N_4(x) = (-x^2L + x^3L)/L^3$$

$N_i(x)$ are the shape functions that blend the nodal displacements over element domain. The plots of these shape functions are given in the Fig. 6.3. The actual deformation pattern of the parent element is the super imposition of these shape functions.

6.5 SHAPE FUNCTIONS: NATURAL CO-ORDINATES

The normalized co-ordinate (also called natural co-ordinate) is defined in terms of local co-ordinates as:

$$z = \frac{2(x - x_1)}{(x_2 - x_1)} - 1 \qquad \text{...(6.21)}$$

Differentiating this with respect to x

$$\frac{dz}{dx} = \frac{2}{(x_2 - x_1)} = \frac{2}{L} \qquad \text{...(6.22)}$$

To obtain the shape function in natural co-ordinates the cubic polynomial function is assumed.

$$V(z) = a_0 + a_1 z + a_2 z^2 + a_3 z^3 \qquad \text{...(6.23)}$$

$$\frac{dV}{dx} = \frac{dV}{dz} \cdot \frac{dz}{dx} = \frac{2}{L}\frac{dV}{dz} \qquad \text{...(6.24)}$$

From the boundary conditions at $z = -1$ and $z = +1$.

$$V(-1) = V_1 = a_0 - a_1 + a_2 - a_3$$

$$\left.\frac{dV}{dz}\right|_{z=-1} = \theta_1 = \frac{2}{L}(a_1 - 2a_2 + 3a_3)$$

$$V(+1) = V_2 = a_0 + a_1 + a_2 + a_3$$

$$\left.\frac{dV}{dz}\right|_{z=+1} = \theta_2 = \frac{2}{L}(a_1 + 2a_2 + 3a_3)$$

The set of equation in the matrix form

$$\begin{Bmatrix} V_1 \\ \theta_1 \\ V_2 \\ \theta_2 \end{Bmatrix} = \begin{bmatrix} 1 & -1 & 1 & -1 \\ 0 & 2/L & -4/L & 6/L \\ 1 & 1 & 1 & 1 \\ 0 & 2/L & 4/L & 6/L \end{bmatrix} \begin{Bmatrix} a_0 \\ a_1 \\ a_2 \\ a_3 \end{Bmatrix} \qquad \text{...(6.25)}$$

The matrix notation of the equation (6.25)

$$\{U\} = [C]\{A\}$$

{A} is the vector of generalized co-ordinates. [C] is the connectivity matrix and $\{U\}$ is the displacement vector. By inverting the matrix [C] the generalized co-ordinates are written as

$$\{A\} = [C]^{-1}\{U\}$$

$$= \frac{1}{4}\begin{bmatrix} 2 & L/2 & 2 & -L/2 \\ -3 & -L/2 & 3 & -L/2 \\ 0 & -L/2 & 0 & L/2 \\ 1 & L/2 & -1 & L/2 \end{bmatrix} \begin{Bmatrix} V_1 \\ \theta_1 \\ V_2 \\ \theta_2 \end{Bmatrix}$$

But

$$V(z) = [1 \;\; z \;\; z^2 \;\; z^3]\{A\}$$

$$= [1 \;\; z \;\; z^2 \;\; z^3]\frac{1}{4}\begin{bmatrix} 2 & L/2 & 2 & -L/2 \\ -3 & -L/2 & 3 & -L/2 \\ 0 & -L/2 & 0 & L/2 \\ 1 & L/2 & -1 & L/2 \end{bmatrix} \begin{Bmatrix} V_1 \\ \theta_1 \\ V_2 \\ \theta_2 \end{Bmatrix}$$

Hence $$V(z) = N_1V_1 + N_2\theta_2 + N_3V^2 + N_4\theta_2 \quad ...(6.26)$$

The shape functions N_i are

$$N_1 = \frac{1}{4}(2 - 3z + z^3) = \frac{1}{4}(1 - z)^2(2 + z)$$

$$N_2 = \frac{L}{8}(1 - z - z^2 + z^3) = \frac{L}{8}(1 + z)^2(1 + z) \quad ...(6.27)$$

$$N_3 = 1/4\ (2 + 3z - z^3) = 1/4\ (1 + z)^2\ (2 - z)$$

$$N_4 = L/8\ (-1 - z + z^2 + z^3) = L/8\ (1 + z)^2\ (z - 1)$$

The plot of the shape functions follow the same pattern as in Fig. 6.3 obtained for the local co-ordinate system.

Thus the expression for the displacement function in terms of shape functions and the displacements at the nodes, is given as

$$V(z) = [N_1(z)\ \ N_2(z)\ \ N_3(z)\ \ N_4(z)] \begin{Bmatrix} V_1 \\ \theta_1 \\ V_2 \\ \theta_2 \end{Bmatrix} \quad ...(6.28)$$

- Element Equations: Stiffness Matrix.

It has been established in mechanics of materials that

$$f_{1y} = +F = EI \left.\frac{d^3V}{dx^3}\right|_{x=0} = EI \frac{d^3}{dx^3}[(N_i(x))\{V\}]$$

Hence

$$f_{1y} = \frac{EI}{L^3}[12\quad 6L\quad -12\quad 6L] \begin{Bmatrix} V_1 \\ \theta_1 \\ V_2 \\ \theta_2 \end{Bmatrix} \quad ...(6.29)$$

$$M_1 = -M = -EI \left.\frac{d^2V}{dx^2}\right|_{x=0} = -EI \frac{d^2}{dx^2}[(N_i)(V)]$$

Hence

$$M_1 = \frac{EI}{L^3}[6L\quad 4L^2\quad -6L\quad 2L^2] \begin{Bmatrix} V_1 \\ \theta_1 \\ V_2 \\ \theta_2 \end{Bmatrix} \quad ...(6.30)$$

$$f_{2y} = -F = EI \left.\frac{d^3V}{dx^3}\right|_{x=L} = -EI \frac{d^3}{dx^3}[(N_i)\{V\}]$$

Hence

$$f_{2y} = \frac{EI}{L^3}[-12\quad -6L\quad 12\quad -6L] \begin{Bmatrix} V_1 \\ \theta_1 \\ V_2 \\ \theta_2 \end{Bmatrix} \quad ...(6.31)$$

$$M_2 = +M = EI \left.\frac{d^2V(x)}{dx^2}\right|_{x=L} = EI \frac{d^2}{dx^2}[N_i(x)\{V\}]$$

Hence

$$M_2 = \frac{EI}{L^3}[6L \quad 2L^2 \quad -6L \quad 4L^2]\begin{Bmatrix} V_1 \\ \theta_1 \\ V_2 \\ \theta_2 \end{Bmatrix} \qquad ...(6.32)$$

The element equations from (6.29) to (6.32) when expressed in the matrix form shows the static equilibrium establishing relation between force and displacements.

$$\begin{Bmatrix} f_{1y} \\ M_1 \\ f_{2y} \\ M_2 \end{Bmatrix} = \frac{EI}{L^3}\begin{bmatrix} 12 & 6L & -12 & 6L \\ 6L & 4L^2 & -6L & 2L^2 \\ -12 & -6L & 12 & -6L \\ 6L & 2L^2 & -6L & 4L^2 \end{bmatrix}\begin{Bmatrix} V_1 \\ \theta_1 \\ V_2 \\ \theta_2 \end{Bmatrix} \qquad ...(6.33)$$

This compares with static equilibrium equation

$$\{F\} = [k^e]\{U\} \qquad ...(6.34)$$

where $[k^e]$ is the elemental stiffness matrix for the Euler-Bernoulli's beam element.

6.5.1 Element Stiffness Matrix: Alternate Method

The radius of curvature of the beam under bending is defined in mechanics of materials. The radius of curvature (curvature)

$$\frac{1}{R} \approx \frac{d^2V(x)}{dx^2} = \frac{d^2z}{dx^2}\cdot\frac{d^2V(z)}{dz^2}$$

$$= \frac{4}{L^2}\frac{d^2V}{dz^2} = \frac{4}{L^2}\frac{d^2}{dz^2}[N_i]^T\{U\}$$

or

$$\frac{1}{R} = \frac{M}{EI} = [L][N]\{U\} \quad \text{where} \quad L = \frac{4}{L^2}\frac{d^2}{dz} \qquad ...(6.35)$$

Hence

$$M = [B]\,EI\,\{U\} \quad \text{where } [B] = [L]\,[N]$$

$$= [B]\,[D]\,\{U\} \ [D] \text{ is the elasticity matrix} \qquad ...(6.36)$$

To estimate the $[B]$ matrix for $1D$ beam element

$$[B] = [L]\,[N]$$

$$= \frac{4}{L^2}\frac{d^2}{dz^2}[N_1 \ N_2 \ N_3 \ N_4]$$

$$= \frac{4}{L^2}\left[\frac{3}{2}z \ \frac{L}{2}\left(\frac{3z-1}{2}\right) - \frac{3z}{2} \ \frac{L}{2}\left(\frac{3z+1}{2}\right)\right]$$

Hence the moment is given by the expression

$$M = \frac{EI}{L^2}[6z \quad L(3z-1) \quad -6z \quad L(3z+1)]\begin{Bmatrix} V_1 \\ \theta_1 \\ V_2 \\ \theta_2 \end{Bmatrix} \qquad ...(6.37)$$

The shear force distribution in the element is obtained

as
$$F = \frac{dM}{dx} = \frac{dM}{dz}\cdot\frac{dz}{dx} = \frac{2}{L}\frac{dM}{dz} \quad ...(6.38)$$

The force/moment and displacement relation is

$$\begin{Bmatrix} f_{1y} \\ M_1 \\ F_{2y} \\ M_2 \end{Bmatrix} = \int_{-1}^{+1} \begin{bmatrix} \frac{dM}{dx} \\ -M \\ -\frac{dM}{dx} \\ +M \end{bmatrix} \begin{Bmatrix} V_1 \\ \theta_1 \\ V_2 \\ \theta_2 \end{Bmatrix} dz \quad ...(6.39)$$

But

$$\frac{dM}{dx} = \frac{2}{L}\frac{dM}{dz} = \frac{2}{L}\left(\frac{EI}{L^2}\right)[6 \quad 3L \quad -6 \quad 3L] \quad ...(6.40)$$

$$\begin{Bmatrix} f_{1y} \\ M_1 \\ F_{2y} \\ M_2 \end{Bmatrix} = \frac{EI}{L^2}\cdot\frac{1}{L}\begin{bmatrix} 12 & 6L & -12 & 6L \\ 6L & 4L^2 & -6L & 2L^2 \\ -12 & -6L & 12 & -6L \\ 6L & 2L^2 & -6L & 4L^2 \end{bmatrix} \begin{Bmatrix} V_1 \\ \theta_1 \\ V_2 \\ \theta_2 \end{Bmatrix} \quad ...(6.41)$$

The compares with the force-displacement relation defining the static equilibrium of the Euler-Bernoull is beam element.

$$\{F\} = [K^e]\,\{U\} \quad ...(6.42)$$

where $[K^e]$ is defined to be the elemental stiffness matrix and $\{F\}$, $\{U\}$ are the force/moment vector and deformation/rotation vector respectively.

6.6 ELEMENT LOAD VECTORS

The possible forces that form the consistent elemental load vectors are:

1. Body force
   (*a*) Due to self-weight
   (*b*) Due to centrifugal loading
2. Traction force due to uniformly distributed load
3. Concentrated point load
4. Moments acting at various sections.

The element load vectors are derived in FEM in the following way:

6.6.1 The Body Force: Beam Element

The body force vector is defined to represent

$$[f^b] = \int_V [N]^T [b]\, dV \quad ...(6.43)$$

For a beam with uniform cross-section, A

$$dV = A\,dx = \frac{AL}{2}dz$$

Hence

$$[f^b] = \frac{AL}{2}\int_{-1}^{+1} [N]^T [b]\, dz \qquad ...(6.44)$$

To evaluate

$$\int_{-1}^{+1} [N]^T dz = \int_{-1}^{+1} \begin{Bmatrix} 1/4\,(2 - 3z + z^3) \\ L/8\,(1 - z - z^2 + z^3) \\ 1/4\,(2 + 3z - z^3) \\ L/8\,(-1 - z + z^2 + z^3) \end{Bmatrix} dz \qquad ...(6.45)$$

$$= [1 \quad L/6 \quad 1 \quad -L/6]^T$$

Hence the load vector due to body force $[b]$ can be expressed as

$$[f^b] = \frac{AL}{2}[b][1 \quad L/6 \quad 1 \quad -L/6]^T \qquad ...(6.46)$$

(*a*) The force due to self weight is defined as the product of mass density and acceleration due to gravity *i.e.*, $[b] = \rho g$.

Hence, the load vector due to self weight

$$[f^b]_w = \frac{A\rho g L}{2}[1 \quad L/6 \quad 1 \quad -L/6]^T \qquad ...(6.47)$$

where ρ = mass density of the material, kg/mm^3

g = acceleration due to gravity, m/sec^2

A = Area of cross section, mm^2

L = length of the element ($= [x_2 - x_1]$), in mm.

(*b*) The force due to centrifugal loading is defined to be the product of mass density and the angular acceleration due to rotation *i.e.*, $[b] = \rho\omega 2 r_{av}$.

But the average radius, $r_{av} = \left(\frac{r_1 + r_2}{2}\right)$ with r_1 and r_2 are the radius of node 1 and node 2 respectively.

Hence, the load vector due to centrifugal loading

$$[f^b]_c = \frac{A\rho\omega^2}{2}\frac{(r_1 + r_2)}{2}\begin{bmatrix} 1 \\ L/6 \\ 1 \\ -L/6 \end{bmatrix} \qquad ...(6.48)$$

ω – angular velocity, rad/sec

r_1 = radius of node 1, mm

r_2 = radius of node 2, mm

6.6.2 The Load Vector: Traction Force

The traction force or the uniformly distributed load is

$$[T] = p(x)$$

And the traction force gives the load vector to be

$$[f^t] = \int_S [N]^T [T]\, dS \qquad ...(6.49)$$

For a unit width of the beam $dS = dx$

In natural co-ordinate, $dx = \frac{1}{2} dz$

Hence $$\{f^t\} = \frac{p(x)L}{2} \int_{-1}^{+1} [N]^T dz \qquad ...(6.50)$$

From equation (6.50) that evaluates the integral

$$\{f^t\} = \frac{p(x)L}{2} \begin{bmatrix} 1 \\ L/6 \\ 1 \\ -L/6 \end{bmatrix} \qquad ...(6.51)$$

6.6.3 Load Vector: Concentrated Load

The set of point loads P_i acting on the given load vector as defined by

$$\{f^P\} = \sum_{i=1}^{n} [N_i]^T \{P_i\} \qquad ...(6.52)$$

Let a point load $P_i = P_1$ be acting at a distance $z = d$ from origin $z = 0$. By the substitution of $z = d$

$$[N_i]^T = \begin{bmatrix} 1/4\,(2 - 3d + d^3) \\ L/8\,(1 - d - d^2 + d^3) \\ 1/4\,(2 + 3d - d^3) \\ L/8\,(-1 - d + d^2 + d^3) \end{bmatrix} \qquad ...(6.53)$$

For example $d = +1$

$$\{f^P\} = P_1\,[0 \;\; 0 \;\; 1 \;\; 0]^T$$

and for $d = -1$

$$\{f^P\} = P_1\,[1 \;\; 0 \;\; 0 \;\; 0]^T$$

6.6.4 Load Vector: Moments

The elemental load vector due to moments M_i acting about various points on the beam is defined to be represented by

$$\{f^m\} = \sum \frac{d}{dx} [N]^T \{M_i\} \qquad ...(6.54)$$

To evaluate the differential of shape functions with respect to x (By chain rule of differentiation)

$$\frac{d}{dx}[N]^T = \frac{dz}{dx} \cdot \frac{d}{dz}[N]^T = \frac{2}{L}\frac{d}{dz}[N]^T$$

$$\frac{d}{dz}[N]^T = \begin{Bmatrix} \frac{d}{dz}[1/4\,(2 - 3z + z^3)] \\ \frac{d}{dz}[L/8\,(1 - z - z^2 + z^3)] \\ \frac{d}{dz}[1/4\,(2 + 3z - z^3)] \\ \frac{d}{dz}[L/8\,(-1 - z + z^2 + z^3)] \end{Bmatrix}$$

$$= \begin{Bmatrix} 1/4\,(-3+3z^2) \\ L/8\,(-1-2z+3z^2) \\ 1/4\,(3-3z^2) \\ L/8\,(-1+2z+3z^2) \end{Bmatrix} \qquad ...(6.55)$$

Let a moment $M = m_1$ act at a distance $z = d$ from the origin $z = 0$. Hence the load vector due to m_1 is given as

$$\{f^m\} = \frac{2m_1}{L}\begin{Bmatrix} 1/4\,(-3+3d^2) \\ L/8\,(-1-2d+3d^2) \\ 1/4\,(3-3d^2) \\ L/8\,(-1+2d+3d^2) \end{Bmatrix} \qquad ...(6.56)$$

Let for example $z = -1 = d$

$$[f^m] = \frac{2m_1}{L}[0 \quad L/2 \quad 0 \quad 0]^T$$

for $z = +1 = d$

$$\{f^m\} = \frac{2m_1}{L}[0 \quad 0 \quad 0 \quad L/2]^T$$

6.7 HERMITE POLYNOMIAL: BEAM ELEMENT

The shape functions can be constructed by various methods for different elements. Inverting the displacement transformation matrix [c] is difficult when the degrees for an element are large in number. Hermite polynomial method of obtaining shape functions is one such method which is useful when the derivatives of displacements are introduced as degrees of freedom. In beam element two degrees of freedom are the transverse displacements and other two are the rotations which are the slopes defined by their derivatives with respect to distance x. Hence Hermite polynomial method to derive the shape functions can be well applied to beam elements.

In a beam element V_1 and V_2 are the transverse displacements and θ_1 and θ_2 are the rotations at two nodes defined for the linear beam element. The Euler-Bernoulli's beam element with two degrees of freedom at each node is as shown in Fig. 6.4. The shape functions for 1D beam elements define the displacement interpolation function as follows:

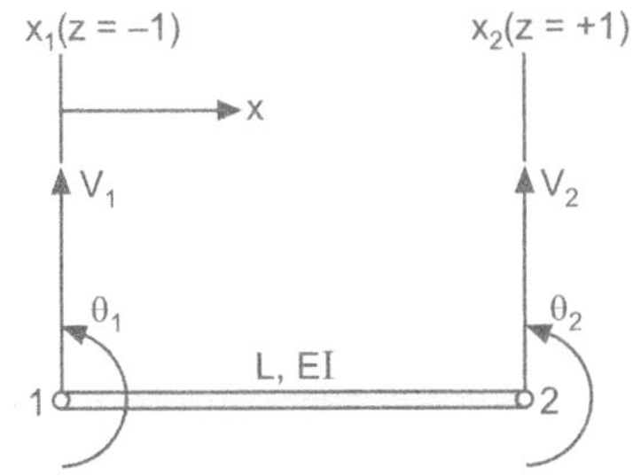

Fig. 6.4. Euler-Bernoulli beam.

$$u(x) = N_1(x)\,V_1 + N_2(x)\theta_1 + N_3(x)\,V_2 + N_4(x)\theta_2 \qquad ...(6.57)$$

where N_1, N_2, N_3 and N_4 are the shape function that are to be estimated using Hermite polynomial.

6.7.1 Shape Function $N_1(x)$

This shape function N_1 corresponds to the displacement V_1 at node 1. The plot of this shape function must assign a value of 1 (unit displacement) at left side (*i.e.,* node 1). The other values displacement V_2 and two rotations θ_1 and θ_2 are assumed zero. The associated shape function is shown in Fig. 6.5.

- Local co-ordinates

Let the cubic polynomial for the shape function N_1

be $$u = N_1 = a_0 + a_1x + a_2x^2 + a_3x^3 \quad ...(6.58)$$

The first derivative:

$$\frac{du}{dx} = a_1 + 2a_2x + 3a_3x^2$$

$$u|_{x=0} = 1 = a_0$$

$$\left.\frac{du}{dx}\right|_{x=0} = 0 = a_1 \quad ...(6.59)$$

$$u|_{x=L} = 0 = a_0 + a_2L^2 + a_3L^3$$

$$\left.\frac{du}{dx}\right|_{x=L} = 0 = 2a_2L + 3a_3L^2$$

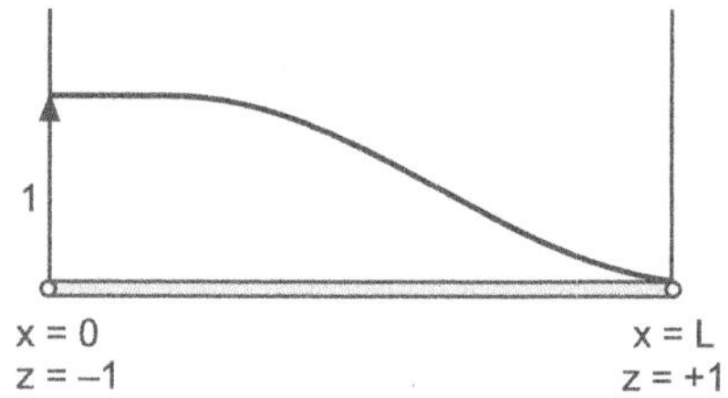

Fig. 6.5. Plot of shape function N_1.

Solving these set of equations simultaneously

$$a_0 = 1, \quad a_1 = 0, \quad a_2 = \frac{-3}{L^2}, \quad a_3 = \frac{2}{L^3}$$

Hence the shape function by substitution in equation (6.58) is

$$N_1 = 1 - \frac{3x^2}{L^2} + \frac{2x^3}{L^3} \quad ...(6.60)$$

- In natural co-ordinates

Let cubic polynomial in natural co-ordinate be

$$u = N_1 = a_0 + a_1z + a_2z^2 + a_3z^3$$

$$\frac{du}{dz} = a_1 + 2a_2z + 3a_3z^2$$

but $$\frac{2}{L}\frac{du}{dz} = \frac{du}{dx}$$

Using the boundary condition at $z = -1$

$$a_0 - a_1 + a_2 - a_3 = 1$$

$$a_1 - 2a_2 + 3a_3 = 0$$

Using the boundary condition $z = +1$

$$a_0 + a_1 + a_2 + a_3 = 0$$

$$a_1 + 2a_2 + 3a_3 = 0$$

The set of simultaneous equations in the matrix form

$$\begin{bmatrix} 1 & -1 & 1 & -1 \\ 0 & 1 & -2 & 3 \\ 1 & 1 & 1 & 1 \\ 0 & 1 & 2 & 3 \end{bmatrix} \begin{Bmatrix} a_0 \\ a_1 \\ a_2 \\ a_3 \end{Bmatrix} = \begin{Bmatrix} 1 \\ 0 \\ 0 \\ 0 \end{Bmatrix}$$

Applying Gauss elimination method of solution

$$a_0 = 1/4 \qquad a_1 = 0 \qquad a_2 = -3/4 \qquad a_3 = 1/2$$

Hence the shape function is

$$N_1 = \left(1/4 - \frac{3}{4}z^2 + \frac{1}{2}z^3\right)$$

This can be written as

$$N_1 = 1/4\,(1 - 3z^2 + 2z^3) \qquad \text{...(6.60}b\text{)}$$

6.7.2 Shape Function N_2

The shape function N_2 corresponds to the rotation θ_1 at node 1. The plot of this shape function must assign unit rotation ($\theta_1 = 1$) at the node 1. The other two displacements (V_1 and V_2) and the rotation at node 2 (θ_2) must assume the value zero. The associated plot of shape function N_2 is shown in Fig. 6.6.

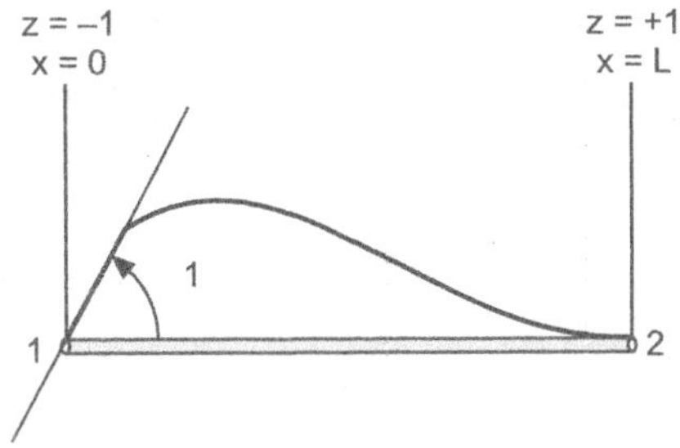

Fig. 6.6. Plot of shape function N_2.

- Local co-ordinates

The polynomial for the displacement

$$u(x) = N_2 = a_0 + a_1x + a_2x^2 + a_3x^3$$

$$\frac{du}{dx} = a_1 + 2a_2x + 3a_3x^2$$

The boundary conditions are

at $x = 0$ $\quad u = 0$

at $x = 0$ $\quad \frac{du}{dx} = 1$

at $x =$ $\quad Lu =$ $\quad 0$

at $x = L$ $\quad \frac{du}{dx} = 0$

This leads to

$$a_0 = 0$$

$$a_1 = 1$$

$$a_1L + a_2L^2 + a_3L^3 = 0$$

$$1 + 2a_2L + 3a_3L^2 = 0$$

By solving these equations simultaneously

$$a_0 = 0, \quad a_1 = 1 \quad a_2 = -2/L \quad a_3 = \frac{1}{L^2}$$

Hence the polynomial, representing shape function N_2

$$N_2 = x - \frac{2x^2}{L} + \frac{x^3}{L^2} \qquad \text{...(6.61)}$$

- Natural co-ordinates

$$u = N_2 = a_0 + a_1z + a_2z^2 + a_3z^3$$

$$\frac{2}{L}\frac{du}{dz} = a_1 + 2a_2z + 3a_3z^2$$

By substituting $z = -1$ and $z = +1$.

$$a_0 - a_1 + a_2 - a_3 = 0$$

$$a_1 - 2a_2 + 3a_3 = L/2$$
$$a_0 + a_1 + a_2 + a_3 = 0$$
$$a_1 + 2a_2 + 3a_3 = 0$$

The matrix form of simultaneous equations are

$$\begin{bmatrix} 1 & -1 & 1 & -1 \\ 0 & 1 & -2 & 3 \\ 1 & 1 & 1 & 1 \\ 0 & 1 & 2 & 3 \end{bmatrix} \begin{Bmatrix} a_0 \\ a_1 \\ a_2 \\ a_3 \end{Bmatrix} = \begin{Bmatrix} 0 \\ L/2 \\ 0 \\ 0 \end{Bmatrix}$$

Solving this by Gauss elimination method.

$$a_0 = L/8 \qquad a_1 = -L/8 \qquad a_2 = -L/8 \qquad a_3 = L/8$$

Now the polynomial for the shape function N_2 is

$$N_2(z) = \frac{L}{8}(1 - z - z^2 + z^3) \qquad \text{...(6.62)}$$

6.7.3 Shape Function, N_3

The shape function N_3 corresponds to the degree of freedom V_2 (the displacement) at node 2. Assigning unit displacement at node 2, the rotations θ_1 and θ_2, and the displacement V_1 are kept zero. The plot of the shape function N_3 is as shown by Fig. 6.7.

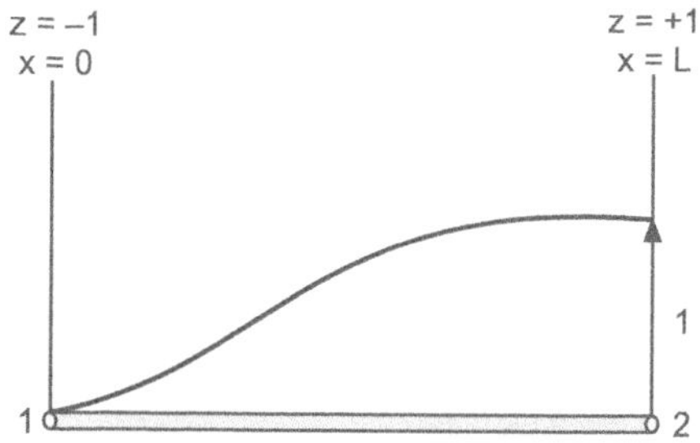

Fig 6.7. Plot of shape function N_3.

The polynomial and its derivative are assumed to a cubic Hermitian polynomial as,

$$u(x) = N_3 = a_0 + a_1x + a_2x^2 + a_3x^3$$

$$\frac{du}{dx} = a_1 + 2a_2x + 3a_3x^2$$

The boundary conditions, when applied leads

at $x = 0 \quad u = 0 \rightarrow a_0 = 0$

at $x = 0 \quad \frac{du}{dx} = 0 \rightarrow a_1 = 0$

at $x = L \quad u = 1 \rightarrow 1 = a_2L^2 + a_3L^3$

at $x = L \quad \frac{du}{dx} = 0 \rightarrow 0 = 2a_2L + 3a_3L^2$

Solving for the generalized co-efficients

$$a_0 = 0, \qquad a_1 = 0 \qquad a_2 = \frac{3}{L^2} \qquad a_3 = \frac{-2}{L^3}$$

The shape function is

$$N_3 = (3x^2/L^2 - 2x^3/L^3) \qquad \text{...(6.63)}$$

- N_3 in natural co-ordinates

The cubic polynomial and its derivative in natural co-ordinate system is

$$u(z) = a_0 + a_1z + a_2z^2 + a_3z^3 = N_3$$

$$\frac{du}{dz} = a_1 + 2a_2z + 3a_3z^2$$

By assigning $z = -1$ and $z = +1$ and applying boundary condition to displacement and its derivative

$$z = -1 \begin{cases} a_0 - a_1 + a_2 - a_3 = 0 \\ a_1 - 2a_2 + 3a_3 = 0 \end{cases}$$

$$z = +1 \begin{cases} a_0 + a_1 + a_2 + a_3 = 1 \\ a_1 + 2a_2 + 3a_3 = 0 \end{cases}$$

The equations is the matrix form

$$\begin{bmatrix} 1 & -1 & 1 & -1 \\ 0 & 1 & -2 & 3 \\ 1 & 1 & 1 & 1 \\ 0 & 1 & 2 & 3 \end{bmatrix} \begin{Bmatrix} a_0 \\ a_1 \\ a_2 \\ a_3 \end{Bmatrix} = \begin{Bmatrix} 0 \\ 0 \\ 1 \\ 0 \end{Bmatrix}$$

Solving the system of equations by Gauss elimination

$$a_0 = 1/2; \quad a_1 = +3/4; \quad a_2 = 0; \quad a_3 = -1/4$$

Hence the shape function N_3 works out to be

$$N_3 = 1/4\,(2 + 3z - z^3) \qquad \text{...(6.64)}$$

6.7.4 The Shape Function N_4

The shape function N_4 corresponds to the rotation θ_2 at node 2. Assigning unit rotation, $\theta_2 = 1$ at node 2 and assuming other degrees of freedom to take zero value the shape function N_4 is obtained.

The plot of the shape function N_4 is depicted in the Fig. 6.8.

- In local co-ordinates

By inputting $x = 0$ and $x = L$ in the cubic polynomial and its derivative

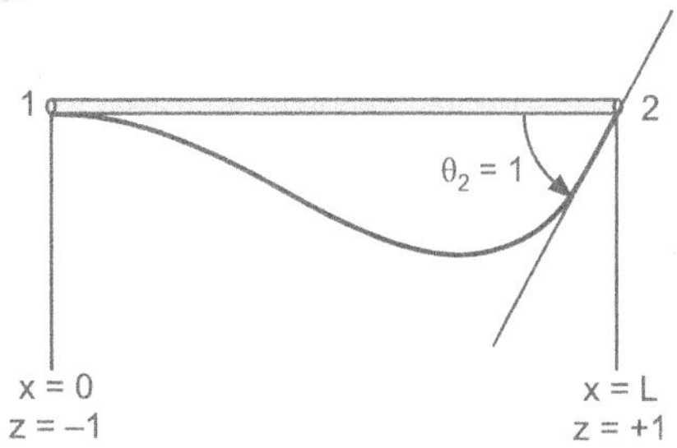

Fig. 6.8. Plot of shape function N_4.

$$u\big|_{x=0} = 0 = a_0$$

$$\frac{du}{dx}\bigg|_{x=0} = 0 = a_1$$

$$u\big|_{x=L} = 0 = a_2L^2 + a_3L^3$$

$$\frac{du}{dx}\bigg|_{x=L} = 1 = 2a_2L + 3a_3L^2$$

Solving the simultaneous equations for a_2 and a_3

$$a_0 = 0; \quad a_1 = 0; \quad a_2 = -\frac{1}{L^2} \quad a_3 = 1/L^3$$

Hence the expression for the shape function N_4

$$N_4 = -\frac{x^2}{L^2} + \frac{x^3}{L^3} \qquad \text{...(6.65)}$$

- In natural co-ordinates

Since
$$\frac{2}{L}\cdot\frac{du}{dz} = \left.\frac{du}{dx}\right|_{x=L} = 1$$
$$\frac{du}{dz} = \frac{L}{2}$$

The cubic polynomial and its derivative in natural co-ordinates are given by
$$u(z) = a_0 + a_1 z + a_2 z^2 + a_3 z^3$$
$$\frac{du}{dz} = a_1 + 2a_2 z + 3a_3 z^2$$

Assigning $z = -1$ and $z = +1$ and applying the boundary condition to these equations
$$u|_{z=-1} = 0 = a_0 - a_1 + a_2 - a_3$$
$$\left.\frac{du}{dz}\right|_{z=-1} = 0 = a_1 - 2a_2 + 3a_3$$
$$u|_{z=+1} = 0 = a_0 + a_1 + a_2 + a_3$$
$$\left.\frac{du}{dz}\right|_{z=+1} = L/2 = a_1 + 2a_2 + 3a_3$$

Expressing these equations in the matrix form
$$\begin{bmatrix} 1 & -1 & 1 & -1 \\ 0 & 1 & -2 & 3 \\ 1 & 1 & 1 & 1 \\ 0 & 1 & 2 & 3 \end{bmatrix} \begin{Bmatrix} a_0 \\ a_1 \\ a_2 \\ a_3 \end{Bmatrix} = \begin{Bmatrix} 0 \\ 0 \\ 0 \\ L/2 \end{Bmatrix}$$

Solve the matrix equation by Gauss elimination
$$a_0 = -\frac{L}{8};\quad a_1 = -\frac{L}{8};\quad a_2 = \frac{L}{8};\quad a_3 = \frac{L}{8}$$

Substituting these values of a_i in the polynomial representing the shape function
$$N_4(z) = \frac{L}{8}(-1 - z + z^2 + z^3) \qquad \text{...(6.66)}$$

Example 6.1. *A cantilever beam loaded at free end by a point load of 120 N and a moment of 120 N-m has a length of 1 m and property EI = 1 unit. Derive the stiffness matrix to the specific case and the unknown displacements and unknown forces. (Refer Fig. 6.9).*

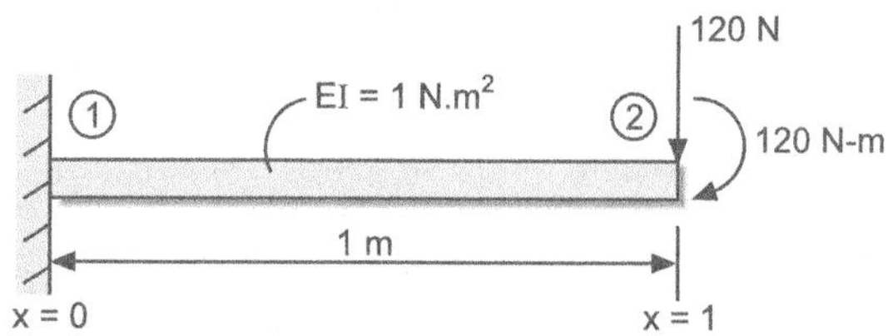

Fig. 6.9

Solution. The displacement function assumed for the beam element is
$$u(x) = a_0 + a_1 x + a_2 x^2 + a_3 x^3 \qquad \text{...}(i)$$
$$u(x=0) = a_0 = 0$$
$$u(x=1) = a_0 + a_1 + a_2 + a_3 = U_{2y}$$
$$\left.\frac{du}{dx}\right|_{x=0} = a_1 = 0$$

$$\left.\frac{du}{dx}\right|_{x=1} = a_1 + 2a_2 + 3a_3 = \theta_2$$

From these set of equation we have two simultaneous equations in two unknowns

$$a_2 + a_3 = u_{2y}$$

$$2a_2 + 3a_3 = \theta_2$$

Solving the same

$$a_3 = (\theta_2 - 2u_{2y})$$

$$a_2 = (3u_{2y} - \theta_2)$$

Substituting the values of a_0 a_1 a_2 and a_3 in (*i*)

$$(3u_{2y} - \theta_2)x^2 + (\theta_2 - 2u_{2y})x^3 = u(x)$$

$$(3x^2 - 2x^3)u_{2y} + (-x^2 + x^3)\theta_2 = u(x)$$

Hence the shape functions are

$$N_1 = 3x^2 - 2x^3$$

$$N_2 = -x^2 + x^3$$

By the definition of stiffness matrix

$$[K] = \int_0^1 [B]^T EI[B]\,dx$$

But

$$[B] = \frac{d^2}{dx^2}[N_1 N_2]$$

$$= [(6 - 12x)\ (-2 + 6x)]$$

Hence the stiffness matrix

$$[K] = EI\int_0^1 \begin{bmatrix}(6-12x)\\(-2+6x)\end{bmatrix}[(6-12x)\,(-2+6x)]\,dx$$

$$= \int_0^1 \begin{bmatrix}(6-12x)^2 & (6-12x)(-2+6x)\\(-2+6x)(6-12x) & (-2+6x)^2\end{bmatrix} dx$$

$$= EI\begin{bmatrix}12 & -6\\-6 & 4\end{bmatrix}$$

But $EI = 1$ unit

From the force displacement relation

$$\{F\} = [K]\{U\}$$

$$\begin{bmatrix}-120\\-120\end{bmatrix} = \begin{bmatrix}12 & -6\\-6 & 4\end{bmatrix}\begin{Bmatrix}u_{2y}\\\theta_2\end{Bmatrix}$$

By Gauss elimination method

$$\theta_2 = \frac{-120}{2} = -60 \text{ rads.}$$

$$u_{2y} = \frac{-120 + 6(-60)}{12} = -40 \text{ m}$$

Hence the unknown displacements are

$$\begin{Bmatrix} u_{2y} \\ \theta_2 \end{Bmatrix} = \begin{Bmatrix} -40 \\ -60 \end{Bmatrix}$$

Example 6.2. *A simply supported beam with a point load of 12 kN is as shown in Fig. 6.10. If the EI = 10^{+3} units and length of the beam is 1 m determine the stiffness matrix specific to the case and the unknown degrees of freedom.*

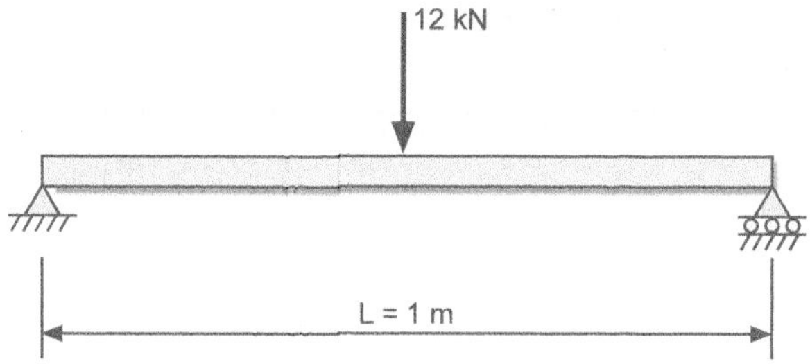

Fig. 6.10.

Solution. The displacement field assumed is

$$u(x) = a_0 + a_1 x + a_2 x^2 + a_3 x^3 \qquad ...(a)$$

The boundary conditions are

at $x = 0$ $\quad u_{1y} = 0 = a_0$

at $x = 0$ $\quad \theta_1 = a_1$

at $x = 1$ $\quad u_{2y} = 0 = a_1 + a_2 + a_3$

at $x = 1$ $\quad \frac{du}{dx} = \theta_2 = a_1 + 2a_2 + 3a_3 = 0$

Solving the equations

$$a_2 + a_3 = -\theta_1$$

$$2a_2 + 3a_3 = \theta_2 - \theta_1$$

$$a_3 = \theta_1 + \theta_2 \qquad a_2 = -(2\theta_1 + \theta_2)$$

Substituting in equation (*a*)

$$u(x) = \theta_1 x - (2\theta_1 + \theta_2)x^2 + (\theta_1 + \theta_2)x^3$$

$$= (x - 2x^2 + x^3)\,\theta_1 + (-x^2 + x^3)\theta_2$$

$$= N_2\theta_1 + N_4\theta_2$$

$$[N] = [(x - 2x^2 + x^3)\ (x^3 - x^2)]$$

Stiffness matrix by definition

$$[K] = EI\int_0^1 [B]^T [B]\, dx$$

$$= EI \int_0^1 \begin{bmatrix} \frac{d^2}{dx^2} N_2 \\ \frac{d^2}{dx^2} N_4 \end{bmatrix} \begin{bmatrix} \frac{d^2}{dx^2} N_2 & \frac{d^2}{dx^2} N_4 \end{bmatrix} dx$$

$$= EI \int_0^1 \begin{bmatrix} \left(\frac{d^2}{dx^2} N_2\right)^2 & \left(\frac{d^2 N_2}{dx^2} \cdot \frac{d^2 N_4}{dx^2}\right) \\ \left(\frac{d^2 N_4}{dx^2} \cdot \frac{d^2 N_2}{dx^2}\right) & \left(\frac{d^2 N_4}{dx^2}\right) \end{bmatrix}$$

$$= EI \int_0^1 \begin{bmatrix} (-4+6x)^2 & (-4+6x)(6x-2) \\ (6x-2)(6x-4) & (6x-2)^2 \end{bmatrix} dx$$

$$= EI \begin{bmatrix} 4 & 2 \\ 2 & 4 \end{bmatrix} = 10^{+3} \begin{bmatrix} 4 & 2 \\ 2 & 4 \end{bmatrix}$$

The force-displacement relation $\{F\} = [K]\{U\}$

$$\begin{Bmatrix} M_1 \\ M_2 \end{Bmatrix} = 10^{+3} \begin{bmatrix} 4 & 2 \\ 2 & 4 \end{bmatrix} \begin{Bmatrix} \theta_1 \\ \theta_2 \end{Bmatrix}$$

$$M_1 = -\frac{PL}{4} = -\frac{12 \times 10^3}{4} = 3 \times 10^3 \text{ N-m}$$

$$M_2 = \frac{PL}{4} = 3 \times 10^3 \text{ N-m}$$

Solving by Gauss elimination

$$6 \times 10^{+3} = 9 \times 10^3$$

$$\theta_2 = 1.5 \text{ rads}$$

$$\theta_1 = \frac{-(2 \times 10^3)\,1.5 - 3 \times 10^3}{4 \times 10^3}$$

$$= -1.5 \text{ rads}$$

The displacement vector

$$\begin{Bmatrix} \theta_1 \\ \theta_2 \end{Bmatrix} = \begin{Bmatrix} -1.5 \\ 1.5 \end{Bmatrix}$$

Example 6.3. *A simply supported beam with a point load in the middle is as shown in Fig. 6.11. If $EI = 10^3$ units and the span is 2 m calculate the unknown forces and displacement considering two elements for the beam.*

Solution. The element stiffness matrix for element # 1

$$[K^{(1)}] = \frac{EI}{L^3} \begin{bmatrix} 12 & +6L & -12 & 6L \\ 6L & 4L^2 & -6L & 2L^2 \\ -12 & -6L & 12 & -6L \\ 6L & 2L^2 & -6L & 4L^2 \end{bmatrix} \begin{Bmatrix} U_{1y} \\ \theta_1 \\ U_{2y} \\ \theta_2 \end{Bmatrix}$$

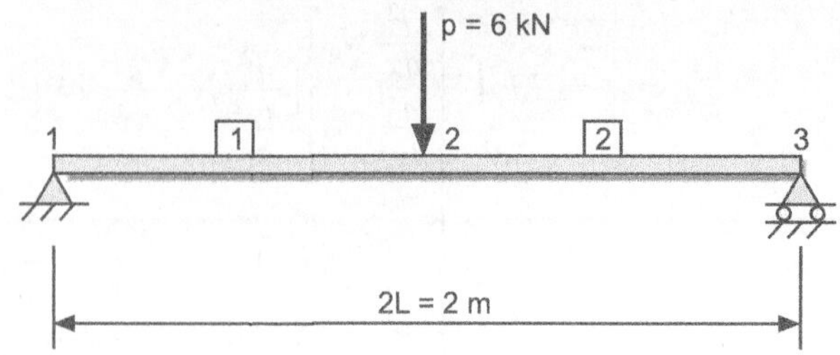

Fig 6.11. Simply supported beam.

For element # 2 the stiffness matrix is

$$[K^{(2)}] = \frac{EI}{L^3}\begin{bmatrix} 12 & 6L & -12 & 6L \\ 6L & 4L^2 & -6L & 2L^2 \\ -12 & -6L & 12 & -6L \\ 6L & 2L^2 & -6L & 4L^2 \end{bmatrix}\begin{Bmatrix} U_{2y} \\ \theta_2 \\ U_{3y} \\ \theta_3 \end{Bmatrix}$$

Assembled equation for the beam taking $EI = 10^3$, $L = 1\text{m}$

$$\begin{Bmatrix} F_1 \\ M_1 \\ F_2 \\ M_2 \\ F_3 \\ M_3 \end{Bmatrix} = 10^3\begin{bmatrix} 12 & 6 & -12 & 6 & 0 & 0 \\ 6 & 4 & -6 & 2 & 0 & 0 \\ -12 & -6 & 24 & +0 & -12 & 6 \\ 6 & 2 & 0 & 8 & -6 & 2 \\ 0 & 0 & -12 & -6 & 12 & -6 \\ 0 & 0 & 6 & 2 & -6 & 4 \end{bmatrix}\begin{Bmatrix} U_{1y} \\ \theta_1 \\ U_{2y} \\ \theta_2 \\ U_{3y} \\ \theta_3 \end{Bmatrix}$$

The displacement boundary conditions are

$$u_{1y} = u_{3y} = \theta_2 = 0.$$

The force boundary conditions are

$$M_1 = -3 \times 10^3 \qquad F_2 = -6 \times 10^3 \qquad M_3 = 3 \times 10^3$$

Hence the force displacement relation is

$$\begin{Bmatrix} -3\times10^3 \\ -6\times10^3 \\ 3\times10^3 \end{Bmatrix} = 10^3\begin{bmatrix} 4 & -6 & 0 \\ -6 & 24 & 6 \\ 0 & 6 & 4 \end{bmatrix}\begin{Bmatrix} \theta_1 \\ u_{2y} \\ \theta_3 \end{Bmatrix}$$

By Gauss elimination method

$$\begin{bmatrix} 4 & -6 & 0 \\ 0 & 10 & 4 \\ 0 & 0 & 8/3 \end{bmatrix}\begin{Bmatrix} \theta_1 \\ u_{2y} \\ \theta_3 \end{Bmatrix} = \begin{Bmatrix} -3 \\ -7 \\ 12 \end{Bmatrix}$$

Hence unknown displacements are

$$\theta_1 = -4.5 \qquad u_{2y} = -2.5 \quad \text{and} \quad \theta_3 = 4.5$$

Unknown forces are F_1, M_2 and F_3

$$\begin{Bmatrix} F_1 \\ M_2 \\ F_3 \end{Bmatrix} = 10^3 \begin{bmatrix} 6 & -12 & 0 \\ 2 & 0 & 2 \\ 0 & -12 & -6 \end{bmatrix} \begin{Bmatrix} -4.5 \\ -2.5 \\ 4.5 \end{Bmatrix}$$

$$F_1 = +6\,(-4.5) - 12(-2.5) = 3 \times 10^3 \text{ N}$$

$$M_2 = 2\,(-4.5) + 2\,(4.5) = 0$$

$$F_3 = -12\,(-4.5) - 6\,(4.5) = 3 \times 10^3 \text{ N}$$

6.8 2D BEAM-COLUMN ELEMENT: LOCAL SYSTEM

The parent beam-column element and the deformed shape of the same are shown in Fig. 6.12. In the deformed element the node 1 translates by $\bar{u}_{x1}$ and $\bar{u}_{y1}$ along $\bar{x}$ and $\bar{y}$ directions respectively. It is subjected rotation $\bar{\theta}_{z1}$ about $\bar{z}$ axis. Similar way the deformations of node 2 are $\bar{u}_{x2}, \bar{u}_{y2}$ and $\bar{\theta}_{z2}$. Hence the 2D beam column element with two nodes has three degree of freedom at each node – totally 6 DOF. The deformation in axial direction reveal the column-effect and the other two represent the beam effect. Such elements are also called frame elements.

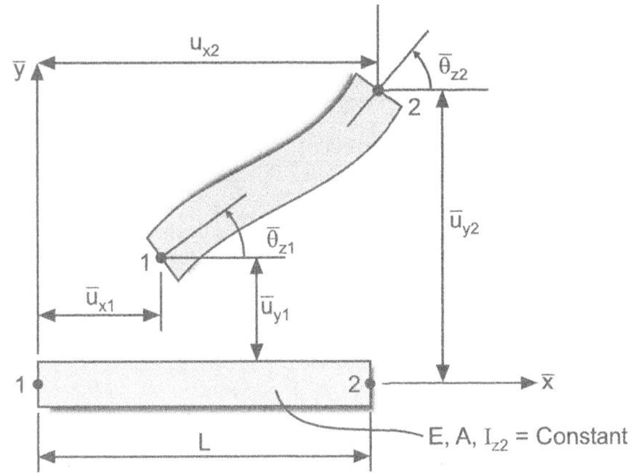

Fig. 6.12. Beam column element in local system.

The deformation vector that defines the degree of freedom for 2D Beam-column element is

$$\{\bar{U}\} = [\bar{u}_{x1} \;\; \bar{u}_{y1} \;\; \bar{\theta}_{z1} \;\; \bar{u}_{x2} \;\; \bar{u}_{y2} \;\; \bar{\theta}_{z2}]^T \qquad \text{...(6.67)}$$

Similarly the force vector that effects the deformations is

$$\{\bar{f}\} = [\bar{f}_{x1} \;\; \bar{f}_{y1} \;\; \bar{m}_{z1} \;\; \bar{f}_{x2} \;\; \bar{f}_{y2} \;\; \bar{m}_{z2}]^T \qquad \text{...(6.68)}$$

The force-displacement relation between force vector and the displacement vector is established through

$$\{\bar{f}\} = [\bar{K}]\{u\} \qquad \text{...(6.69)}$$

$[\bar{k}]$ is the elemental stiffness matrix in local co-ordinate system. The stiffness matrix can be the combination of bar contribution and Euler-Bernoulli beam contribution.

$$[\bar{K}] = [\bar{K}]_{bar} + [\bar{K}]_{beam} \qquad \text{...(6.70)}$$

where

$$[\bar{K}]_{bar} = \frac{EA}{L}\begin{bmatrix} 1 & 0 & 0 & -1 & 0 & 0 \\ & 0 & 0 & 0 & 0 & 0 \\ & & 0 & 0 & 0 & 0 \\ & & & 1 & 0 & 0 \\ & \text{symmetric} & & & 0 & 0 \\ & & & & & 0 \end{bmatrix} \quad ...(6.71)$$

(Bar contribution)

and

$$[\bar{K}]_{beam} = \frac{EI}{L^3}\begin{bmatrix} 0 & 0 & 0 & 0 & 0 & 0 \\ & 12 & 6L & 0 & -12 & 6L \\ & & 4L^2 & 0 & -6L & 2L^2 \\ & & & 0 & 0 & 0 \\ & \text{symmetric} & & & 12 & -6L \\ & & & & & 4L^2 \end{bmatrix} \quad ...(6.72)$$

(Beam contribution)

It may be noted that the stiffness matrix is a 6 × 6 square matrix which is also symmetric. The elements of stiffness matrix are dependent on the geometric properties like A (area), L (Length) and I (area moment of inertia), and the mechanical property of materials E (Young's Modulus). It is clear that stiffness matrix is independent of force.

6.9 2D BEAM-COLUMN ELEMENT: GLOBAL SYSTEM

When the local co-ordinate system is rotated about the global system, it is needed to carry out a transformation for all the element variables.

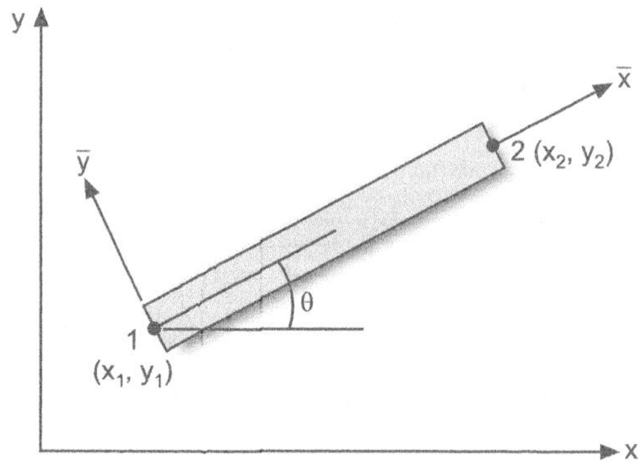

Fig. 6.13. Beam Column Element-Global System.

Let the deformation vector in the global system of co-ordinates be

$$\{u\} = \{u_{x1}\ u_{y1}\ \theta_{z1}\ u_{x2}\ u_{y2}\ \theta_{z2}\}^T$$

Similarly the force vector be

$$\{f\} = [f_{x1}\ f_{y1}\ m_{z1}\ f_{x2}\ f_{y2}\ m_{z2}]^T$$

By transformation

$$\{\bar{u}\} = [T]\{u\} \text{ – for deformation vector}$$

$$\{\bar{f}\} = [T][f] \text{ – for force vector}$$

By the force displacement relation in local system

$$\{\bar{f}\} = [\bar{K}]\{\bar{u}\}$$

or

$$[T]\,[f] = [\bar{K}][T]\{u\}$$

Hence

$$[f] = [T]^T[\bar{K}][T]\{u\} \quad ...(6.73)$$

For orthogonal system $[T]^{-1} - [T]^T$.

Hence the stiffness matrix in Global system Fig. 6.13 is written as

$$[K] = [T]^T\,[\bar{K}]\,[T]$$

where

$$[T] = \begin{bmatrix} l & m & 0 & 0 & 0 & 0 \\ -m & l & 0 & 0 & 0 & 0 \\ 0 & 0 & 1 & 0 & 0 & 0 \\ 0 & 0 & 0 & l & m & 0 \\ 0 & 0 & 0 & -m & l & 0 \\ 0 & 0 & 0 & 0 & 0 & 1 \end{bmatrix} \quad ...(6.74)$$

where

$$l = \cos\theta = \frac{(x_2 - x_1)}{L}$$

$$m = \sin\theta = \frac{(y_2 - y_1)}{L}$$

and the length of element $L = [(x_2 - x_1)^2 + (y_2 - y_1)^2]^{1/2}$ (x_1, y_1) and (x_2, y_2) are the co-ordinates of the nodes 1 and 2 respectively. l and m are the direction cosines and L is the length of the element.

Example 6.4. *Determine the displacements of the following structure given that $EI = 3000$ kN.m^2 and $AE = 6 \times 10^5$ kN. (Refer Fig. 6.14).*

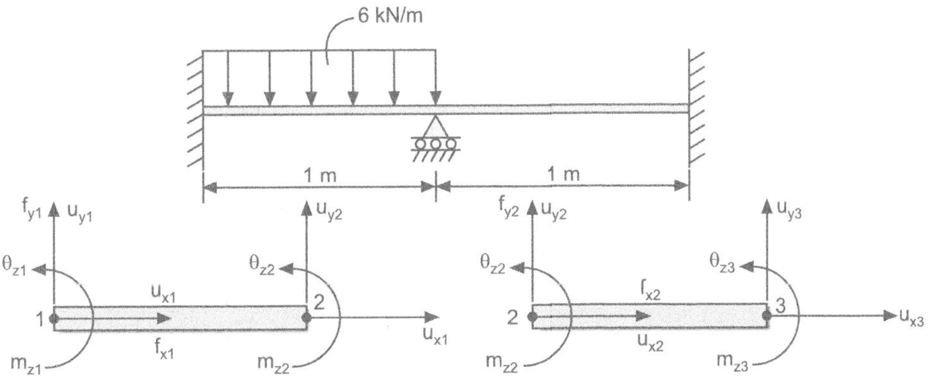

Fig. 6.14. Fixed beam structure.

Solution.

The displacement boundary condition specify that $u_{x1} = u_{y1} = \theta_{z1} = u_{x3} = u_{y3} = \theta_{z3} = u_{y2} = 0$

The forces pertaining to unknown displacements u_{x2} and θ_{z2} are $f_{x2} = 0$ and $m_{z2} = \dfrac{qL^2}{12}$.

We know $q = 6$ kN/m $L = 1$ m.

Hence $$M_{z2} = 0.5 \text{ kN/m}.$$

The force displacement relation after elimination of the boundary conditions lead to the following stiffness matrix,

$$[K] = \underset{\text{Element \#1}}{\begin{bmatrix} \frac{AE}{L} & 0 \\ 0 & \frac{4EI}{L} \end{bmatrix}} + \underset{\text{Element \#2}}{\begin{bmatrix} \frac{AE}{L} & 0 \\ 0 & \frac{4EI}{L} \end{bmatrix}}$$

Or it can also be written as

$$[K]\{U\} = \begin{bmatrix} \frac{2AE}{L} & 0 \\ 0 & \frac{8EI}{L} \end{bmatrix} \begin{Bmatrix} ux_2 \\ \theta_{z2} \end{Bmatrix} = \begin{bmatrix} 0 \\ 0.5 \end{bmatrix}$$

Hence by manipulation

$$\theta_{z2} = \frac{0.5L}{8EI} = \frac{0.5(1)}{24 \times 10^3}$$

$$u_{x2} = 0$$

It conforms that the displacement in x-direction does not exists.

Example 6.5. *A structural frame with two element with the sectional and material properties, $EI = 6 \times 10^3$ N.m^2 and $AE = 6 \times 10^6$ N is as shown in Fig. 6.15. Determine unknown displacements and unknown reactions for the structure.*

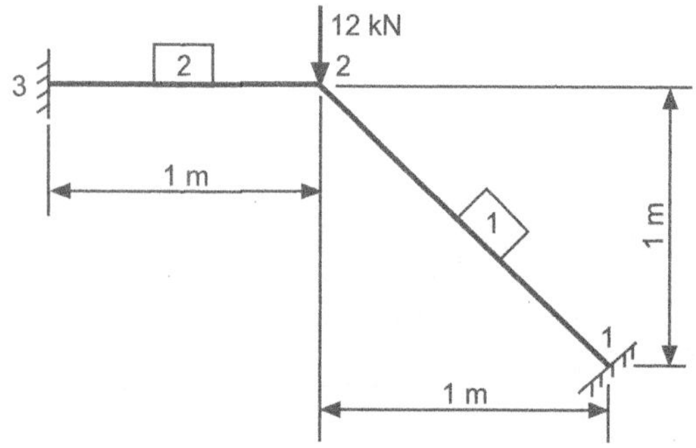

Fig. 6.15. A structural frame.

Solution.

Connectivity table

Elements	*Nodes*
1	1, 2
2	2, 3

Co-ordinates and direction cosines

Elements	*Nodes*	*Co-ordinates*	*length*	*l*	*m*	*lm*	m^2
1	1	(2, 0)	$\sqrt{2}$	$\frac{1}{\sqrt{2}}$	$-\frac{1}{\sqrt{2}}$	$-\frac{1}{\sqrt{2}}$	$\frac{1}{2}$
	2	(1, 1)					
2	2	(1, 1)	1	1	0	1	0
	3	(0, 1)					

The boundary conditions for the elements # 1 are $U_{x1} = U_{y1} = \theta_{z1} = 0$. U_{x2}, U_{y2} and θ_{z2} are unknowns. Hence the elemental stiffness matrix (for the unknown degress of freedom) in local co-ordinates system are

$$[\bar{K}_{12}] = \begin{bmatrix} \frac{AE}{L_1} & 0 & 0 \\ 0 & \frac{12EI}{L_1^3} & -\frac{6EI}{L_1^3} \\ 0 & -\frac{6EI}{L_1^3} & \frac{4EI}{L_1^3} \end{bmatrix}$$

The transformation matrix is

$$T = \begin{bmatrix} l & m & 0 \\ -m & l & 0 \\ 0 & 0 & 1 \end{bmatrix} = \begin{bmatrix} 1/\sqrt{2} & -1/\sqrt{2} & 0 \\ 1/\sqrt{2} & 1/\sqrt{2} & 0 \\ 0 & 0 & 1 \end{bmatrix}$$

The elemental stiffness matrix for element # 1 in the global co-ordinate system is.

$$[K_{12}] = \begin{bmatrix} 1/\sqrt{2} & 1/\sqrt{2} & 0 \\ -1/\sqrt{2} & 1/\sqrt{2} & 0 \\ 0 & 0 & 1 \end{bmatrix} [\bar{K}_{12}] \begin{bmatrix} 1/\sqrt{2} & -1/\sqrt{2} & 0 \\ 1/\sqrt{2} & 1/\sqrt{2} & 0 \\ 0 & 0 & 1 \end{bmatrix}$$

$$= \begin{bmatrix} \left(\frac{AE}{2L_1} - \frac{6EI}{L_1^3}\right) & \left(-\frac{AE}{2L_1} + \frac{6EI}{L_1^3}\right) & -\frac{6EI}{\sqrt{2}L_1^2} \\ \left(-\frac{AE}{2L_1} + \frac{6EI}{L_1^3}\right) & \left(\frac{AE}{2L_1} + \frac{6EI}{L_1^3}\right) & -\frac{6EI}{\sqrt{2}L_1^2} \\ \left(-\frac{6EI}{\sqrt{2}L_1^2}\right) & -\frac{6EI}{\sqrt{2}L_1^2} & \frac{4EI}{L_1} \end{bmatrix}$$

By the substitution of the numerical values for EI and AE in the above matrix.

$$[K_{12}] = 6 \times 10^3 \begin{bmatrix} 351.4 & -351.4 & -2.12 \\ -351.4 & 355.7 & -2.12 \\ -2.12 & -2.12 & 2.83 \end{bmatrix}$$

In the global co-ordinate the stiffness matrix for element # after the elimination of the DOF that are all zero.

$$[K_{23}] = \begin{bmatrix} \frac{AE}{L_2} & 0 & 0 \\ 0 & \frac{12EI}{L_2^3} & \frac{6EI}{L_2^2} \\ 0 & \frac{6EI}{L_2^2} & \frac{4EI}{L_2^2} \end{bmatrix}$$

On the numerical substitution of *EI*, *AE* and $l_2 = 1$ m

$$[K_{23}] = 6\times10^3\begin{bmatrix} 10^3 & 0 & 0 \\ 0 & 12 & 6 \\ 0 & 6 & 4 \end{bmatrix}$$

The assembled global stiffness matrix is

$$[K] = [K_{12}] + [K_{23}]$$

$$= 6\times10^3\begin{bmatrix} 1351.4 & -351.4 & -2.12 \\ -351.4 & 367.7 & 3.88 \\ -2.12 & 3.88 & 6.83 \end{bmatrix}$$

The force-displacement relation is

$$[K]\{U\} = \{F\}$$

$$6\times10^3\begin{bmatrix} 1351.4 & -351.4 & -2.12 \\ -351.4 & 367.7 & 3.88 \\ -2.12 & 3.88 & 6.83 \end{bmatrix}\begin{Bmatrix} u_{x2} \\ u_{y2} \\ \theta_{z2} \end{Bmatrix} = \begin{Bmatrix} 0 \\ 12\times10^3 \\ 0 \end{Bmatrix}$$

With the application of Gauss elimination method

$$\begin{bmatrix} 1351.4 & -351.4 & -2.12 \\ 0 & 1062.7 & 12.8 \\ 0 & 0 & 2166.7 \end{bmatrix}\begin{Bmatrix} u_{x2} \\ u_{y2} \\ \theta_{z2} \end{Bmatrix} = \begin{Bmatrix} 0 \\ 7.69 \\ -7.69 \end{Bmatrix}$$

$$\theta_{z2} = \frac{-7.69}{2166.7} = -0.0035 \text{ rad.}$$

$$u_{y2} = 0.00719 \text{ m}$$

$$u_{x2} = 0.00187 \text{ m}$$

Hence the unknown displacements are

$$\{u\} = \begin{Bmatrix} u_{x2} \\ u_{y2} \\ \theta_{z2} \end{Bmatrix} = \begin{Bmatrix} 0.00187 \\ 0.00719 \\ -0.0035 \end{Bmatrix}$$

POINTS TO BE REMEMBER

- Euler-Bernoulli's beam element has two nodes.
- 1D Beam element has two degree of freedom at each node: transverse translation and rotation.

- The displacement function used has been the third degree polynomial for the 1D beam with four degrees of freedom.
- 1D two noded beam element has four degrees of freedom.
- The expressions for shape functions of 1D beam element have the degree three.
  There are four shape functions for the 1D beam element.
- The stiffness matrix representing 1D beam element is 4 × 4 square matrix.
- The kroneckor delta property of shape functions is not applicable to beam elements.
- The displacement transformation method of obtaining shape functions requires the inversion of 4 × 4 matrix (connectivity matrix).
- Hermite polynomial method is the easier method to obtain the shape functions, of 1D beam element.
- Hermitean method employs third degree polynomial for the shape functions.
- Beam-bar combination in FEM is called frame element.
- The factors in the stiffness matrix of a 1D beam element are: 1. Young's modulus 2. Moment of inertia 3. The length of the beam.

EXERCISE

1. Give the expressions for the forces and moments for Euler Bernoulli's beam element using the beam bending theory.
2. Discuss the continuity, compatibility and completeness requirements of beam element.
3. Obtain the shape functions by displacement approach using third degree displacement function.
4. Plot the shape functions fo Euler Bernoulli's beam element.
5. Obtain the shape functions of beam element in natural co-ordinates.
6. Obtain the shape functions of 1D beam element in global co-ordinates using Hermite polynomial method.
7. Using the Hermetian method obtain the shape functions in natural co-ordinates.
8. Obtain the elemental stiffness matrix for the Euler Bernoulli's beam element.
9. Obtain the expression for load vector in beam element due to self weight.
10. Obtain the matrix expression for load vector in beam element due to traction force (UDL).
11. Obtain the expression in matrix form the load vector due to concentrated load and the moment acting at a point on the beam.
12. Obtain the stiffness matrix for the 1D frame element.
13. Obtain the transformation matrix to transform the stiffness matrix in local cartesian system to global co-ordinate system.
14. Obtain the load vector in the matrix form for the beam element acted upon by the centrifugal force.

Chapter 7

Finite Element Formulation of 2-Dimensional Elements

अचार्यात् पादमादत्ते।
पादं शिष्यः स्वमेधया॥
पादं सब्रह्मचारिभ्यः।
पादं कालक्रमेण हि॥ (महाभारत्)

[The disciple acquires part of his education from the teacher, a part by self study, another part by discussion with class-mates and the remaining part by self-experience with time.]

7.1 TRIANGULAR ELEMENT: THREE NODED

To represent the displacement variation at the element level 2-D element level and to develop the shape functions or the linear interpolation functions consider a triangular element with three nodes. Let the vertices or the nodes at the corner be denoted by the lebels '1', '2' and '3'. Let the X-Y co-ordinates of these nodes be (x_i, y_i) for $i = 1, 2$ and 3 as shown in the Fig. 7.1. At this instant let us consider one degree of freedom at each node of the element. Hence total degrees of freedom works out to be three.

Let N_i for $i = 1, 2, 3$ be the shape functions that interpolate the displacements at the nodes to generalize the displacements distribution in the element. The shape functions used for interpolation have the following properties:

1. N_i is linear in x and y variables.
2. N_i has the value of 'o' at node j for $i \neq j$ and values 1 at node i.

The distribution of displacement in terms of shape functions and degrees of freedom at nodes 1, 2 and 3, is expressed as

$$u(x, y) = N_1u_1 + N_2u_2 + N_3u_3 \quad ...(7.1)$$

The assumed expression for the shape functions $N_i(x, y)$ be

$$N_i(x, y) = a_i + b_i\, x + c_i y \quad ...(7.2)$$

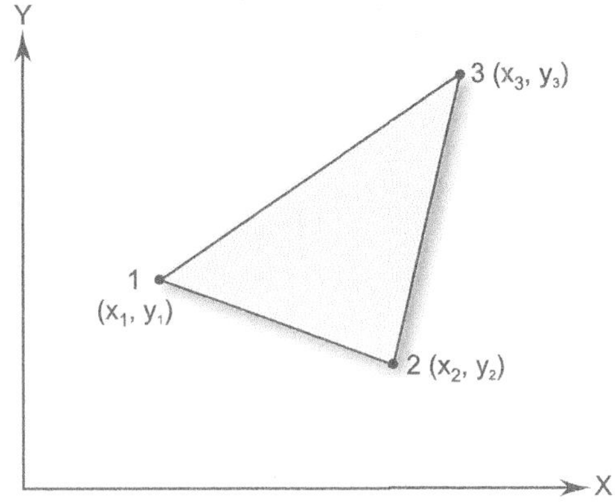

Fig. 7.1. Triangular element in global system.

7.2 SHAPE FUNCTIONS: TRIANGULAR ELEMENT

a_i, b_i and c_i are the constant co-efficients to be determined from the co-ordinates of the nodes and the property 2 of the shape functions. With this we arrive at the equations.

$$\begin{aligned} 1 &= a_1 + b_1\, x_1 + c_1\, y_1 \\ 0 &= a_1 + b_1\, x_2 + c_1\, y_2 \\ 0 &= a_1 + b_1\, x_3 + c_1\, y_3 \end{aligned}$$

$$\begin{aligned} 0 &= a_2 + b_2\, x_1 + c_2\, y_1 \\ 1 &= a_2 + b_2\, x_2 + c_2\, y_2 \\ 0 &= a_2 + b_2\, x_3 + c_2\, y_3 \end{aligned}$$

$$\begin{aligned} 0 &= a_3 + b_3\, x_1 + c_3\, y_1 \\ 0 &= a_3 + b_3\, x_2 + c_3\, y_2 \\ 1 &= a_3 + b_3\, x_3 + c_3\, y_3 \end{aligned}$$

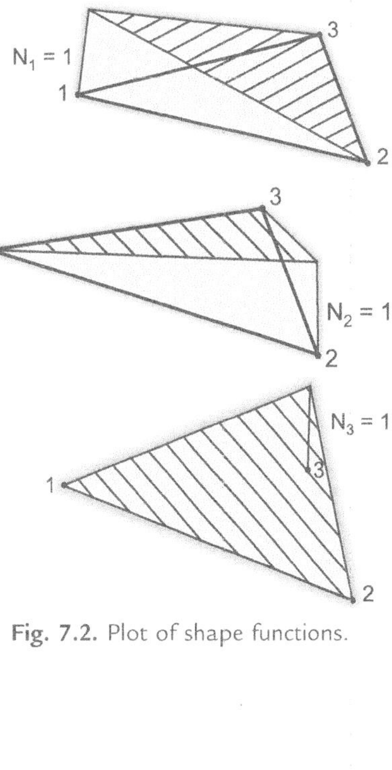

Fig. 7.2. Plot of shape functions.

These equations can be written in the form

$$\begin{bmatrix} 1 \\ 0 \\ 0 \end{bmatrix} = [C]\begin{Bmatrix} a_1 \\ b_1 \\ c_1 \end{Bmatrix}, \quad \begin{bmatrix} 0 \\ 1 \\ 0 \end{bmatrix} = [C]\begin{Bmatrix} a_2 \\ b_2 \\ c_2 \end{Bmatrix}, \quad \begin{bmatrix} 0 \\ 0 \\ 1 \end{bmatrix} = [C]\begin{Bmatrix} a_3 \\ b_3 \\ c_3 \end{Bmatrix}$$

where the connectivity matrix

$$[C] = \begin{bmatrix} 1 & x_1 & y_1 \\ 1 & x_2 & y_2 \\ 1 & x_3 & y_3 \end{bmatrix} \quad \text{...(7.3)}$$

$$N_i(x, y) = [1 \quad x \quad y]\begin{Bmatrix} a_i \\ b_i \\ c_i \end{Bmatrix}$$

$$[N_1, N_2, N_3] = [1 \quad x \quad y]\begin{bmatrix} a_1 & b_1 & c_1 \\ a_2 & b_2 & c_2 \\ a_3 & b_3 & c_3 \end{bmatrix}^T$$

$$= [1 \quad x \quad y][C]^{-1}\begin{bmatrix} 1 & 0 & 0 \\ 0 & 1 & 0 \\ 0 & 0 & 1 \end{bmatrix}$$

The inverse of the non-singular matrix is the transpose of co-factor matrix divided by the determinant of the matrix.

Hence

$$[C]^{-1} = \frac{1}{\det [C]}\begin{bmatrix} (x_2y_3 - x_3y_2) & (y_2 - y_3) & (x_3 - x_2) \\ (x_3y_1 - x_1y_3) & (y_3 - y_1) & (x_1 - x_3) \\ (x_1y_2 - x_2y_1) & (y_1 - y_2) & (x_2 - x_1) \end{bmatrix}^T \quad \text{...(7.4)}$$

But

det $[C] = 2A$ where A is the area of the triangle. We write the shape functions as

$$N_i(x, y) = [1 \quad x \quad y]\frac{1}{2A}\begin{bmatrix} (x_2y_3 - x_3y_2) & -(x_1y_3 - x_3y_1) & (x_1y_2 - x_2y_1) \\ (y_2 - y_3) & (y_3 - y_1) & (y_1 - y_2) \\ (x_3 - x_2) & (x_1 - x_3) & (x_2 - x_1) \end{bmatrix} \quad \text{...(7.5)}$$

$$\begin{bmatrix} 1 & 0 & 0 \\ 0 & 1 & 0 \\ 0 & 0 & 1 \end{bmatrix}$$

Hence the formulae for the shape functions are

$$N_1 = \frac{1}{2A}[(x_2y_3 - x_3y_2) + (y_2 - y_3)x + (x_3 - x_2)y] \quad \text{...(7.6)}$$

$$N_2 = \frac{1}{2A}[-(x_1y_3 - x_3y_1) + (y_3 - y_1)x + (x_1 - x_3)y] \quad \text{...(7.7)}$$

$$N_3 = \frac{1}{2A}[(x_1y_2 - x_2y_1) + (y_1 - y_2)x + (x_2 - x_1)y] \quad \text{...(7.8)}$$

It may be noted that by substituting the coordinates of node 1 *i.e.*, (x_1, y_1) in the variable (x, y) we get $N_1 = 1 \quad N_2 = 0 \quad N_3 = 0$

Similarly for node 2, *i.e.*, (x_2, y_2)

$$N_1 = 0 \qquad N_2 = 1 \qquad N_3 = 0$$

Also for the node 3 with $(x, y) = (x_3, y_3)$ we get

$$N_1 = 0 \qquad N_2 = 0 \qquad N_3 = 1$$

Further it is quite evident that the property

$$\sum_{i=1}^{3} Ni = 1$$

7.3 TRIANGULAR ELEMENT: 2-DIMENSIONAL

Let the element displacement vector for the 2-*D* triangular element be denoted by $\{u\}$ whose components at each node be $\{u_{ix}\}$ and $\{u_{iy}\}$ for $i = 1$, 2 and 3. This clearly indicates that the three nodded element has totally six degrees of freedom with 2 *DOF* at each node. The displacement field is expressed as

$$\begin{aligned} u(x, y) &= N_1\, u_{1x} + N_2\, u_{2x} + N_3\, u_{3x} \\ v(x, y) &= N_1\, u_{1y} + N_2 u_{2y} + N_3\, u_{3y} \end{aligned} \qquad ...(7.9)$$

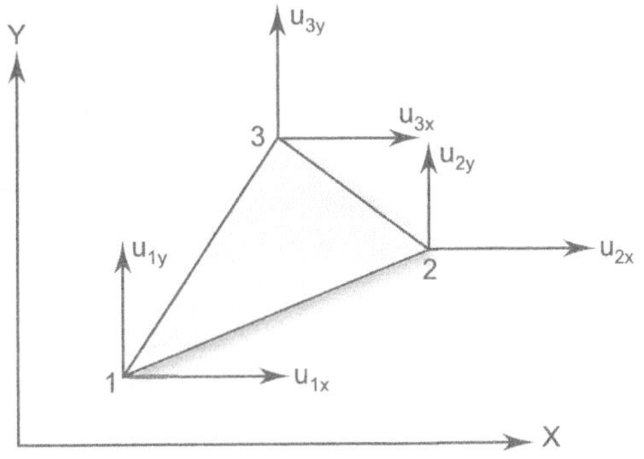

Fig. 7.3. 2-*D* triangle 2-DOF at node.

The degress of freedom as shown by the Fig. 7.3 can hence be related as

$$\{u\} = \{N\}\,\{d\} \qquad ...(7.10)$$

where $\{d\} = \{u_{1x} \;\; u_{1y} \;\; u_{2x} \;\; u_{2y} \;\; u_{3x} \;\; u_{3y}\}$

and $[N]$ is the array of 2×6 dimensions expressed in terms of shape functions in the form

$$[N] = \begin{bmatrix} N_1 & 0 & N_2 & 0 & N_3 & 0 \\ 0 & N_1 & 0 & N_2 & 0 & N_3 \end{bmatrix} \qquad ...(7.11)$$

7.4 STRAIN-DISPLACEMENT MATRIX

The two dimensional strain distribution in an element is expressed to be of the following form

$$\{\in\} = \begin{Bmatrix} \in_x \\ \in_y \\ \gamma_{xy} \end{Bmatrix} = \begin{Bmatrix} \dfrac{\partial u_x}{\partial x} \\ \dfrac{\partial u_y}{\partial y} \\ \dfrac{\partial u_x}{\partial y} + \dfrac{\partial u_y}{\partial x} \end{Bmatrix} \qquad ...(7.12)$$

$$= \begin{bmatrix} \frac{\partial}{\partial x} & 0 \\ 0 & \frac{\partial}{\partial y} \\ \frac{\partial}{\partial y} & \frac{\partial}{\partial x} \end{bmatrix} \begin{Bmatrix} u_x \\ u_y \end{Bmatrix} = [\partial]\{u\}$$

But we have

$$\{u\} = [N]\{d\}$$
$$= [\partial][N]\{d\}$$
$$= [B]\{d\}$$

It may be recalled that for a triangular element (with 3 nodes) with one degree-freedom at each node.

$$[N]^T = [N_1\ N_2\ N_3] = [1\ \ x\ \ y]\,[C]^{-1}$$

$$= [1\ \ x\ \ y]\begin{bmatrix} a_1 & a_2 & a_3 \\ b_1 & b_2 & b_3 \\ c_1 & c_2 & c_3 \end{bmatrix} \quad ...(7.13)$$

with the formulation of shape functions the strain displacement, $[B]$ matrix can be expressed as :

$$[B] = \begin{bmatrix} \frac{\partial N_1}{\partial x} & 0 & \frac{\partial N_2}{\partial x} & 0 & \frac{\partial N_3}{\partial x} & 0 \\ 0 & \frac{\partial N_1}{\partial y} & 0 & \frac{\partial N_2}{\partial y} & 0 & \frac{\partial N_3}{\partial y} \\ \frac{\partial N_1}{\partial y} & \frac{\partial N_1}{\partial x} & \frac{\partial N_2}{\partial y} & \frac{\partial N_2}{\partial x} & \frac{\partial N_3}{\partial y} & \frac{\partial N_3}{\partial x} \end{bmatrix} \quad ...(7.14)$$

Using the equation (7.13) for N_i and substituting in equation (7.14)

$$[B] = \begin{bmatrix} b_1 & 0 & b_2 & 0 & b_3 & 0 \\ 0 & c_1 & 0 & c_2 & 0 & c_3 \\ c_1 & b_1 & c_2 & b_2 & c_3 & b_3 \end{bmatrix}$$

where

$$\left.\begin{aligned} b_1 &= \frac{1}{2A}(y_2 - y_3)\ ;\ c_1 = \frac{1}{2A}(x_3 - x_2) \\ b_2 &= \frac{1}{2A}(y_3 - y_1)\ ;\ c_2 = \frac{1}{2A}(x_1 - x_3) \\ b_3 &= \frac{1}{2A}(y_1 - y_2)\ ;\ c_3 = \frac{1}{2A}(x_2 - x_1) \end{aligned}\right\} \quad ...(7.15)$$

7.5 STIFFNESS MATRIX: PLANE STRESS CONDITION

The finite element procedure: The stiffness matrix is expressed in terms of the strain-displacement matrix $[B]$ and the elasticity matrix $[D]$, as given by

$$[K] = \int_V [B]^T [D][B]\,dV \quad ...(7.16)$$

Even though the $[B]$ matrix is same for the triangular element the $[D]$ matrix (elasticity matrix) is specific to the case like plane stress condition of state.

In the plane stress condition, the elasticity matrix is expressed in the form

$$[D] = \frac{E}{(1-\mu^2)}\begin{bmatrix} 1 & \mu & 0 \\ \mu & 1 & 0 \\ 0 & 0 & \frac{1-\mu}{2} \end{bmatrix} \quad ...(7.17)$$

where E = is the Young's modulus

and μ is the Poisson's ratio

Hence the stiffness matrix is written as

$$[K] = \int [B]^T \frac{E}{(1-\mu^2)}\begin{bmatrix} 1 & \mu & 0 \\ \mu & 1 & 0 \\ 0 & 0 & \left(\frac{1-\mu}{2}\right) \end{bmatrix}[B]\, dV \quad ...(7.18)$$

This is written symbollically as

$$[K] = \frac{E}{(1-\mu^2)}\begin{bmatrix} k_{11} & k_{12} & k_{13} & k_{14} & k_{15} & k_{16} \\ & k_{22} & k_{23} & k_{24} & k_{25} & k_{26} \\ & & k_{33} & k_{34} & k_{35} & k_{36} \\ & & & k_{44} & k_{45} & k_{46} \\ & \text{symmetry} & & & k_{55} & k_{56} \\ & & & & & k_{66} \end{bmatrix} \quad ...(7.19)$$

The elements of the stiffness matrix k_{ij} can be written to be

$$K_{11} = \int_V \left[b_1^2 + \left(\frac{1-\mu}{2}\right)c_1^2\right] dV$$

$$K_{12} = \int_V b_1c_1\left(\frac{1+\mu}{2}\right) dV$$

$$K_{13} = \int_V \left[b_1b_2 + c_1c_2\left(\frac{1-\mu}{2}\right)\right] dV$$

$$K_{14} = \int_V \left[\mu b_1c_2 + \left(\frac{1-\mu}{2}\right) b_2c_1\right] dV$$

$$K_{15} = \int_V \left[b_1b_3 + c_1c_3\left(\frac{1-\mu}{2}\right)\right] dV$$

$$K_{16} = \int_V \left[\mu b_1c_3 + \left(\frac{1-\mu}{2}\right)c_1b_3\right] dV$$

$$K_{22} = \int_V \left[c_1^2 + \left(\frac{1-\mu}{2}\right)b_1^2\right] dV$$

$$K_{23} = \int_V \left[\mu b_2 c_1 + \left(\frac{1-\mu}{2}\right)b_1 c_2\right] dV$$

$$K_{24} = \int_V \left[c_1 c_2 + \left(\frac{1-\mu}{2}\right)b_1 b_2\right] dV$$

$$K_{25} = \int_V \left[\mu b_3 c_1 + \left(\frac{1-\mu}{2}\right)b_1 c_3\right] dV$$

$$K_{26} = \int_V \left[c_1 c_3 + \left(\frac{1-\mu}{2}\right)b_1 b_3\right] dV$$

$$K_{33} = \int_V \left[b_2^2 + c_2^2\left(\frac{1-\mu}{2}\right)\right] dV$$

$$K_{34} = \int_V \left[b_2 c_2\left(\frac{1+\mu}{2}\right)\right] dV$$

$$K_{35} = \int_V \left[b_2 b_3 + \left(\frac{1-\mu}{2}\right)c_2 c_3\right] dV$$

$$K_{36} = \int_V \left[\mu b_2 c_3 + \left(\frac{1-\mu}{2}\right)b_3 c_2\right] dV$$

$$K_{44} = \int_V \left[c_2^2 + b_2^2\left(\frac{1-\mu}{2}\right)\right] dV$$

$$K_{45} = \int_V \left[\mu b_3 c_2 + \left(\frac{1-\mu}{2}\right)b_2 c_3\right] dV$$

$$K_{46} = \int_V \left[c_2 c_3 + b_2 b_3\left(\frac{1-\mu}{2}\right)\right] dV$$

$$K_{55} = \int_V \left[b_3^2 + \left(\frac{1-\mu}{2}\right)c_3^2\right] dV$$

$$K_{56} = \int_V \left[b_3 c_3\left(\frac{1+\mu}{2}\right)\right] dV$$

$$K_{66} = \int \left[c_3^2 + \left(\frac{1-\mu}{2}\right)b_3^2\right] dV$$

7.6 PLANE STRAIN CONDITION: TRIANGULAR ELEMENT

The strain-displacement $[B]$ matrix is same for this case also but the elasticity matrix for the plane strain condition is

$$[D] = \frac{E}{(1+\mu)(1-2\mu)}\begin{bmatrix}(1-\mu) & \mu & 0 \\ \mu & (1-\mu) & 0 \\ 0 & 0 & \dfrac{(1-2\mu)}{2}\end{bmatrix} \qquad ...(7.20)$$

And the stiffness matrix can be obtained by the manipulation of the following formulation which is left as an exercise

$$[K] = \int [B]^T[D][B]dV$$

Example 7.1. *3-noded right angle triangle*

Figure 7.4 shows a 3-noded right angle triangular element with one degree of freedom at each node. The strain is constant in this 2-D element. The problem is to find the shape functions and the stiffness matrix assuming a linear displacement field for the element.

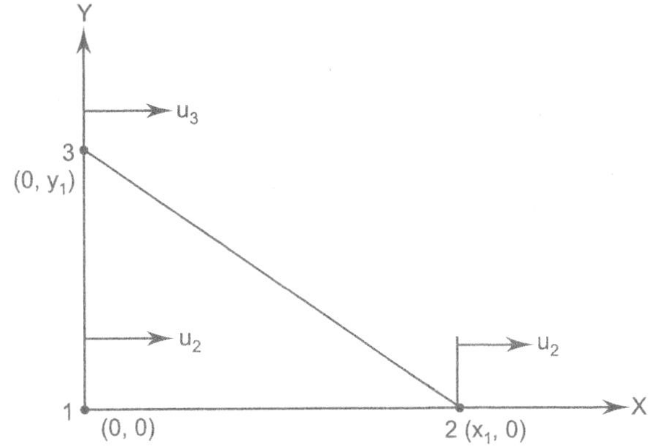

Fig. 7.4. 3-noded right angle triangle element.

Solution. The linear displacement field assumed for the 2-D triangular element with three DOF can be expressed as

$$u(x, y) = a_0 + a_1 x + a_2 y \qquad ...(7.21)$$

In the matrix form

$$\{u\} = [1 \quad x \quad y]\begin{Bmatrix}a_0 \\ a_1 \\ a_2\end{Bmatrix} = [P]\{A\} \qquad ...(7.22)$$

u_1, u_2 and u_3 are the displacement boundary conditions available at the three nodes with $i = 1, 2, 3$. The condition at the three nodes with co-ordinates.

node # 1 → $(0, 0)$

node # 2 → $(x_1, 0)$

node # 3 → $(0, y_1)$

At node 1 with $x = 0$ $y = 0$; $u = u_1$

At node 2, $x = x_1$, $y = 0$ and $u = u_2$

At node 3, $x = 0$, $y = y_1$ and $u = u_3$

From these set of boundary conditions we obtain the following set of equations

$$\begin{aligned} u_1 &= a_0 \\ u_2 &= a_0 + a_1 x_1 \\ u_3 &= a_0 + a_2 y_1 \end{aligned} \qquad ...(7.23)$$

The equations (7.13) in the matrix form

$$\begin{Bmatrix} u_1 \\ u_2 \\ u_3 \end{Bmatrix} = \begin{bmatrix} 1 & 0 & 0 \\ 1 & x_1 & 0 \\ 1 & 0 & y_1 \end{bmatrix} \begin{Bmatrix} a_0 \\ a_1 \\ a_2 \end{Bmatrix}$$

$$\begin{aligned} \{u\} &= [C]\{A\} \\ \{A\} &= [C]^{-1}\{u\} \\ u(x, y) &= [P]\{A\} \\ &= [P][C]^{-1}\{u\} \end{aligned} \qquad ...(7.24)$$

But

$$[C]^{-1} = \frac{1}{x_1 y_1} \begin{bmatrix} x_1 y_1 & 0 & 0 \\ -y_1 & y_1 & 0 \\ -x_1 & 0 & x_1 \end{bmatrix} \qquad ...(7.25)$$

Hence the shape functions are

$$N_i = [N_1 \quad N_2 \quad N_3]^T = [P][C]^{-1}$$

$$= [1 \quad x \quad y] \frac{1}{x_1 y_1} \begin{bmatrix} x_1 y_1 & 0 & 0 \\ -y_1 & y_1 & 0 \\ -x_1 & 0 & x_1 \end{bmatrix} \qquad ...(7.26)$$

To write the shape functions

$$\begin{aligned} N_1 &= \frac{1}{x_1 y_1}(x_1 y_1 - y_1 x - x_1 y) \\ N_2 &= \frac{1}{x_1 y_1}(y_1 x) \\ N_3 &= \frac{1}{x_1 y_1}(x_1 y) \end{aligned} \qquad ...(7.27)$$

Properties of shape functions are confirmed with the following manner

1. $u(x, y) = N_1(x, y)\, u_1 + N_2(x, y)\, u_2 + N_3(x, y)\, u_3$
   where u_1, u_2 and u_3 are the nodal properties
2. $\Sigma N_i = N_1 + N_2 + N_3 = 1$
3. at node # 1 $(x, y) = (0, 0)$ we have $N_1 = 1$ $\quad N_2 = 0 \quad N_3 = 0$
   at node # 2 $(x, y) = (x_1, 0)$ we have $N_1 = 0$ $\quad N_2 = 1 \quad N_3 = 0$
   at node # 3 $(x, y) = (0, y_1)$ we have $N_1 = 0$ $\quad N_2 = 0 \quad N_3 = 1$

Stiffness matrix: plane stress

The two dimensional strain relations are

$$\epsilon_x = \frac{\partial u}{\partial x} \qquad \epsilon_y = \frac{\partial v}{\partial y} \qquad \gamma_{xy} = \frac{\partial u}{\partial y} + \frac{\partial v}{\partial x} \qquad ...(7.28)$$

But

$$u = \Sigma N_i\, u_i$$

$$v = \Sigma N_i V_i \quad \text{Here all } V_i \text{ are zero}$$

Hence

$$\epsilon_x = \left(-\frac{1}{x_1}\right)u_1 + \left(\frac{1}{x_1}\right)u_2 + (0)\,u_3$$

$$\epsilon_y = (0)u_1 + (0)u_2 + (0)u_3 \qquad ...(7.29)$$

$$\gamma_{xy} = \left(-\frac{1}{y_1}\right)u_1 + (0)\,u_2 + \frac{1}{y_1}(u_3)$$

In the matrix form the strain-displacement relation takes the shape to suggest the relation for the [B] matrix in the following way.

For a plane strain case

$$[D] = \frac{E}{(1+\mu)(1-2\mu)}\begin{bmatrix}(1-\mu) & \mu & 0\\ \mu & (1-\mu) & 0\\ 0 & 0 & \dfrac{(1-2\mu)}{2}\end{bmatrix} \qquad ...(7.30)$$

The stiffness matrix by definition

$$[K] = \int_V [B]^T[D][B]dV \qquad ...(7.31)$$

But

$$\int_V dV = At = \left(\frac{x_1 y_1}{2}\right)t \qquad ...(7.32)$$

Hence stiffness matrix for a plane strain case

$$[K] = \frac{At}{x_1y_1}\begin{bmatrix}-y_1 & 0 & -x_1\\ y_1 & 0 & 0\\ 0 & 0 & x_1\end{bmatrix}\frac{E}{(1+\mu)(1-2\mu)}\begin{bmatrix}(1-\mu) & \mu & 0\\ \mu & (1-\mu) & 0\\ 0 & 0 & \dfrac{(1-2\mu)}{2}\end{bmatrix}\frac{1}{x_1y_1}\begin{bmatrix}-y_1 & y_1 & 0\\ 0 & 0 & 0\\ -x_1 & 0 & x_1\end{bmatrix}$$

$$= \frac{Et}{2x_1y_1(1+\mu)(1-2\mu)}\begin{bmatrix}(1-\mu)y_1^2 + x_1^2\left(\dfrac{1-2\mu}{2}\right) & -\mu y_1^2 & -x_1^2\left(\dfrac{1-2\mu}{2}\right)\\ -\mu y_1^2 & \mu y_1^2 & 0\\ -x_1^2\left(\dfrac{1-2\mu}{2}\right) & 0 & x_1^2\left(\dfrac{1-2\mu}{2}\right)\end{bmatrix} \qquad ...(7.33)$$

7.7 RECTANGULAR ELEMENT

The purpose of this section is to determine the displacement function $u(x, y)$ for the quadrilateral (rectangular) element as shown in the Fig. 7.5. The element is configured with four nodes; 1, 2, 3 and 4 whose co-ordinates are (x_i, y_i) for $i = 1$ to 4. The requirements, the function should satisfy are:

1. $u(x, y)$ assumes the values u_1, u_2, u_3 and u_4 at the respective nodal points.
2. $u(x, y)$ varies linearly along the edges of the element.

Meeting these requirements the displacement function is expanded in a polynomial in terms of x and y. Let $u(x, y)$ be written in the form.

$$u(x, y) = a + bx + cy + dxy \quad ...(7.34)$$

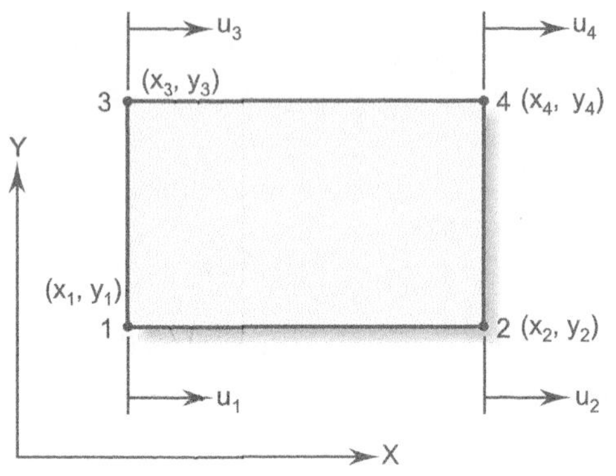

Fig. 7.5. Rectangular element with 4 DOF.

The co-efficients a, b, c and d are to be determined satisfying the two requirements as mentioned above with substitution of the conditions at four nodes, we arrive at the set of four equations, the unknowns being the generalized co-efficients a, b, c and d.

$$\begin{aligned} u_1 &= a + bx_1 + cy_1 + dx_1y_1 \\ u_2 &= a + bx_2 + cy_2 + dx_2y_2 \\ u_3 &= a + bx_3 + cy_3 + dx_3y_3 \\ u_4 &= a + bx_4 + cy_4 + dx_4y_4 \end{aligned} \quad ...(7.35)$$

The simultaneous equations (7.35) can be written in the matrix form

$$\begin{Bmatrix} u_1 \\ u_2 \\ u_3 \\ u_4 \end{Bmatrix} = \begin{bmatrix} 1 & x_1 & y_1 & x_1y_1 \\ 1 & x_2 & y_2 & x_2y_2 \\ 1 & x_3 & y_3 & x_3y_3 \\ 1 & x_4 & y_4 & x_4y_4 \end{bmatrix} \begin{Bmatrix} a \\ b \\ c \\ d \end{Bmatrix} \quad ...(7.36)$$

$$\{u\} = [C]\{A\}$$

The co-efficient matrix

$$\begin{aligned} \{A\} &= [C]^{-1}\{u\} \\ u(x, y) &= [1\ x\ y\ xy]\{A\} \\ &= [1\ x\ y\ xy][C]^{-1}\{u\} \\ &= [N]^T\{u\} \end{aligned}$$

where N_i are the four shape functions that satisfy the equation

$$u(x, y) = N_1u_1 + N_2u_2 + N_3u_3 + N_4u_4$$

It may be noted that the shape functions are to satisfy the following property

$$[C]^T [N] = \begin{bmatrix} 1 & 1 & 1 & 1 \\ x_1 & x_2 & x_3 & x_4 \\ y_1 & y_2 & y_3 & y_4 \\ x_1y_1 & x_2y_2 & x_3y_3 & x_4y_4 \end{bmatrix} \begin{bmatrix} N_1 \\ N_2 \\ N_3 \\ N_4 \end{bmatrix} = \begin{Bmatrix} 1 \\ x \\ y \\ xy \end{Bmatrix} \qquad ...(7.37)$$

7.8 SHAPE FUNCTIONS: SQUARE ELEMENT

A square element of length of sides equal to 'b' is considered with origin at the centroid of the element. The co-ordinates of the four nodes 1, 2, 3 and 4 are ($-b/2$, $-b/2$), ($b/2$, $-b/2$), ($-b/2$, $b/2$) and ($b/2$, $b/2$). The representation of the square element is as shown by Fig. 7.6. At each node one displacement in x direction are considered totalling to 4 *DOF*.

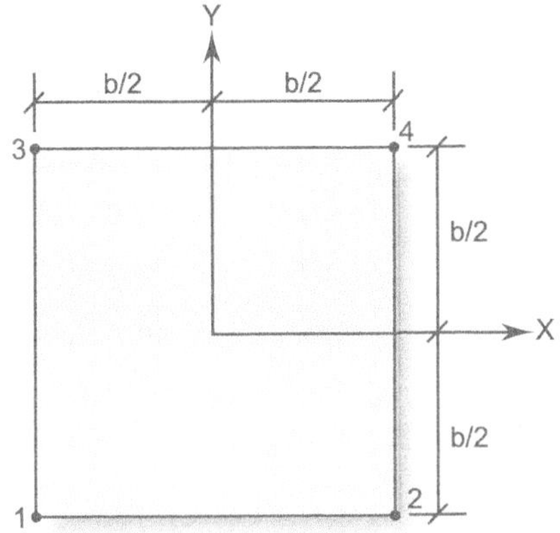

Fig. 7.6. Square element.

At node # 1, the condition is

$$u = u_1 \text{ with } (x, y) = \left(-\frac{b}{2}, -\frac{b}{2}\right)$$

Hence

$$u_1 = a_0 - \frac{b}{2}a_1 - \frac{b}{2}a_2 + \frac{b^2}{4}a_3 \qquad ...(7.38)$$

At node # 2, the condition is

$$u = u_2 \text{ with } (x, y) = \left(\frac{b}{2}, -\frac{b}{2}\right)$$

Hence

$$u_2 = a_0 + \frac{b}{2}a_1 - \frac{b}{2}a_2 - \frac{b^2}{4}a_3 \qquad ...(7.39)$$

At node # 3, the condition is

$$u = u_3 \text{ with } (x, y) = \left(-\frac{b}{2}, \frac{b}{2}\right)$$

Hence

$$u_3 = a_0 - \frac{b}{2}a_1 + \frac{b}{2}a_2 - \frac{b^2}{4}a_3 \qquad ...(7.40)$$

At node # 4, the condition is

$$u = u_4 \text{ with } (x, y) = \left(\frac{b}{2}, \frac{b}{2}\right)$$

Hence
$$u_4 = a_0 + \frac{b}{2}a_1 + \frac{b}{2}a_2 + \frac{b^2}{4}a_3 \quad ...(7.41)$$

The four equations (7.38) to (7.41) can be written in the matrix form as

$$\begin{Bmatrix} u_1 \\ u_2 \\ u_3 \\ u_4 \end{Bmatrix} = \begin{bmatrix} 1 & -b/2 & -b/2 & b^2/4 \\ 1 & b/2 & -b/2 & -b^2/4 \\ 1 & -b/2 & b/2 & -b^2/4 \\ 1 & b/2 & b/2 & b^2/4 \end{bmatrix} \begin{Bmatrix} a_0 \\ a_1 \\ a_2 \\ a_3 \end{Bmatrix} \quad ...(7.42)$$

$$\{u\} = [e]\{A\}$$
$$\{A\} = [C]^{-1}\{u\}$$

$$[C]^{-1} = \frac{1}{b}\begin{bmatrix} b/4 & b/4 & b/4 & b/4 \\ -1/2 & 1/2 & -1/2 & 1/2 \\ -1/2 & -1/2 & 1/2 & 1/2 \\ 1/b & -1/b & -1/b & 1/b \end{bmatrix} \quad ...(7.43)$$

The polynomial assumed for the linear square 4-noded element is

$$u(x, y) = a_0 + a_1x + a_2y + a_3xy$$
$$= [1\ x\ y\ xy]\{A\}$$
$$= [1\ x\ y\ xy][C]^{-1}\{u\}$$

Hence the shape functions are

$$[N_i]^T = [1\ x\ y\ xy][C]^{-1}$$

Hence the shape functions are

$$N_1 = \left(\frac{1}{4} - \frac{x}{2b} - \frac{y}{2b} + \frac{xy}{b^2}\right)$$

$$N_2 = \left(\frac{1}{4} + \frac{x}{2b} - \frac{y}{2b} - \frac{xy}{b^2}\right) \quad ...(7.44)$$

$$N_3 = \left(\frac{1}{4} - \frac{x}{2b} + \frac{y}{2b} - \frac{xy}{b^2}\right)$$

$$N_4 = \left(\frac{1}{4} + \frac{x}{2b} + \frac{y}{2b} + \frac{xy}{b^2}\right)$$

The properties that can be noted are

1. $\sum_{i=1}^{4} N_i = 1$

2. $N_i(x_j, y_j) = \begin{cases} 1 & \text{for } i = j \\ 0 & \text{for } i \neq j \end{cases}$

7.9 THE STIFFNESS MATRIX

From the definition of stiffness matrix in the *FEM* procedures

$$[K] = \int_V [B]^T [D][B] dV \qquad ...(7.45)$$

From the two dimensional strain relations

$$[\in] = \begin{Bmatrix} \in_x \\ \in_y \\ \gamma_{xy} \end{Bmatrix} = \begin{bmatrix} \frac{\partial}{\partial x} & 0 \\ 0 & \frac{\partial}{\partial y} \\ \frac{\partial}{\partial y} & \frac{\partial}{\partial x} \end{bmatrix} \begin{Bmatrix} u \\ V \end{Bmatrix} = [L]\,[N]\,\{U\} = [B]\,\{U\} \qquad ...(7.46)$$

$\{u\}$ is the displacements in x direction

$\{V\}$ is the displacements in y direction.

where $[B]$ is the strain displacement matrix defined by the shape functions to be expressed as the following manner.

The strain displacement matrix

$$[B] = \begin{bmatrix} \frac{\partial N_1}{\partial x} & 0 & \frac{\partial N_2}{\partial x} & 0 & \frac{\partial N_3}{\partial x} & 0 & \frac{\partial N_4}{\partial x} & 0 \\ 0 & \frac{\partial N_1}{\partial y} & 0 & \frac{\partial N_2}{\partial y} & 0 & \frac{\partial N_3}{\partial y} & 0 & \frac{\partial N_4}{\partial y} \\ \frac{\partial N_1}{\partial y} & \frac{\partial N_1}{\partial x} & \frac{\partial N_2}{\partial y} & \frac{\partial N_2}{\partial x} & \frac{\partial N_3}{\partial x} & \frac{\partial N_3}{\partial x} & \frac{\partial N_4}{\partial y} & \frac{\partial N_4}{\partial x} \end{bmatrix}$$

$$= \begin{bmatrix} -\frac{1}{2b}+\frac{y}{b^2} & 0 & \frac{1}{2b}-\frac{y}{b^2} & 0 & -\frac{1}{2b}-\frac{y}{b^2} & 0 & \frac{1}{2b}+\frac{y}{b^2} & 0 \\ 0 & -\frac{1}{2b}+\frac{x}{b^2} & 0 & -\frac{1}{2b}-\frac{x}{b^2} & 0 & \frac{1}{2b}-\frac{x}{b^2} & 0 & \frac{1}{2b}+\frac{x}{b^2} \\ \left(-\frac{1}{2b}+\frac{x}{b^2}\right) & \left(-\frac{1}{2b}+\frac{y}{b^2}\right) & \left(-\frac{1}{2b}-\frac{x}{b^2}\right) & \left(\frac{1}{2b}-\frac{y}{b^2}\right) & \left(\frac{1}{2b}-\frac{x}{b^2}\right) & \left(-\frac{1}{2b}-\frac{y}{b^2}\right) & \left(\frac{1}{2b}+\frac{x}{b^2}\right) & \left(\frac{1}{2b}+\frac{y}{b^2}\right) \end{bmatrix}$$

...(7.47)

Elasticity matrix : Plane stress

$$[D] = \frac{E}{(1-\mu^2)} \begin{bmatrix} 1 & \mu & 0 \\ \mu & 1 & 0 \\ 0 & 0 & \left(\frac{1-\mu}{2}\right) \end{bmatrix}$$

7.9.1 Plane Strain Case

$$[D] = \frac{E}{(1+\mu)(1-2\mu)} \begin{bmatrix} (1-\mu) & \mu & 0 \\ \mu & (1-\mu) & 0 \\ 0 & 0 & \left(\frac{1-2\mu}{2}\right) \end{bmatrix}$$

By substituting the $[D]$ matrix for the plane stress and plane strain cases in the equation ( )

the elemental stiffness matrix is obtained for the 2–D element of uniform thickness, t.

$$\int_V dV = t \times \text{area.}$$

7.10 ISOPARAMETRIC FORMULATION

One way to improve the accuracy of discretization of a curved boundary is to increase the number of triangular elements or the quadrilateral elements to represent curved boundary. But it is more convenient to improve the accuracy by using different element with curved edge to match the boundary of the given region.

These elements can be developed by the method so called "isoparametric formulation". The procedure of this method is based on a mapping or transformation of a relatively simple element into desired shape. The Fig. 7.7 shows the mapping of a simple right angled triangular element into arbitrary triangular element.

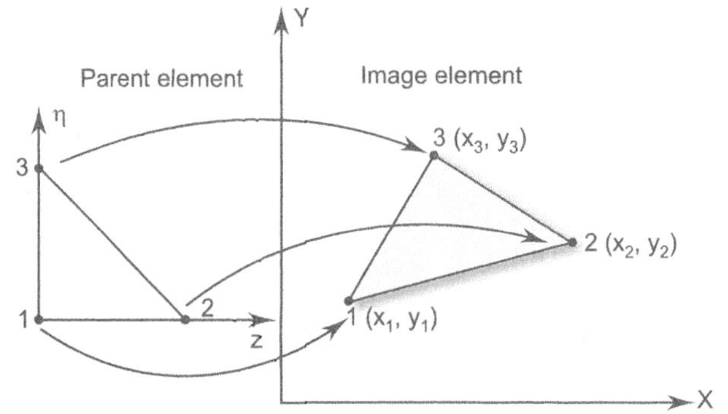

Fig. 7.7. Mapping of triangular element.

The right angled triangle element is usually called the 'parent element', and the arbitrary triangle is called the image element. In this particular case z and η are called the local co-ordinates and x and y are called the global co-ordinates the local co-ordinates (z, η) are related to global system (x, y) by parametric equations, with include the following steps.

1. Develop the shape functions for the parent element.
2. Use these shape functions to represent the dependent field variable over the image element which is called the 'isoparemetric' element.

Step 1: Development of shape function

Consider a right angled triangular element in the local natural co-ordinate system (z, η). To find the shape functions assume a linear polynomial in (ζ) and (η).

$$u(\zeta, \eta) = a_0 + a_1\zeta + a_2\eta_1 \qquad ...(7.48)$$

At node # 1, $(\zeta, \eta) = (0, 0)$ and $u = u_1$

At node # 2, $(\zeta, \eta) = (1, 0)$ and $u = u_2$

At node # 3, $(\zeta, \eta) = (0, 1)$ and $u = u_3$

These set of conditions lead to set of equations

$$u_1 = a_0$$

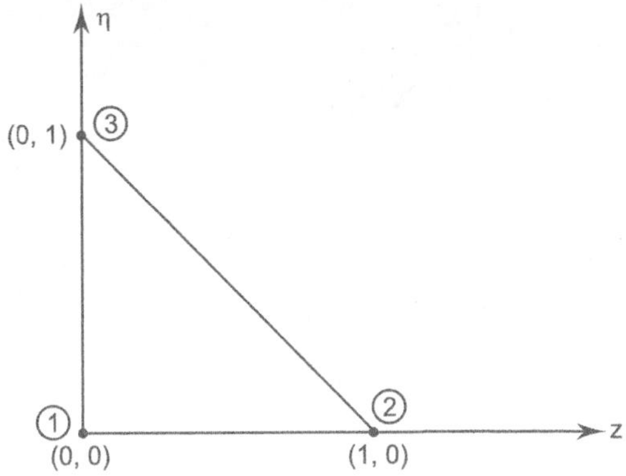

Fig. 7.8. Parent element.

$$u_2 = a_0 + a_1$$
$$u_3 = a_0 + a_2$$

By manipulation of these equations

$$a_0 = u_1$$
$$a_1 = (u_2 - u_1)$$
$$a_2 = (u_3 - u_1)$$

Substituting these constants in the polynomial assumed

$$u(\zeta, \eta) = u_1 + (u_2 - u_1)\,\zeta + (u_3 - u_1)\eta$$
$$= (1 - \zeta - \eta)\,u_1 + \zeta\,u_2 + \eta\,u_3 \qquad \text{...(7.49)}$$

Comparing this with the equation

$$u(\zeta\ \eta) = \Sigma\,N(\zeta, \eta)\,u_i$$

The shape functions are

$$N_1 = (1 - \zeta - \eta)$$
$$N_2 = \zeta \qquad \text{...(7.50)}$$
$$N_3 = \eta$$

The properties observed are

$$\sum_{i=1}^{3} N_i = 1 \quad \text{and} \quad N_i(\zeta_j\ \eta_j) = \begin{cases} 0 & i \neq j \\ 1 & i = j \end{cases}$$

Step 2: Mapping

Consider the mapping of the right angled triangular element (parent element) into arbitrary triangular element. The expression for mapping is given by

Hence the mapped points are

$$x(\zeta, \eta) = \sum_{i=1}^{3} N_i x_i \qquad \text{...(7.51)}$$

$$= N_1x_1 + N_2x_2 + N_3x_3$$
$$= (1 - \zeta - \eta)x_1 + x_2 + \eta x_3$$

Similarly

$$y(\zeta, \eta) = \sum_{i=1}^{3} N_i y_i \qquad ...(7.52)$$

$$= (1 - \zeta - \eta)\, y_1 + \zeta y_2 + \eta y_3$$

By inverting these expressions it is possible to express ζ and η in terms of x and y.

$$(x - x_1) = (x_2 - x_1)\, \zeta + (x_3 - x_1)\eta$$

$$(y - y_1) = (y_2 - y_1)\, \zeta + (y_3 - y)\eta$$

Solving these equations of ζ and η

$$\zeta = \frac{(x - x_1)(y_3 - y_1) - (y - y_1)(x_3 - x_1)}{(y_3 - y_1)(x_2 - x_1) - (y_2 - y_1)(x_3 - x_1)} \qquad ...(7.53)$$

$$\eta = \frac{(x - x_1)(y_2 - y_1) - (y - y_1)(x_2 - x_1)}{(x_3 - x_1)(y_2 - y_1) - (y_3 - y_1)(x_2 - x_1)} \qquad ...(7.54)$$

7.11 QUADRILATERAL ELEMENT: ISOPARAMETRIC FORMULATION

The parent element assumed is a square element with four nodes. The local natural co-ordinate system has the origin located in the centre of the square. The co-ordinates of the square is assumed to be $(\zeta_i\ \eta_i)$ which is (–1, – 1) for node 1, (1, – 1) for node 2 (– 1, 1) for node 3 and (1, 1) for node 4. The parent element is mapped with the quadrilateral element in the global co-ordinate system as shown in Fig. 7.9.

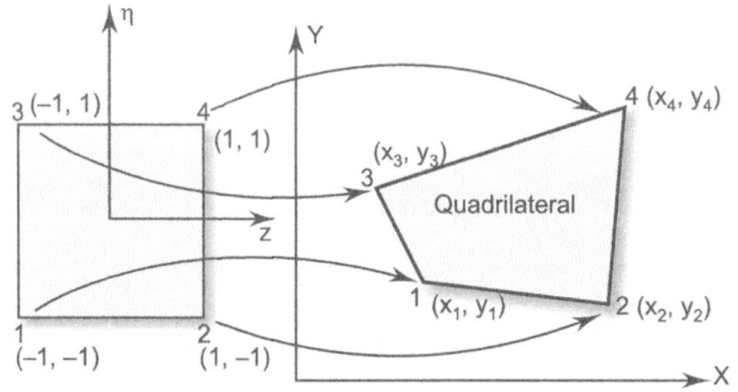

Fig. 7.9. Quadrilateral element.

The displacement function assumed is

$$u(\zeta, \eta) = a_0 + a_1\zeta + a_2\eta + a_3\zeta\eta \qquad ...(7.55)$$

Conditions

$$u = u_1 \text{ at } (-1, -1)$$

$$u = u_2 \text{ at } (1, -1)$$

$$u = u_3 \text{ at } (-1, 1)$$

$$u = u_4 \text{ at } (1, 1)$$

This leads to the set of equations

$$\begin{Bmatrix} u_1 \\ u_2 \\ u_3 \\ u_4 \end{Bmatrix} = \begin{bmatrix} 1 & -1 & -1 & 1 \\ 1 & 1 & -1 & -1 \\ 1 & -1 & 1 & -1 \\ 1 & 1 & 1 & 1 \end{bmatrix} \begin{Bmatrix} a_0 \\ a_1 \\ a_2 \\ a_3 \end{Bmatrix}$$

$$\{u\} = [C]\{A\}$$

$$[C]^{-1} = \frac{1}{4}\begin{bmatrix} 1 & 1 & 1 & 1 \\ -1 & 1 & -1 & 1 \\ -1 & -1 & 1 & 1 \\ 1 & -1 & -1 & 1 \end{bmatrix}$$

Hence the polynomial

$$u\{\zeta, \eta\} = [1, \zeta, \eta, \zeta\eta]\,[C]^{-1}\,\{u\} \quad \text{...(7.56)}$$

The shape functions are given by

$$N_1 = \frac{1}{4}[1-\zeta-\eta+\zeta\eta] = \frac{1}{4}(1-\zeta)(1-\eta)$$

$$N_2 = \frac{1}{4}[1+\zeta-\eta-\zeta\eta] = \frac{1}{4}(1+\zeta)(1-\eta)$$

$$N_3 = \frac{1}{4}[1-\zeta+\eta-\zeta\eta] = \frac{1}{4}(1-\zeta)(1+\eta)$$

$$N_4 = \frac{1}{4}[1+\zeta+\eta+\zeta\eta] = \frac{1}{4}(1+\zeta)(1+\eta) \quad \text{...(7.57)}$$

Hence by mapping

$$x(\zeta, \eta) = \sum_{i=1}^{4} N_i x_i$$

$$y(\zeta, \eta) = \sum_{i=1}^{4} N_i y_i$$

7.12 PLOT OF SHAPE FUNCTIONS: RECTANGULAR ELEMENT

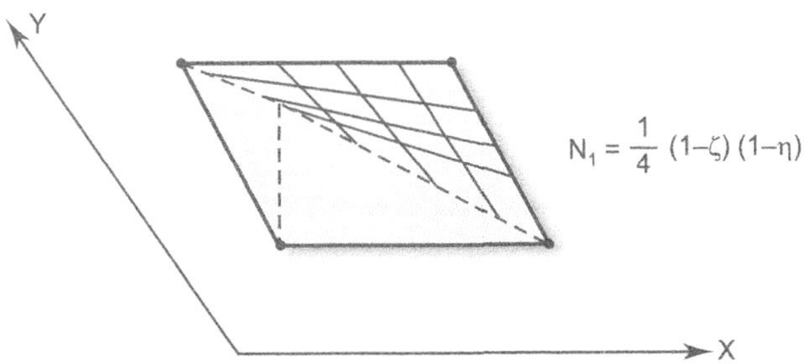

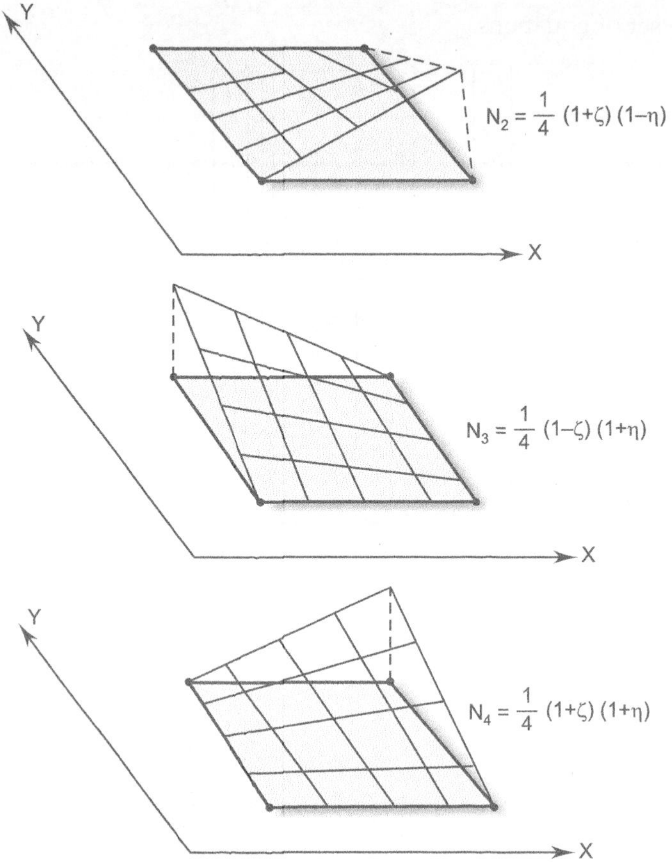

Fig. 7.10. Plot of shape function: rectangular element.

Derivatives of shape functions

$$
\begin{aligned}
\frac{\partial N_1}{\partial z} &= -\frac{1}{4}(1-\eta) & \frac{\partial N_1}{\partial \eta} &= -\frac{1}{4}(1-\zeta) \\
\frac{\partial N_2}{\partial z} &= \frac{1}{4}(1-\eta) & \frac{\partial N_2}{\partial \eta} &= -\frac{1}{4}(1+\zeta) \\
\frac{\partial N_3}{\partial z} &= -\frac{1}{4}(1+\eta) & \frac{\partial N_3}{\partial \eta} &= \frac{1}{4}(1-\zeta) \\
\frac{\partial N_4}{\partial z} &= \frac{1}{4}(1+\eta) & \frac{\partial N_4}{\partial \eta} &= \frac{1}{4}(1+\zeta)
\end{aligned}
\qquad \text{...(7.58)}
$$

7.13 TRANSFORMATION: JACOBIAN

To transform from local domain to the global domain we can make use of the shape functions derived in natural co-ordinate system. In the global co-ordinate system the x and y can be expressed as

$$x = x(\zeta\eta) = \sum_{i=1}^{4} N_i(\zeta\,\eta)\,x_i$$

$$y = y(\zeta\eta) = \sum_{i=1}^{4} N_i(\zeta, \eta)\, y_i$$

The global variables functions of the shape functions in the parent domain, which is a non-dimensional formulation. The element equation that gives the stiffness matrix needs the derivatives of shape functions, equations (7.58) transformed to the global co-ordinate system.

By chain rule of derivatives, yields

$$\frac{\partial N_i}{\partial z} = \frac{\partial N_i}{\partial x}\cdot\frac{\partial x}{\partial z} + \frac{\partial N_i}{\partial y}\cdot\frac{\partial y}{\partial z} \qquad ...(7.59)$$

$$\frac{\partial N_i}{\partial z} = \frac{\partial N_i}{\partial x}\cdot\frac{\partial x}{\partial \eta} + \frac{\partial N_i}{\partial y}\cdot\frac{\partial y}{\partial \eta} \qquad ..(7.60)$$

In the matrix form for the equations (7.59) and (7.60)

$$\begin{bmatrix} \frac{\partial N_i}{\partial \zeta} \\ \frac{\partial N_i}{\partial \eta} \end{bmatrix} = \begin{bmatrix} \frac{\partial x}{\partial \zeta} & \frac{\partial y}{\partial \zeta} \\ \frac{\partial y}{\partial \eta} & \frac{\partial y}{\partial \eta} \end{bmatrix} \begin{bmatrix} \frac{\partial N_i}{\partial x} \\ \frac{\partial N_i}{\partial y} \end{bmatrix} = [J]^T \begin{Bmatrix} \frac{\partial N_i}{\partial x} \\ \frac{\partial N_i}{\partial y} \end{Bmatrix} \qquad ...(7.61)$$

The $[J]$ matrix is called the Jacobian matrix given by

$$[J] = \begin{bmatrix} \frac{\partial x}{\partial \zeta} & \frac{\partial x}{\partial \eta} \\ \frac{\partial y}{\partial \zeta} & \frac{\partial y}{\partial \eta} \end{bmatrix} = \text{Jacobian Matrix.} \qquad ...(7.62)$$

Using the transformation, equation (7.59–7.60) into global domain is evaluted using equation (7.62) as :

$$[J] = \begin{bmatrix} x_i \frac{\partial N_i}{\partial \zeta} & x_i \frac{\partial N_i}{\partial \eta} \\ y_i \frac{\partial N_i}{\partial \zeta} & y_i \frac{\partial N_i}{\partial \eta} \end{bmatrix} \qquad ...(7.63)$$

The shape functions and their partial derivatives are derived and known in the parent domain. Also known are the global co-ordinates of the nodes, (x_i, y_i). Hence the derivatives of the shape functions in the global co-ordinate system are given by

$$\begin{bmatrix} \frac{\partial N_i}{\partial x} \\ \frac{\partial N_i}{\partial y} \end{bmatrix} = [J^T]^{-1} \begin{bmatrix} \frac{\partial N_i}{\partial \zeta} \\ \frac{\partial N_i}{\partial \eta} \end{bmatrix}$$

7.14 STIFFNESS MATRIX: GLOBAL DOMAIN

The element stiffness matrix in the global domain is given by

$$[K] = \int_0^x \int_0^y [B]^T [D][B]\, dx\, dy \qquad ...(7.64)$$

We know that $dA = dx\, dy = |J|\, d\zeta\, d\eta$

$|J|$ is the the determinant of $[J]$ which is often called the Jacobian.

Introducing the strain-displacement matrix in the global domain.

$$[B] = [\bar{J}^T]^{-1}\begin{bmatrix} \frac{\partial N_1}{\partial \zeta} & \frac{\partial N_2}{\partial \zeta} & \frac{\partial N_3}{\partial \zeta} & \frac{\partial N_4}{\partial \zeta} \\ \frac{\partial N_1}{\partial \eta} & \frac{\partial N_2}{\partial \eta} & \frac{\partial N_3}{\partial \eta} & \frac{\partial N_4}{\partial \eta} \end{bmatrix} \qquad ...(7.65)$$

$$= [\bar{J}^T]^{-1}[\bar{B}]$$

The derivatives of the shape functions are in the parent domain in the above $[\bar{B}]$ matrix which gets transformed into global domain by the introduction of the Jacobian matrix.

Hence the element stiffness matrix

$$[K] = \int_{-1}^{+1}\int_{-1}^{+1} [B]^T[D][B]\,|\,J\,|\,d\zeta\,d\eta \qquad ...(7.66)$$

From equations (7.65) and (7.66)

$$[K] = \int_{-1}^{+1}\int_{-1}^{+1} [\bar{J}^{-1}][\bar{B}][D][\bar{J}^{-1}][\bar{B}]\,|\,J\,|\,d\zeta\,d\eta \qquad ...(7.67)$$

7.15 ELEMENT LOAD VECTOR: BODY FORCE

It has been defined to represent the load the vector due to body force, as

$$f_b = \int_V N^T[b]\,dV \qquad ...(7.68)$$

For the 2-D element (triangular-CST element) as shown in Fig. 7.11, the thickness is assumed to be constant (*i.e.*, t) the element load vector is given as :

$$f_b = t\int_A N^T[b]\,dA$$

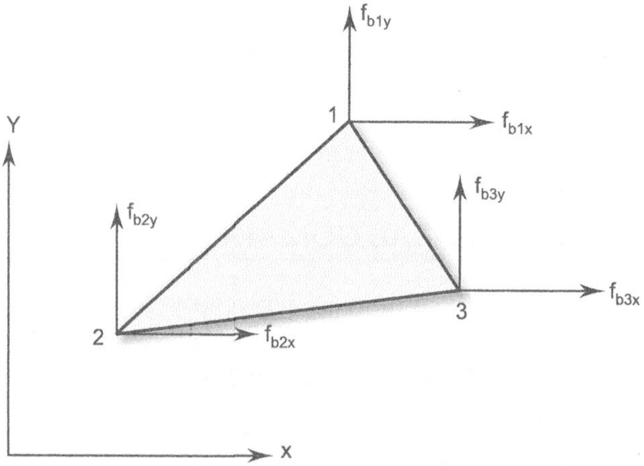

Fig. 7.11. Load vector: body force.

Hence the elemental load vector due to body force in the CST element is written as

$$\{f_b\} = \begin{Bmatrix} f_{b1x} \\ f_{b1y} \\ f_{b2x} \\ f_{b2y} \\ f_{b3x} \\ f_{b3y} \end{Bmatrix} = \begin{bmatrix} t\int N_1[b]_x dA \\ t\int N_1[b]_y dA \\ t\int N_2[b]_x dA \\ t\int N_2[b]_y dA \\ t\int N_3[b]_x dA \\ t\int N_3[b]_y dA \end{bmatrix} \qquad ...(7.69)$$

7.16 ELEMENTAL LOAD VECTOR: TRACTION FORCE

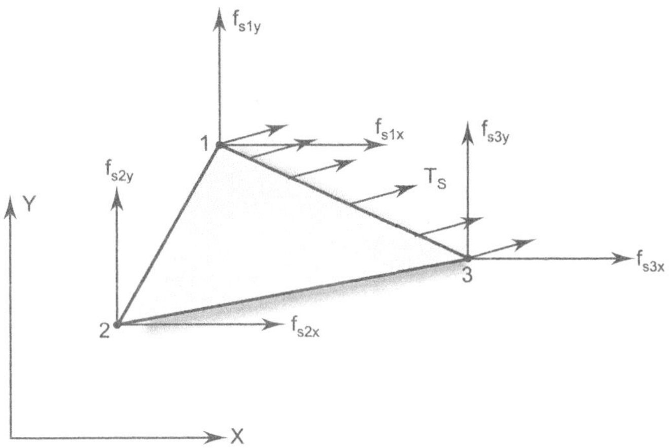

Fig. 7.12. Load vector : Traction.

From the definition of the load vector due to the surface traction force on the CST element as shown in the Fig. (7.12)

$$\{f_s\} = \int_S N^T T_s \, dS \qquad ...(7.70)$$

For the CST element of constant thickness t

$$\{f_s\} = t \int_{L1-3} N^T|_{\text{along } 1-3} T_s \, dS \qquad ...(7.71)$$

$$\{f_s\} = \begin{bmatrix} t\int N_1^T|_{1-3} T_{sx} dx \\ t\int N_1^T|_{1-3} T_{sy} dy \\ t\int N_3|_{1-3} T_{sx} dx \\ t\int N_3|_{1-3} T_{sy} dy \end{bmatrix} = \begin{Bmatrix} f_{s1x} \\ f_{s1y} \\ f_{s3x} \\ f_{s3y} \end{Bmatrix} \qquad ...(7.72)$$

7.17 TRIANGULAR ELEMENT: ISOPARAMETRIC REPRESENTATION

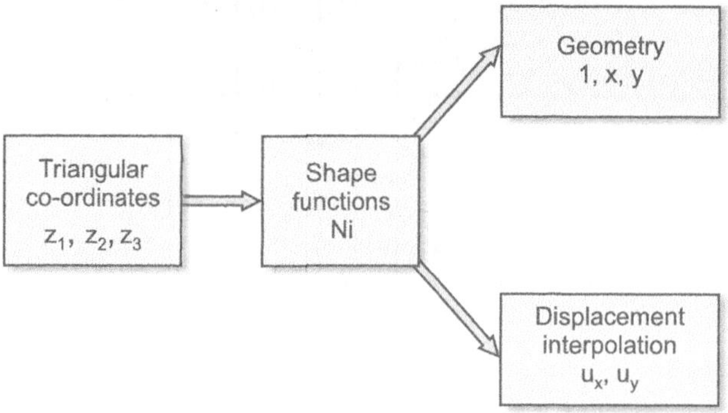

Fig. 7.13. Block diagram.

Linear Triangular Element : Observations

- In constant strain triangular (CST) element the same strain exists inside each element. Hence CST element is used in areas where the strain gradient is small.
- It is suggested not to use CST in the region of stress concentration, crucial areas in the structure such as edges of a hole and corners. Avoid using CST element near the load and support points.
- This element is less accurate than quadrilateral element.
- This is recommended for quick and preliminary FE analysis of 2-D problems.
- Used by most mesh generators for complex geometry generally mesh transition areas like fine to coarse mesh.
- CST elements are not recommended for general analysis purposes as large number of elements are required for reasonable accuracy.

7.18 CST ELEMENT IN AREA CO-ORDINATES

Consider an arbitrary triangular element with corner nodes having the co-ordinates (x_i, y_i) where $i = 1, 2$ and 3. At any point inside the triangle divide the considered element into three triangles of area A_1, A_2 and A_3 and shown by the Fig. (7.14).

If A is the total area of the triangular element then the area co-ordinates of the point $p(x, y)$ inside the triangles, are defined to be :

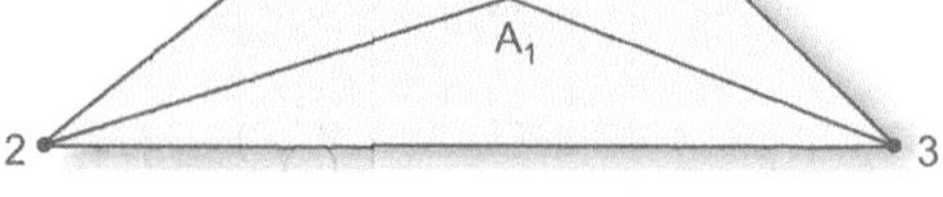

Fig. 7.14. Area co-ordinates.

$$L_1 = \frac{A_1}{A}$$

$$L_2 = \frac{A_2}{A} \qquad \text{...(7.73)}$$

$$L_3 = \frac{A_3}{A}$$

It may be noted that

$$\Sigma L_i = 1 \qquad \text{...(7.74)}$$

Further L_i is defined with respect to (x, y) as

$$L_i = \frac{a_i + b_i x + c_i y}{2A} \qquad \text{...(7.75)}$$

It may be further noted that

$$L_1 x_1 + L_2 x_2 + L_3 x_3 = x$$

and

$$L_1 y_1 + L_2 y_2 + L_3 y_3 = y \qquad \text{...(7.76)}$$

It is also evident that

$$L_1 = N_1 \qquad L_2 = N_2 \qquad L_3 = N_3$$

where N_i are the shape functions of the element that interpolates the distribution of the properties inside the elements in terms of the properties at the corner node :

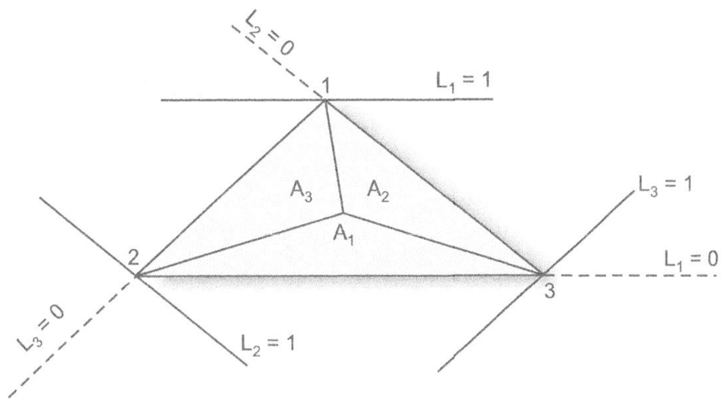

Fig. 7.15. Shape function configuration.

The area co-ordinates in Fig. (7.15) depict the Kronecker delta property

$$L_i(x_j, y_j) = \begin{cases} 1 & \text{for } i = j \\ 0 & \text{for } i \neq j \end{cases} \qquad \text{...(7.77)}$$

For

$$A = A_1 + A_2 + A_3$$

$$L_1 + L_2 + L_3 = \frac{A_1 + A_2 + A_3}{A} = 1$$

Example 7.2. *A rectangular plate is loaded and supported as shown in the Fig. 7.16. The thickness of the lamina is 0.5 cm. The Young's Modulus of the material is 30×10^6 N/cm^2. If the Poisson's ratio is 0.25*

(a) *Compute the unknown nodal displacements*

(b) *Compute stresses in two elements.*

Solution. The given problem is a case of plane-stress problem of 2-*D* finite element. The rectangular plate is divided into two triangular element. The unknown displacements are displacements in *x*-direction for node 1 and 2 and *y*-displacement for node 2.

The element equation requires the computation of elasticity, [*D*] matrix, and [*B*] matrix, also the nodal forces.

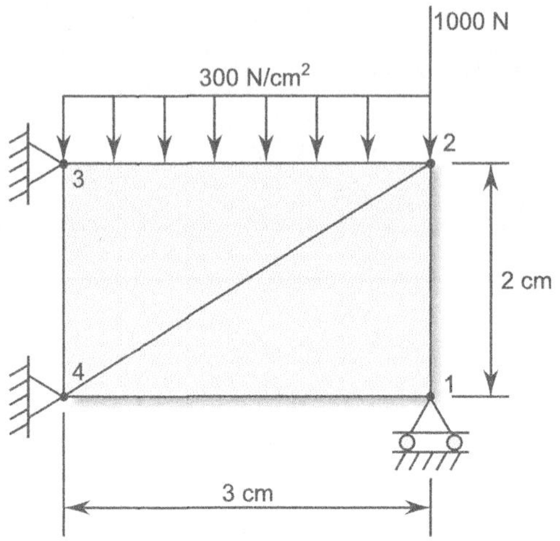

Fig. 7.16. Problem-Triangular element.

- Elasticity matrix

$$[D] = \frac{E}{(1-\mu^2)}\begin{bmatrix} 1 & \mu & 0 \\ \mu & 1 & 0 \\ 0 & 0 & \left(\frac{1-\mu}{2}\right) \end{bmatrix}$$

with the substitution of E and $\mu = 0.25$

$$[D] = 10^7 \times \begin{bmatrix} 3.2 & 0.8 & 0 \\ 0.8 & 3.2 & 0 \\ 0 & 0 & 1.2 \end{bmatrix}$$

Node element connectivity table.

Element	*node i*	*node j*	*node k*	*Area*
1	1	2	4	3 cm^2
2	3	4	2	3 cm^2

Nodal co-ordinates

Node Number	*x–*	*y–*
1	3	0
2	3	2
3	0	2
4	0	0

- Strain-displacement Matrix [B]

$$[B] = \frac{1}{2A}\begin{bmatrix} b_1 & 0 & b_2 & 0 & b_3 & 0 \\ 0 & c_1 & 0 & c_2 & 0 & c_3 \\ c_1 & b_1 & c_2 & b_2 & c_3 & b_3 \end{bmatrix}$$

$$b_1 = y_2 - y_3 \qquad b_2 = y_3 - y_1 \qquad b_3 = y_1 - y_2$$

$$c_1 = x_3 - x_2 \qquad c_2 = x_1 - x_3 \qquad c_3 = x_2 - x_1$$

For element # 1

$$x_1 = 3 \qquad x_2 = 3 \qquad x_3 = 0$$

$$y_1 = 0 \qquad y_2 = 2 \qquad y_3 = 0$$

Hence

$$b_1 = 2 \qquad b_2 = 0 \qquad b_3 = -2$$

$$c_1 = -3 \qquad c_2 = 3 \qquad c_3 = 0$$

Hence

$$[B]^{(1)} = \frac{1}{2\times 3}\begin{bmatrix} 2 & 0 & 0 & 0 & -2 & 0 \\ 0 & -3 & 0 & 3 & 0 & 0 \\ -3 & 2 & 3 & 0 & 0 & -2 \end{bmatrix}$$

Similarly for element # 2

$$x_1 = 0 \qquad x_2 = 0 \qquad x_3 = 3$$

$$y_1 = 2 \qquad y_2 = 0 \qquad y_3 = 2$$

$$b_1 = -2 \qquad b_2 = 0 \qquad b_3 = 2$$

$$c_1 = 3 \qquad c_2 = -3 \qquad c_3 = 0$$

The strain displacement matrix is

$$[B]^{(2)} = \frac{1}{2\times 3}\begin{bmatrix} -2 & 0 & 0 & 0 & 2 & 0 \\ 0 & 3 & 0 & -3 & 0 & 0 \\ 3 & -2 & -3 & 0 & 0 & 2 \end{bmatrix}$$

From the element equation the element stiffness matrix is defined to be

$$[K]^{(e)} = At\,[B]^T\,[D]\,[B]$$

For element 1

$$[K]^{(1)} = 3 \times 0.5\,[B]^{(1)T}\,[D]\,[B]^{(1)}$$

$$[K]^{(1)} = 10^7 \times \begin{array}{c} \begin{matrix} u_1 & v_1 & u_2 & v_2 & u_4 & v_4 \end{matrix} \\ \begin{bmatrix} 0.9833 & -0.5 & -0.45 & 0.2 & -0.533 & 0.3 \\ & 1.4 & 0.3 & -1.2 & 0.2 & -0.2 \\ & & 0.45 & 0 & 0 & -0.3 \\ & \text{sym.} & & 1.2 & -0.2 & 0 \\ & & & & 0.533 & 0.2 \\ & & & & & 1.4 \end{bmatrix} \end{array}$$

Similarly for the element # 2

$$[K]^2 = \text{At}\,[B^{(2)}]^T\,[D]\,[B^{(2)}]$$

$$
= 10^7 \times \begin{array}{c} \begin{array}{cccccc} u_3 & v_3 & u_4 & v_4 & u_3 & v_2 \end{array} \\ \begin{bmatrix} 0.9833 & -0.5 & -0.45 & 0.2 & -0.533 & 0.3 \\ & 1.4 & 0.3 & -1.2 & 0.2 & -0.2 \\ & & 0.45 & 0 & 0 & -0.3 \\ & & & 1.2 & -0.2 & 0 \\ & & & & 0.533 & 0.2 \\ & & & & & 1.4 \end{bmatrix} \end{array}
$$

Considering the boundary conditions at the supports

$$u_3 = v_3 = u_4 = v_4 = v_1 = 0$$

The unknown degrees of freedom are u_1, u_2 and v_2. Assemble the global stiffness matrix corresponding to non-zero degrees of freedom.

The global stiffness matrix

$$
[K] = 10^7 \begin{bmatrix} 0.983 & -0.45 & 0.2 \\ -0.45 & 0.983 & 0 \\ 0.2 & 0 & 1.4 \end{bmatrix} \begin{array}{c} u_1 \\ u_2 \\ v_2 \end{array}
$$

$$
\begin{array}{ccc} u_1 & u_2 & v_2 \end{array}
$$

Consistent Nodal Loads

$$
\{f\} = \begin{Bmatrix} f_{1x} \\ f_{2x} \\ f_{2y} \end{Bmatrix} = \begin{Bmatrix} 0 \\ 0 \\ -1000 + f_{s2y} \end{Bmatrix}
$$

$$
f_{s2y} = \int_0^3 N_3|_{3-2} [T]\, t\, dx
$$

$$
N_3 \Big|_{\text{along } (3-2)} = \frac{x}{3}
$$

$$
= (-300)(0.5) \int_0^3 \left(\frac{x}{3}\right) dx
$$

$$
= (-50) \frac{x^2}{2} \Big|_0^3 = -225\ N
$$

Hence $\qquad F_{2y} = -1225N$

The force-displacement relation is

$$[K]\,\{u\} = \{f\}$$

$$
10^7 \times \begin{bmatrix} 0.983 & -0.45 & 0.2 \\ -0.45 & 0.983 & 0 \\ 0.2 & 0 & 1.4 \end{bmatrix} \begin{Bmatrix} u_1 \\ u_2 \\ v_2 \end{Bmatrix} = \begin{Bmatrix} 0 \\ 0 \\ -1225 \end{Bmatrix}
$$

Elemental stress $\qquad \sigma^{(1)} = [D]\,[B]^{(1)}\{d\}^{(1)}$

$$
\{d\}^{(1)} = [u_1 \quad v_1 \quad u_2 \quad v_2 \quad u_4 \quad v_4]
$$

The force displacement relation is solved to obtain

$$\begin{Bmatrix} u_1 \\ v_1 \\ u_2 \\ v_2 \\ u_3 \\ v_3 \end{Bmatrix} = \begin{Bmatrix} 0.2337 \\ 0 \\ 0.1069 \\ -0.9084 \\ 0 \\ 0 \end{Bmatrix} \times 10^{-4} \text{ cm}$$

Hence the stress in element # 1

$$\sigma^{(1)} = \begin{Bmatrix} -114.1 \\ -1391.1 \\ -761.0 \end{Bmatrix} \text{N/cm}^2$$

Similarly for the element # 2

$$\sigma^{(2)} = [D]\,[B]^{(2)}\,\{d\}^{(2)}$$

$$\{d\}^{(2)} = [\,u_3\; v_3,\; u_4\; v_4,\; u_2\; v_2]$$

$$= [0 \quad 0 \quad 0 \quad 0 \quad 0.1069, -0.9084] \times 10^{-4} \text{ cm}$$

Hence the stress in element # 2

$$\sigma^{(2)} = \begin{bmatrix} 114.1 \\ 28.52 \\ -363.35 \end{bmatrix} \text{N/cm}^2$$

Example 7.3. *For the CST element shown in Fig. (7.17) compute the vector of nodal loads due to the surface traction force.*

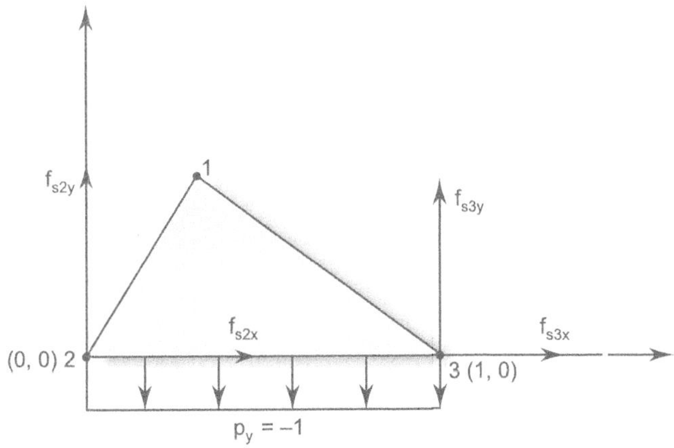

Fig. 7.17. CST element with traction.

Solution. By definition of the load vector due to surface traction

$$\{f_s\} = \int_S N^T T_s dS$$

$$[T_s] = \begin{Bmatrix} P_x \\ P_y \end{Bmatrix} = \begin{Bmatrix} 0 \\ -1 \end{Bmatrix}$$

Hence

$$f_{s2x} = f_{s3x} = 0$$

$$f_{s2y} = t\int_0^1 N_2|_{(2-3)}\, p_y dx$$

$$f_{s3y} = t\int_0^1 N_3|_{(2-3)}\, p_y dx$$

$$N_2|_{\text{along }(2-3)} = \frac{1}{2A}(a_2 + b_2x + c_2y)$$

$$\text{Along } (2-3)\ y = 0 \qquad \text{Hence} \quad N_2 = \frac{a_2 + b_2x}{2A}$$

$$N_2|_{2-3} = \frac{(x_3y_1 - x_1y_3) + (y_3 - y_1)x}{2A}$$

$$= \frac{y_1(1-x)}{2A}$$

$$2A = \begin{vmatrix} 1 & x_1 & y_1 \\ 1 & x_2 & y_2 \\ 1 & x_3 & y_3 \end{vmatrix} = \begin{vmatrix} 1 & x_1 & y_1 \\ 1 & 0 & 0 \\ 1 & 1 & 0 \end{vmatrix} = y_1$$

$$N_2|_{2-3} = (1-x)$$

Hence

$$f_{s2y} = t\int_0^1 (1-x)(-1)\,dx$$

$$= t\left(\frac{x^2}{2} - x\right)\Big|_0^1$$

$$= -\frac{t}{2}$$

To compute $N_3|_{\text{along }(2-3)}$

$$N_3 = \frac{a_3 + b_3x + c_3y}{2A} = \frac{a_3 + b_3x}{2A}$$

$$= \frac{(x_2y_1 - x_1y_2) + (y_1 - y_2)x}{y_1} \qquad (x_2 = 1,\ x_1 = 0 \text{ and } y_2 = 0)$$

$$= \frac{y_1(1+x)}{y_1} = (1+x)$$

$$f_{s3y} = t\int_0^1 (1+x)(-1)\,dx = -\frac{3}{2}t$$

POINTS TO BE REMEMBER

- The triangular and quadrilateral elements are the 2-dimensional elements in FEM.
- The triangular element with three nodes has six degrees of freedom.
- The displacement function assumed for three nodded triangular element is linear in x and y co-ordinates.
- The linear triangular element has three shape functions that are linear.
- The strains are constant in a CST element.
- The x direction strain, the y direction strain and the shear strains are constant in a CST element, since the displacement function assumed is linear.
- A linear quadrilateral element has four nodes.
- The displacement function assumed for the quadrilateral element is non-linear.
- The CST elements are not preferred to quadrilateral element in general practical problem. The number of CST elements required are more.
- The summation of the area co-ordinates is equal to one.
- CST elements are less accurate then quadrilateral element hence not recommended in the critical areas and stress concentration regions of the structure.

EXERCISE

1. Derive the expressions for the shape functions for three nodded triangular element with one degree of freedom at each node.
2. Obtain $[B]$ (strain-displacement) matrix for the three nodded triangular element.
3. Obtain the shape functions for a 3-nodded right angled triangular element with base 'b' and height 'h' in cartesian system.
4. Obtain the shape functions in natural co-ordinate system, for the 3-nodded triangular element with one degree freedom at each node.
5. Obtain the stiffness matrix for a CST element.
6. Obtain the stiffness matrix for the triangular element with plane stress condition.
7. Explain the finite element formulation of a rectangular element.
8. Obtain the shape functions of the square element in local cartesian co-ordinate system.
9. Derive the $[B]$ matrix for a square element.
10. Obtain the isoparametric formulation of triangular element with one degree of freedom at each node.
11. Obtain the shape functions of square element in natural co-ordinate system.
12. Explain the isoparametric formulation of a rectangular element.

Chapter 8

HIGHER ORDER ELEMENTS AND CONVERGENCE CRITERIA

तैलात् रक्षेत् जलात् रक्षेत्।
रक्षेत् शिथिल बन्धनात्॥
पर हस्त गतात् रक्षेत्।
एवं वदति पुस्तकम्॥

[The book says – protect from oil, protect from water and should protect from pages becoming loose. Then keep it away from others hands.]

8.1 HIGHER ORDER ELEMENTS

In order to derive the shape functions it was assumed that the displacement field is a polynomial of any degree for all cases considered. It has been previously shown that for the two noded bar element with one degree of freedom at each node it is convenient to assume a linear polynomial with two generalized co-ordinates in the derivation of the shape function. But this has the limitation on the accuracy of result obtained when a problem of real world physics is considered. Hence an attempt was made to formulate higher order elements in one-dimension and two dimension, also three dimension as well.

The polynomials considered for 1-*D* and 2-*D* cases are.

1-*D* $$u(x) = a_0 + a_1x + a_2x^2 + \dots a_nx^n \qquad ...(8.1)$$

2-*D* $$u(x, y) = a_0 + a_1x + a_2y + a_3xy + \dots\dots \qquad ...(8.2)$$

The co-efficients a_i represent generalized co-ordinates that is equal to the number of degrees-of-freedom for the considered element.

8.2 3-NODED BAR ELEMENT: SHAPE FUNCTION

Consider a three noded bar (truss) element in a local Cartesian co-ordinate system. Let us consider one degree of freedom at each node. Let the length of the element be $2l$ and the mid-node be considered at a distance l from the end nodes. A is the uniform cross-section of the element and Young's modulus E is the mechanical property of the material. The configuration of higher order element is as shown in Fig. 8.1. The displacement of the respective nodes in the axial direction be u_1, u_2 and u_3. The on-going section highlights the formulation of the shape function using the displacement transformation approach, and the approximate higher order displacement field is predicted. From the definition of the strain and using the displacement field in terms of shape functions, the strain displacement matrix $[B]$ is obtained.

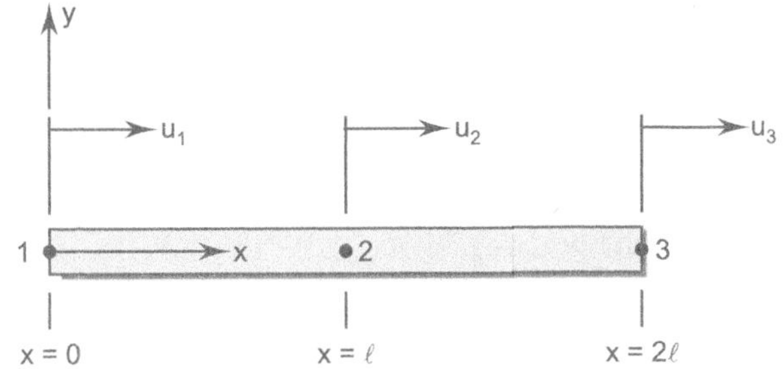

Fig. 8.1. Three noded bar element.

The second order (quadratic) displacement field be represented by the function written as

$$u(x) = a_0 + a_1 x + a_2 x^2 \quad \text{...(8.3)}$$

In the matrix form the equation (8.3) is

$$u(x) = [1 \;\; x \;\; x^2]\begin{Bmatrix} a_0 \\ a_1 \\ a_2 \end{Bmatrix}$$

Let u_1, u_2 and u_3 be the displacements of the nodes 1, 2 and 3 respectively. The co-ordinates of node 1, 2 and 3 are (0, 0), (l, 0), ($2l$, 0) respectively.

Using these boundary conditions the generalized co-ordinate a_i's are evaluated. The specific displacement (Dirichlet) boundary conditions are

$$\begin{aligned} u &= u_1 \quad \text{at } x = 0 \\ u &= u_2 \quad \text{at } x = l \\ u &= u_3 \quad \text{at } x = 2l \end{aligned} \quad \text{...(8.4)}$$

With the substitution of the conditions (8.4) in equation (8.3) the following system of equations are arrived at

$$\begin{aligned} u_1 &= a_0 \\ u_2 &= a_0 + a_1 l + a_2 l^2 \\ u_3 &= a_0 + 2a_1 l + 4a_2 l^2 \end{aligned} \quad \text{...(8.5)}$$

Arranging these equations (8.5) in the matrix form

$$\begin{bmatrix} 1 & 0 & 0 \\ 1 & l & l^2 \\ 1 & 2l & 4l^2 \end{bmatrix}\begin{Bmatrix} a_0 \\ a_1 \\ a_2 \end{Bmatrix} = \begin{Bmatrix} u_1 \\ u_2 \\ u_3 \end{Bmatrix}$$

This is of the form

$$[C]\{A\} = \{u\} \quad \text{...(8.6)}$$

Further the connectivity matrix is inverted to form the following:

$$[C]^{-1} = \frac{1}{2l^3}\begin{bmatrix} 2l^3 & 0 & 0 \\ -3l^2 & 4l^2 & -l^2 \\ l & -2l & l \end{bmatrix} \quad \text{...(8.7)}$$

To obtain the shape functions the parameter matrix $[P] = [1\ x\ x^2]$ is multiplied with inverse of the connectivity matrix $[C]$.

$$[N] = [P]\,[C]^{-1}$$

$$= [1 \quad x \quad x^2]\frac{1}{2l^3}\begin{bmatrix} 2l^3 & 0 & 0 \\ -3l^2 & 4l^2 & -l^2 \\ l & -2l & l \end{bmatrix}$$

Hence the three shape functions are

$$N_1 = \frac{1}{2l^3}(2l^3 - 3l^2x + lx^2)$$

$$N_2 = \frac{1}{2l^3}(4l^2x - 2lx^2) \quad \text{...(8.8)}$$

$$N_3 = \frac{1}{2l^3}(-l^2x + lx^2)$$

It may be verified that the property of the shape function is satisfied as

$$\sum_{i=1}^{n} N_i = 1 \quad \text{...(8.9)}$$

The displacement function in terms of shape functions and the nodal displacements is

$$u(x) = N_1u_1 + N_2u_2 + N_3u_3$$

Observations:

- At node 1, $x = 0$

  then
  $$N_1 = 1$$
  $$N_2 = 0$$
  $$N_3 = 0$$

  The plot of $N_1 = \frac{1}{2l^3}[2l^3 - 3l^2x + lx^2]$ is shown by curve 1 in Fig. 8.2.

- At node 2, $x = l$

  then
  $$N_1 = 0$$
  $$N_2 = 1$$
  $$N_3 = 0$$

  The plot of the shape function $N_2 = \frac{1}{2l^3}[4l^2x - 2lx^2]$ is shown by curve 2 in Fig. 8.2.

- At node 3, $x = 2l$

then
$$N_1 = 0$$
$$N_2 = 0$$
$$N_3 = 1$$

The plot of the shape function $N_3 = \frac{1}{2l^3}[-l^2x + lx^2]$ is shown by the curve 3 in Fig. 8.2.

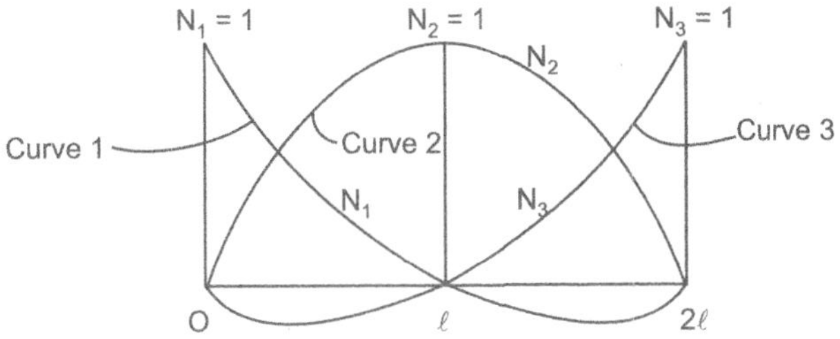

Fig. 8.2. Plot of the shape functions.

The displacement in any point in the element is given by

$$u(x) = N_1u_1 + N_2u_2 + N_3u_3 = \sum_{i=1}^{n} N_iu_i \qquad ...(8.10)$$

From the strain displacement relationship the [B] matrix is given by

$$[B] = \left[\frac{dN_1}{dx}, \frac{dN_2}{dx}, \frac{dN_3}{dx}\right]$$

$$= \left[\left\{\frac{1}{2l^3}(-3l^2 + 2lx)\right\}, \{1/2l^3(4l^2 - 4lx)\}, \{1/2l^3(-l^2 + 2lx)\}\right] \qquad ....(8.11)$$

8.3 3-NODED BAR: STIFFNESS MATRIX

To obtain the stiffness matrix in local Cartesian co-ordinate system for the 3-noded bar element, we resort to the definition of element stiffness matrix

$$[K] = \int_0^{2l} [B]^T[D][B]\,A\,dx$$

With the manipulation the elemental stiffness matrix is

$$[K] = AE\int_0^{2l}\begin{bmatrix} \left(\frac{dN_1}{dx}\right)^2 & \left(\frac{dN_1}{dx}\right)\left(\frac{dN_2}{dx}\right) & \left(\frac{dN_1}{dx}\right)\left(\frac{dN_3}{dx}\right) \\ \left(\frac{dN_2}{dx}\right)\left(\frac{dN_1}{dx}\right) & \left(\frac{dN_2}{dx}\right) & \left(\frac{dN_2}{dx}\right)\left(\frac{dN_3}{dx}\right) \\ \left(\frac{dN_3}{dx}\right)\left(\frac{dN_1}{dx}\right) & \left(\frac{dN_3}{dx}\right)\left(\frac{dN_2}{dx}\right) & \left(\frac{dN_3}{dx}\right)^2 \end{bmatrix} dx \qquad ...(8.12)$$

This can be well represented with the equivalent form.

$$[K] = \begin{bmatrix} k_{11} & k_{12} & k_{13} \\ k_{21} & k_{22} & k_{23} \\ k_{31} & k_{33} & k_{33} \end{bmatrix} \qquad ...(8.13)$$

In the above expression the derivatives of the shape functions are substituted as

$$\begin{aligned}\frac{dN_1}{dx} &= \frac{1}{2l^3}(-3l^2 + 2lx)\\ \frac{dN_2}{dx} &= \frac{1}{2l^3}(4l^2 - 4lx)\\ \frac{dN_3}{dx} &= \frac{1}{2l^3}(-l^2 + 2lx)\end{aligned} \qquad \text{...(8.14)}$$

To evaluate the elements of the stiffness matrix

$$k_{11} = AE\int_0^{2l}\left(\frac{dN_1}{dx}\right)^2 dx$$

$$= \frac{AE}{4l^6}\int_0^{2l}(-3l^2 + 2lx)^2 dx$$

$$= \frac{7}{6}\frac{AE}{l}$$

$$k_{22} = AE\int_0^{2l}\left(\frac{dN_2}{dx}\right)^2 dx$$

$$= \frac{AE}{4l^6}\int_0^{2l}(4l^2 - 4lx)^2 dx$$

$$= \frac{8}{3}\frac{AE}{l}$$

$$k_{33} = AE\int_0^{2l}\left(\frac{dN_3}{dx}\right)^2 dx$$

$$= \frac{AE}{4l^6}\int_0^{2l}(-l^2 + 2lx)^2 dx$$

$$= \frac{7}{6}\frac{AE}{l}$$

From the symmetry with respect to forward diagonal

$$k_{12} = k_{21} \qquad k_{13} = k_{31} \qquad \text{and} \qquad k_{23} = k_{32}$$

$$k_{12} = AE\int_0^{2l}\left(\frac{dN_1}{dx}\right)\left(\frac{dN_2}{dx}\right)dx$$

$$= \frac{AE}{4l^6}\int_0^{2l}(-3l^2 + 2lx)(4l^2 - 4lx)\,dx$$

$$= -4/3 \frac{AE}{l}$$

$$k_{13} = AE \int_0^{2l} \left(\frac{dN_1}{dx}\right)\left(\frac{dN_3}{dx}\right)dx$$

$$= \frac{AE}{4l^6} \int_0^{2l} (-3l^2 + 2lx)(-l^2 + 2lx)\,dx$$

$$= 1/6 \frac{AE}{l}$$

$$k_{23} = AE \int_0^{2l} \left(\frac{dN_2}{dx}\right)\left(\frac{dN_3}{dx}\right)dx$$

$$= \frac{AE}{4l^6} \int_0^{2l} (4l^2 - 4lx)(-l^2 + 2xl)\,dx$$

$$= -\frac{4}{3}\frac{AE}{l}$$

The elements of the stiffness matrix can be written as

$$[K]^e = \frac{AE}{6l}\begin{bmatrix} 7 & -8 & 1 \\ -8 & 16 & -8 \\ 1 & -8 & 7 \end{bmatrix} \qquad \text{...(8.15)}$$

Properties: The stiffness matrix is a symmetric matrix hence it is singular. Addition of elements of any row or any column is zero. All the elements of the diagonal are positive. The stiffness is the function of geometry and the mechanical property of material.

8.4 SHAPE FUNCTION: NATURAL CO-ORDINATES

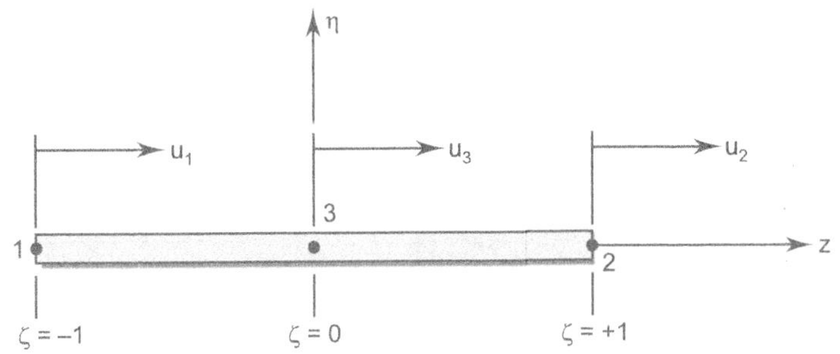

Fig. 8.3. 3-noded bar in natural co-ordinates.

Consider a 3-noded bar element with one degree of freedom at each node. The natural co-ordinate gets defined in terms of cartesian system to be

$$\zeta = \frac{2(x - x_1)}{(x_2 - x_1)} - 1$$

when $x = x_1$; $\zeta = -1$ when $x = x_2$; $\zeta = +1$.

The second order polynomial in natural co-ordinate to represent the quadratic bar element can be written as

$$u(\zeta) = a_0 + a_1\zeta + a_2\zeta^2 \quad \text{...(8.16)}$$

The displacement conditions are

$$\begin{aligned} u &= u_1 \quad \text{at } \zeta = -1 \\ u &= u_2 \quad \text{at } \zeta = +1 \\ u &= u_3 \quad \text{at } \zeta = 0 \end{aligned} \quad \text{...(8.17)}$$

Inserting these conditions in the quadratic equation (8.16) results in the set of equations.

$$\begin{aligned} a_0 - a_1 + a_2 &= u_1 \\ a_0 + a_1 + a_2 &= u_2 \\ a_0 &= u_3 \end{aligned}$$

On manipulation

$$\begin{aligned} a_0 &= u_3 \\ a_1 &= \frac{u_2 - u_1}{2} \\ a_2 &= \frac{u_1 - 2u_3 + u_2}{2} \end{aligned} \quad \text{...(8.18)}$$

With the substitution of the generalized co-ordinates a_i in equation (8.16)

$$u(\zeta) = u_3 + \frac{(u_2 - u_1)}{2}\zeta + \left(\frac{u_1 - 2u_3 + u_2}{2}\right)\zeta^2 \quad \text{...(8.19)}$$

Rearranging the terms in this equation leads to

$$u(\zeta) = \left[\frac{\zeta(\zeta-1)}{2}\right]u_1 + \left[\frac{\zeta(\zeta+1)}{2}\right]u_2 + [(1-\zeta)(1+\zeta)]\,u_3 \quad \text{...(8.20)}$$

On comparison with the equation

$$u = \sum N_i u_i$$

The shape functions are obtained to be

$$\begin{aligned} N_1 &= \frac{\zeta(\zeta-1)}{2} \\ N_2 &= \frac{\zeta(\zeta+1)}{2} \\ N_3 &= (1-\zeta)(1+\zeta) \end{aligned} \quad \text{...(8.21)}$$

The shape functions satisfy the property $\Sigma N_i = 1$.

8.5 STIFFNESS MATRIX: NATURAL CO-ORDINATES

Using the shape functions in natural co-ordinate the following procedure is adapted to formulate the stiffness matrix. From the definition of strain, the strain-displacement relation is obtained to be

$$\in \;=\; [B]\,\{u\} \qquad ...(8.22)$$

But the strain-displacement transformation matrix [B] is written as

$$[B] \;=\; \left[\frac{dN_1}{dx}, \frac{dN_2}{dx}, \frac{dN_3}{dx}\right] \qquad ...(8.23)$$

$$\frac{dN_i}{dx} \;=\; \frac{dNi}{d\zeta}\cdot\frac{dz}{dx} = \frac{2}{L_e}\frac{dNi}{d\zeta}$$

where $L_e = 2l$ = length of the bar element.

To derive the stiffness matrix we have the relation

$$\int_{-1}^{+1} [B]^T [E] [B]\, dV \qquad ...(8.24)$$

for a bar of uniform cross-sectional area A, $dV = Adx$.

But $$dx \;=\; \frac{L_e}{2} d\zeta$$

To evaluate k_{11}:

$$k_{11} \;=\; AE \int_{-1}^{+1} \left(\frac{dN_1}{dx}\right)^2 dx$$

$$=\; AE \int_{-1}^{+1} \left(\frac{2}{L_e}\frac{dN_1}{d\zeta}\right)^2 \frac{L_e}{2} d\zeta$$

$$=\; \frac{2AE}{L_e} \int_{-1}^{+1} \left(\zeta - \frac{1}{2}\right)^2 d\zeta$$

$$=\; \frac{2AE}{(2l)}(7/6) = \frac{7AE}{6l}$$

To evaluate $k_{12} = k_{21}$:

$$k_{12} \;=\; AE \int_{0}^{2l} \left(\frac{dN_1}{dx}\right)\left(\frac{dN_2}{dx}\right) dx$$

$$=\; \frac{2AE}{L_e} \int_{-1}^{+1} \left(\zeta - \frac{1}{2}\right)\left(\zeta + \frac{1}{2}\right) d\zeta$$

$$=\; \frac{2AE}{(2l)}(1/6) = \frac{AE}{6l}$$

To evaluate $k_{13} = k_{31}$:

$$k_{13} = AE \int_0^{2l} \left(\frac{dN_1}{dx}\right)\left(\frac{dN_3}{dx}\right) dx$$

$$= \frac{2AE}{L_e} \int_{-1}^{+1} \left(\zeta + \frac{1}{2}\right)(-2\zeta)\, d\zeta$$

$$= \frac{2AE}{(2l)}\left(-\frac{8}{6}\right) = -\frac{8AE}{6l}$$

To evaluate k_{22}:

$$k_{22} = AE \int_0^{2l} \left(\frac{dN_2}{dx}\right)^2 dx$$

$$= \frac{2AE}{L_e} \int_{-1}^{+1} \left(\zeta + \frac{1}{2}\right)^2 d\zeta$$

$$= \frac{2AE}{2l}\left(\frac{7}{6}\right) = \frac{7AE}{6l}$$

To evaluate $k_{23} = k_{32}$:

$$k_{23} = AE \int_0^{2l} \left(\frac{dN_2}{dx}\right)\left(\frac{dN_3}{dx}\right) dx$$

$$= \frac{2AE}{L_e} \int_{-1}^{+1} \left(\zeta + \frac{1}{2}\right)(-2\zeta)\, d\zeta$$

$$= \frac{2AE}{(2l)}(-4/3) = -\frac{8AE}{6l}$$

To evaluate k_{33}:

$$k_{33} = AE \int_0^{2l} \left(\frac{dN_3}{dx}\right)^2 dx$$

$$= \frac{2AE}{L_e} \int_{-1}^{+1} (-2\zeta)^2 d\zeta$$

$$= \frac{2AE}{(2l)}\left(\frac{8}{3}\right) = \frac{16AE}{6l}$$

Arranging the stiffness co-efficients in the matrix form

$$[K]^e = \frac{AE}{6l}\begin{bmatrix} 7 & 1 & -8 \\ 1 & 7 & -8 \\ -8 & -8 & 16 \end{bmatrix} \quad ...(8.25)$$

This is the stiffness matrix for 3-noded bar element. It may be noted that the order of the matrix is same as the number of degrees of freedom assumed for the higher order element. The matrix confirms the properties of the stiffness matrix. The elements of the matrix are dependent on the geometrical and mechanical properties of materials.

Example 1. A bar of uniform cross-section Fig. 8.4 of 100 mm^2 and length 150 mm when subjected to a tensile load deforms 0.2 mm and 0.3 mm at the end nodes and 0.5 mm in the middle. Find the strain at distances 50 mm and 100 mm in the bar (from left end) and plot the displacement along the length.

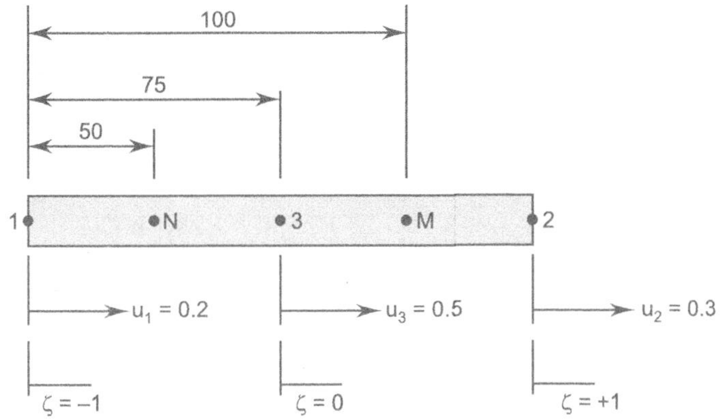

Fig. 8.4. Problem on 3-noded bar.

Solution. The natural co-ordinate by definition

$$\zeta = \frac{2(x - x_1)}{(x_2 - x_1)} - 1$$

where $x_1 = 0.0$ and $x_2 = 150$

But $x_m = 100$ and $x_n = 50$

Hence

$$\zeta_m = \frac{2(100 - 0)}{(150 - 0)} - 1 = \frac{1}{3}$$

$$\zeta_n = \frac{2(50 - 0)}{(150 - 0)} - 1 = \frac{-1}{3}$$

Given that $u_1 = 0.2 \quad u_2 = 0.3 \quad u_3 = 0.5$

The displacement field in terms of shape functions

$$u(\zeta) = N_1 u_1 + N_2 u_2 + N_3 u_3$$

$$N_1^m = \frac{(\zeta_m - 1)}{2}\zeta_m = -\frac{1}{9}$$

$$N_2^m = \frac{(\zeta_m + 1)}{2}\zeta_m = \frac{2}{9}$$

$$N_3^m = (1 - \zeta_m)(1 + \zeta_m) = \frac{8}{9}$$

Similarly

$$N_1^n = \frac{(\zeta_n - 1)}{2}\zeta_n = -\frac{2}{9}$$

$$N_2^n = \frac{(\zeta_n + 1)}{2}\zeta_n = -\frac{1}{9}$$

$$N_3^n = (1 - \zeta_n)(1 + \zeta_n) = \frac{8}{9}$$

Hence

$$u_m = -1/9(0.2) + 2/9(0.3) + 8/9(0.5) = \frac{4.4}{9}$$

$$u_n = 2/9(0.2) + (-1/9)(0.3) + 8/9(0.5) = \frac{4.1}{9}$$

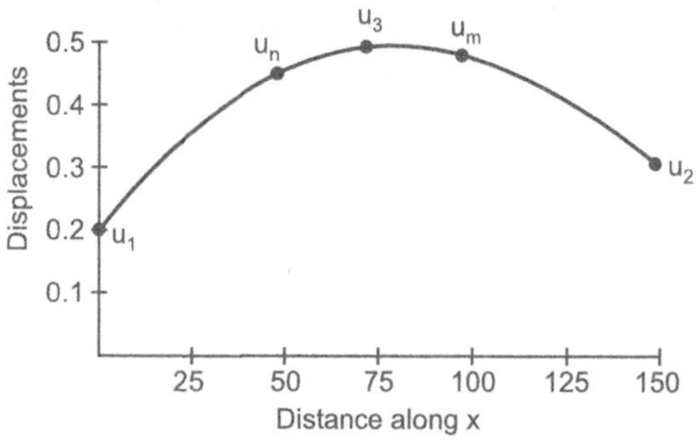

Fig. 8.5. Displacement plot.

The strain distribution

$$\in = \frac{du}{dx} = \frac{du}{dz}\cdot\frac{dz}{dx} = \frac{2}{L}\frac{du}{d\zeta}$$

But

$$u = \Sigma N_i u_i$$

$$\in = \frac{2}{L}\left[\frac{dN_1}{d\zeta}u_1 + \frac{dN_2}{d\zeta}u_2 + \frac{dN_3}{d\zeta}u_3\right]$$

$$\frac{dN_1}{d\zeta} = \zeta - \frac{1}{2}, \quad \frac{dN_2}{d\zeta} = \zeta + \frac{1}{2} \quad \text{and} \quad \frac{dN_3}{d\zeta} = -2\zeta$$

strain at point M

$$\in_m = \left[-\frac{1}{6}(0.2) + \frac{5}{6}(0.3) - 2/3\,(0.5)\right]\frac{2}{150} = -\frac{0.7}{450}$$

strain at point N

$$\in_n = \left[-5/6(0.2) + 1/6(0.3) + \frac{2}{3}(0.5)\right]\frac{2}{150} = \frac{1.3}{450}$$

At point N the strain is tensile and at point M the strain is compressive (*i.e.*, –ve)

8.6 LAGRANGE POLYNOMIAL

The polynomial functions that yield function value at finite number of points, are called the Lagrange's Polynomials. The shape functions have been derived for the bar element. Linear, quadratic displacement polynomials were used for the 2-DOF system (*i.e.*, two noded bar and 3-DOF system (*i.e.*, 3-noded bar) respectively. Using Lagrange's polynomial approach it is possible to formulate the shape functions for any order element without having gone into derive them by conventional methods.

Lagrange's polynomials ensure the continuity of the used function (displacement function) at the end of the element. But the continuity of the derivative cannot be ensured at the end of the element irrespective of the order of the polynomials when Lagrange polynomial are used to represent the shape function.

8.7 LINEAR ELEMENT: LAGRANGE POLYNOMIAL

Displacement approximation

$$u(x) = a_0 + a_1(x) \quad \text{...(8.26)}$$

The function is linear with the order of x equal to one.

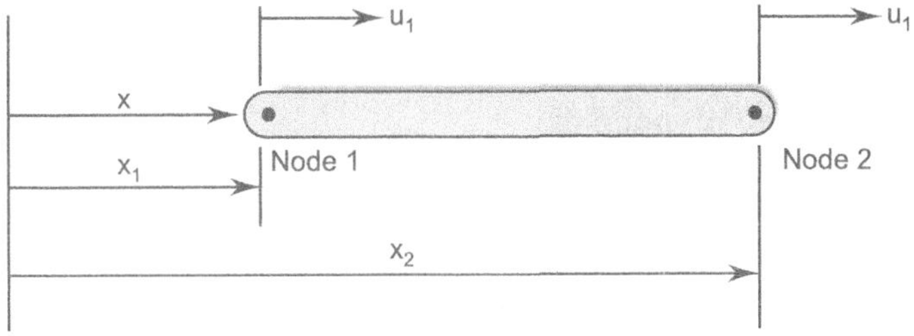

Fig. 8.6. Bar element with 2-nodes.

The degrees of freedom are two, hence we are supposed to use two shape functions.

The condition available:

$$u(x_1) = u_1 \quad \text{and} \quad u(x_2) = u_2$$

The shape function

$$N_1 = \left[\frac{x - x_2}{x_1 - x_2}\right] \quad \text{...(8.27)}$$

and

$$N_2 = \left[\frac{x - x_1}{x_2 - x_1}\right]$$

Displacement function

$$u(x) = N_1 u_1 + N_2 u_2$$

$$= \sum_{i=1}^{n} N_i(x)\, u_i$$

when N_i are obtained by the direct application of Lagrange polynomial they are represented by Lagrange function $L_i(x)$.

Hence

$$L_1 = \left[\frac{x - x_2}{x_1 - x_2}\right] \qquad ...(8.28)$$

$$L_2 = \left[\frac{x - x_1}{x_2 - x_1}\right]$$

with the use of L_i

$$u(x) = \sum_{i=1}^{2} L_i(x)\, u_i \qquad ...(8.29)$$

Lagrange Polynomial: One Dimensional

A bar element is one dimensional. The order of linear element is one. The order of quadratic and cubic elements are 2 and 3 respectively. The Lagrange's Polynomial in one dimension of the order 'n' can be defined as:

$$L_i(x) = \prod_{\substack{j=1 \\ i \neq j}}^{n} \left[\frac{(x - x_j)}{(x_i - x_j)}\right] \quad \text{for } i = 1 \text{ to } (n + 1) \qquad ...(8.30)$$

where i is the node under consideration and j are the nodes in the element other than i in the x-direction. n is the order of the element. There exist $(n + 1)$ Lagrange functions to represent the displacement function.

Linear Element: Lagrange Polynomials

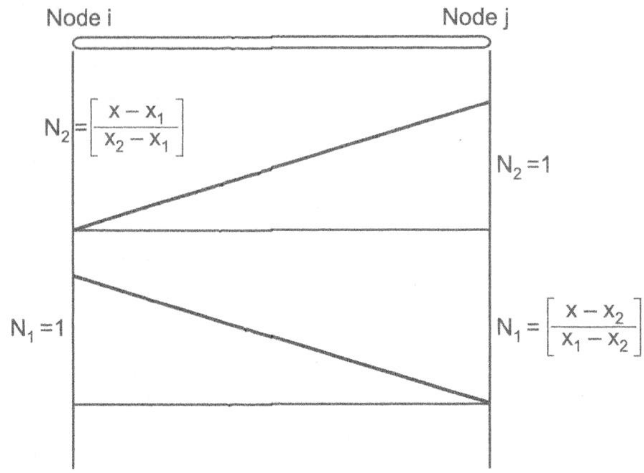

Fig. 8.7. Plot of Lagrange functions.

To write L_1

The node under consideration is 1 (i = 1).

The nodes in the x-direction other than node 1 is node 2 *i.e.*, j = 2. From equation (8.30)

$$L_1(x) = \left(\frac{x - x_2}{x_1 - x_2}\right) \qquad ...(8.31)$$

To write L_2

Now the considered node is 2 (*i.e.*, i = 2). The other node is 1 (*i.e.*, j = 1).

From equation (8.30)

$$L_2 = \left[\frac{x - x_1}{x_2 - x_1}\right] \quad ...(8.32)$$

The displacement function is

$$u(x) = \left[\frac{x - x_2}{x_1 - x_2}\right] u_1 + \left[\frac{x - x_1}{x_2 - x_1}\right] u_2 \quad ...(8.33)$$

The plot of Lagrange functions L_i is as shown in Fig. 8.7. To verify the plot

when $x = x_1$ $L_1 = 1$ and $L_2 = 0$

when $x = x_2$ $L_1 = 0$ and $L_2 = 1.$

8.8 LAGRANGE POLYNOMIAL: QUADRATIC ELEMENT

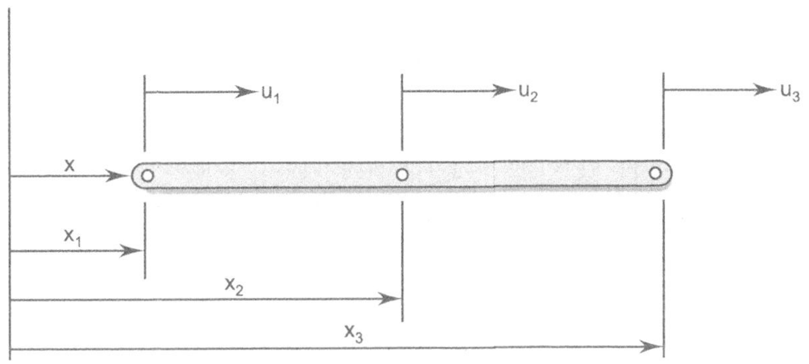

Fig. 8.8. Quadratic bar element.

Displacement Function in generalized co-ordinate

$$u(x) = a_0 + a_1 x + a_2 x^2 \quad ...(8.34)$$

In terms of Lagranges Functions

$$u(x) = \sum_{i=1}^{n} L_i(x) u_i \quad ...(8.35)$$

To write L_1

Node under consideration $i = 1$.

Node next to node 1 are node 2 and node 3. Hence $j = 2, 3$

$$L_1 = \frac{(x - x_2)(x - x_3)}{(x_1 - x_2)(x_1 - x_3)} \quad ...(8.36)$$

To write L_2

Node under consideration is 2 *i.e.*, $i = 2$

Nodes next to node 2 are node 1 and 3, *i.e.*, $j = 1, 3$

Hence the Lagrange function from equation (8.30)

$$L_2 = \frac{(x - x_1)(x - x_3)}{(x_2 - x_1)(x_2 - x_3)} \quad ...(8.37)$$

To write L_3

Similar to the previous two L_1 and L_2 here $i = 3$ and $j = 1, 2$

Hence from equation (8.30)

$$L_3 = \frac{(x - x_1)(x - x_2)}{(x_3 - x_1)(x_3 - x_2)} \quad ...(8.38)$$

To plot the functions

$$L_1 = \frac{(x - x_2)(x - x_3)}{(x_1 - x_2)(x_1 - x_3)}$$

$$L_2 = \frac{(x - x_1)(x - x_3)}{(x_2 - x_1)(x_2 - x_3)}$$

$$L_3 = \frac{(x - x_1)(x - x_2)}{(x_3 - x_1)(x_3 - x_2)}$$

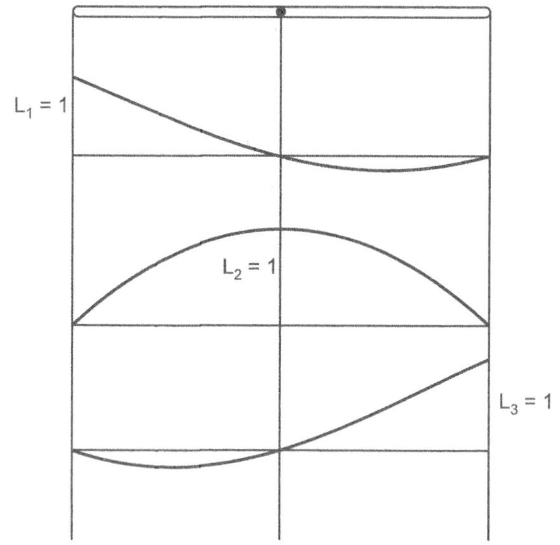

Fig. 8.9. Plot of Lagrange functions.

8.9 CUBIC ELEMENT: BAR ELEMENT

Displacement function

$$u(x) = a_0 + a_1x + a_2x^2 + a_3x^3 \quad ...(8.39)$$

Conditions

$$u(x_1) = u_1 \quad u(x_2) = u_2 \quad u(x_3) = u_3 \quad u(x_4) = u_4$$

Shape function $\quad i = 1$ and $j = 2, 3, 4$

$$L_1 = \frac{(x - x_2)(x - x_3)(x - x_4)}{(x_1 - x_2)(x_1 - x_3)(x_1 - x_4)}$$

$$i = 2; j = 1, 3, 4$$

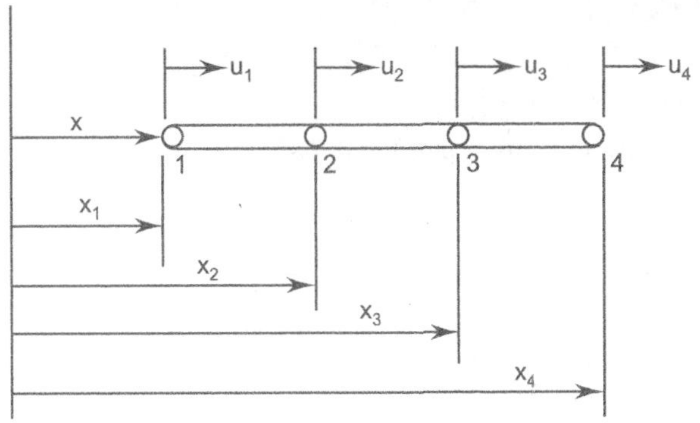

Fig. 8.10. 4 noded bar element.

$$L_2 = \frac{(x-x_1)(x-x_3)(x-x_4)}{(x_2-x_1)(x_2-x_3)(x_2-x_4)}$$

$$i = 3; \quad j = 1, 2, 4$$

$$L_3 = \frac{(x-x_1)(x-x_2)(x-x_4)}{(x_3-x_1)(x_3-x_2)(x_3-x_4)}$$

$$i = 4 \text{ and } j = 1, 2, 3$$

$$L_4 = \frac{(x-x_1)(x-x_2)(x-x_3)}{(x_4-x_1)(x_4-x_2)(x_4-x_3)}.$$

8.10 LAGRANGE'S POLYNOMIAL: NATURAL CO-ORDINATES

The Lagrange's Polynomial in natural co-ordinate for the nth order element is defined to be

$$L_i(\zeta) = \prod_{\substack{j=1 \\ i \neq j}}^{n} \frac{(\zeta-\zeta_j)}{(\zeta_i-\zeta_j)} \quad \text{for } i = 1 \text{ to } n+1 \qquad ...(8.40)$$

Linear element

At node 1:

$$i = 1 \quad \text{and} \quad j = 2$$

$$\zeta_i = -1 \qquad \zeta_j = +1$$

$$L_1(\zeta) = \frac{(\zeta-1)}{(-1-1)} = \frac{(1-\zeta)}{2}$$

At node 2:

$$i = 2 \quad \text{and} \quad j = 1$$

$$\zeta_i = +1 \qquad \zeta_j = -1$$

$$L_2(\zeta) = \frac{(\zeta+1)}{(1+1)} = \frac{(1+\zeta)}{2}$$

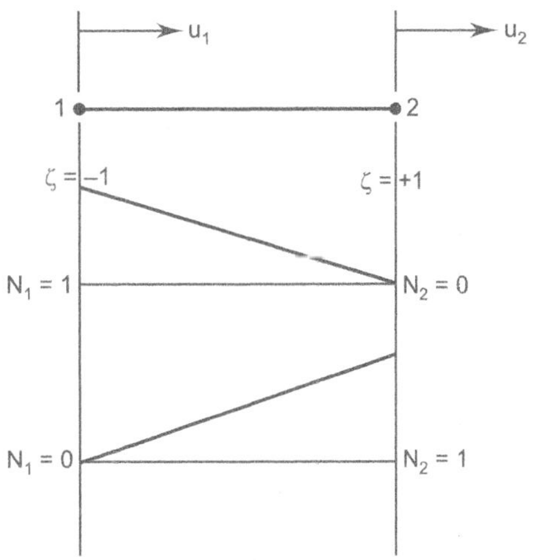

Fig. 8.11. Shape function.

Property

1. at node 1 $\zeta = -1$ $L_1 = 1$ and $L_2 = 0$

   at node 2 $\zeta = +1$ $L_1 = 0$ and $L_2 = 1$

2. $\Sigma L_i = \left(\frac{1-\zeta}{2}\right) + \left(\frac{1+\zeta}{2}\right) = 1.$

The order of the displacement, $n = 1$.

Number of Lagrange's Polynomials = $n + 1 = 2$.

8.11 QUADRATIC ELEMENT

Order of polynomial = $n = 2$.

Number of Lagrange functions, $n + 1 = 3$.

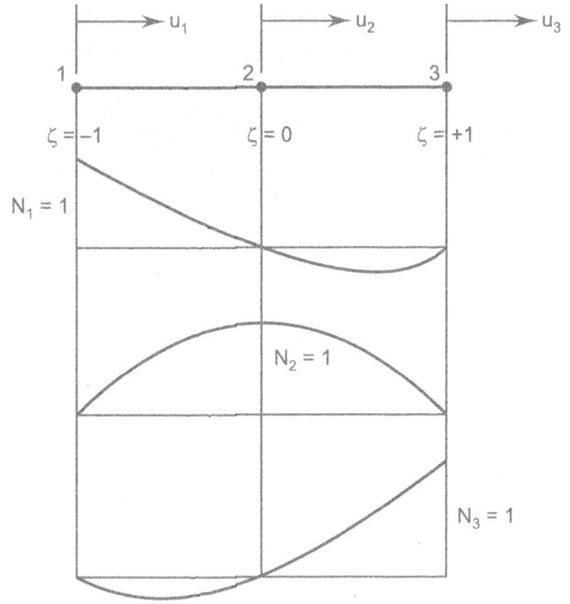

Fig. 8.12. Quadratic element.

At node 1:

$$i = 1 \quad \text{and} \quad j = 2, 3$$

$$\zeta_i = -1 \quad \zeta_j = 0, +1$$

$$L_1 = \frac{(\zeta - 0)(\zeta - 1)}{(-1-0)(-1-1)}$$

$$= \frac{\zeta(\zeta - 1)}{2} \qquad \text{...(8.41)}$$

At node 2:

$$i = 2; \qquad j = 1, 3$$

$$\zeta_i = 0 \qquad \zeta_j = -1, +1$$

$$L_2 = \frac{(\zeta+1)(\zeta-1)}{(0+1)(0-1)} = (1-\zeta)(1+\zeta) \qquad \text{...(8.42)}$$

At node 3:

$$i = 3; \quad j = 1, 2$$

$$\zeta_i = +1 \quad \text{and} \quad \zeta_j = -1,\, 0$$

$$L_3 = \frac{(\zeta+1)(\zeta-0)}{(1+1)(1-0)} = \frac{\zeta(\zeta+1)}{2} \qquad \text{...(8.43)}$$

The dispacement function is

$$u(\zeta) = L_1(\zeta)u_1 + L_2(\zeta)u_2 + L_3(\zeta)u_3 \qquad \text{...(8.44)}$$

when

$$\zeta = -1 \quad N_1 = 1 \quad N_2 = 0 \quad N_3 = 0$$
$$\zeta = 0 \quad N_1 = 0 \quad N_2 = 1 \quad N_3 = 0$$
$$\zeta = +1 \quad N_1 = 0 \quad N_2 = 0 \quad N_3 = 1$$

8.12 CUBIC ELEMENT: NATURAL CO-ORDINATES

One dimensional cubic element is the third order bar element with four nodes having one degree of freedom at each node. Hence four Lagrange's interpolation functions are required to describe the displacement distribution in the element. The Fig. 8.13 shows the configuration of the element and the plot of the Lagrange functions.

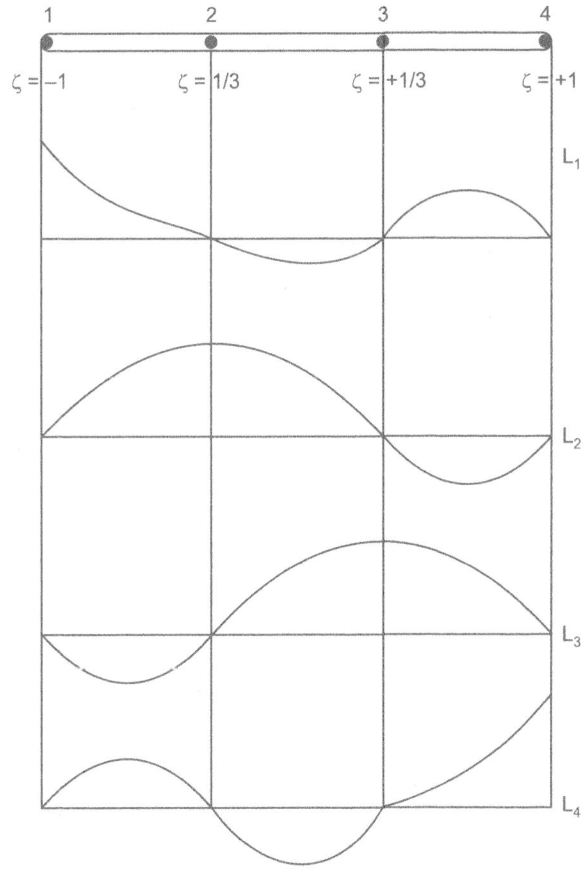

Fig. 8.13. Cubic element.

At node 1,

$$i = 1 \qquad j = 2, 3, 4$$

$$\zeta_i = -1 \qquad \zeta_j = -\,1/3,\ 1/3,\ 1$$

Hence

$$L_1 = \frac{\left(\zeta+\frac{1}{3}\right)\left(\zeta-\frac{1}{3}\right)(\zeta-1)}{\left(-1+\frac{1}{3}\right)\left(-1-\frac{1}{3}\right)(-1-1)}$$

$$= -\frac{9}{16}\left(\zeta+\frac{1}{3}\right)\left(\zeta-\frac{1}{3}\right)(\zeta-1)$$

At node 2,

$$i = 2, \quad j = 1, 3, 4$$

$$\zeta_i = -1/3 \quad \zeta_j = -1, 1/3, +1$$

Hence

$$L_1 = \frac{\left(\zeta-\frac{1}{3}\right)(\zeta+1)(\zeta-1)}{\left(-\frac{1}{3}-\frac{1}{3}\right)\left(-\frac{1}{3}+1\right)\left(-\frac{1}{3}-1\right)}$$

$$= -\frac{27}{16}(\zeta+1)(\zeta-1/3)(\zeta-1)$$

At node 3,

$$i = 3, \quad j = 1, 2, 4$$

$$\zeta_i = +\frac{1}{3} \quad \zeta_j = -1, -1/3, +1$$

Hence

$$L_3 = \frac{(\zeta+1)\left(\zeta+\frac{1}{3}\right)(\zeta-1)}{\left(\frac{1}{3}+1\right)\left(\frac{1}{3}+\frac{1}{3}\right)\left(\frac{1}{3}-1\right)} = -\frac{27}{16}(\zeta+1)\left(\zeta+\frac{1}{3}\right)(\zeta-1)$$

At node 4,

$$i = 4, \quad j = 1, 2, 3$$

$$\zeta_i = +1 \quad \zeta_j = -1, -\frac{1}{3}, +\frac{1}{3}$$

Hence

$$L_4 = \frac{(\zeta+1)\left(\zeta+\frac{1}{3}\right)\left(\zeta-\frac{1}{3}\right)}{(1+1)\left(1+\frac{1}{3}\right)\left(1-\frac{1}{3}\right)} = \frac{9}{16}(\zeta+1)\left(\zeta+\frac{1}{3}\right)\left(\zeta-\frac{1}{3}\right)$$

It may be verified that the property

$$\sum_{i=1}^{4} L_i(\zeta) = 1 \text{ is satisfied in the element.}$$

8.13 2-D ELEMENTS: LAGRANGE'S POLYNOMIAL

The Lagrange's polynomial of order 'n' in cartesian co-ordinates for a two dimensional element is given by

$$L_i = \prod_{\substack{j=1 \\ k=1 \\ i \neq j,k}}^{\substack{k=m \\ j=n}} \frac{(x-x_j)}{(x_i-x_j)} \frac{(y-y_k)}{(y_i-y_k)} \qquad \text{...(8.45)}$$

The Lagrange's Polynomial of order 'n' in natural co-ordinates for a two dimensional element is given by

$$L_i(\xi, \eta) = \prod_{\substack{j=1 \\ k=1 \\ i \neq j, k}}^{\substack{k=m \\ j=n}} \frac{(\xi - \xi_j)(\eta - \eta_k)}{(\xi_i - \xi_j)(\eta_i - \eta_k)} \qquad \text{...(8.46)}$$

8.14 HIGHER ORDER ELEMENT: 2-DIMENSIONAL

A triangular element with three corner nodes and three mid-side nodes is quadratically higher order two dimensional element. A quadrilateral with four corner nodes and four mid-side nodes is also higher order element that is quadratic. Let us find out the shape functions of such elements using Lagrange's polynomial method.

Triangular element: Lagrange's polynomial.

Assume a right angle triangular element with six nodes as shown in Fig. 8.14. The element is identified in natural co-ordinate system.

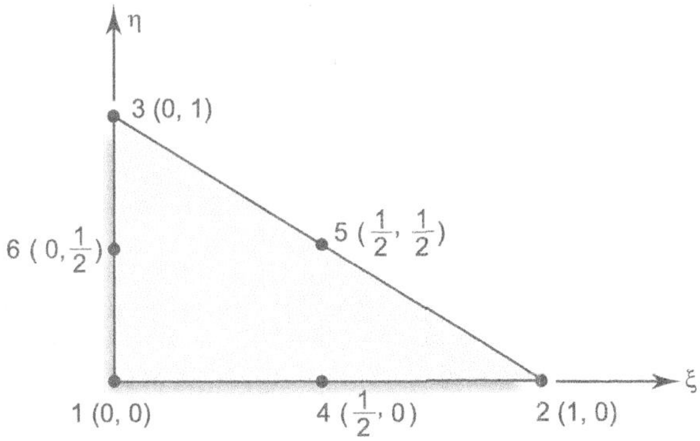

Fig. 8.14. Higher order triangular element.

The Lagrange's polynomial is described by

$$L_i(z, \eta) = \prod_{\substack{j=1 \\ k=1}}^{n, m} \frac{(\xi - \xi_j)(\eta - \eta_k)}{(\xi_i - \xi_j)(\eta_i - \eta_k)}$$

Node i	j nodes	k nodes
1	4, 2	6, 3
2	1, 4	nil
3	nil	6, 1
4	1, 2	5
5	6	4
6	5	1, 3

The Lagrange's polynomial representing the shape functions are listed as follows:

$$L_1 = \frac{\left(\xi - \frac{1}{2}\right)(\xi - 1)\left(\eta - \frac{1}{2}\right)(\eta - 1)}{\left(0 - \frac{1}{2}\right)(0 - 1)\left(0 - \frac{1}{2}\right)(0 - 1)}$$

$$= (2\xi - 1)\ (\xi - 1)\ (2\eta - 1)\ (\eta - 1)$$

$$L_2 = \frac{(\xi - 0)\left(\xi - \frac{1}{2}\right)}{(1 - 0)\left(1 - \frac{1}{2}\right)} = \zeta(2\zeta - 1)$$

$$L_3 = \frac{\left(\eta - \frac{1}{2}\right)(\eta - 0)}{\left(1 - \frac{1}{2}\right)(1 - 0)} = \eta(2\eta - 1)$$

$$L_4 = \frac{(\xi - 1)(\xi - 0)\left(\eta - \frac{1}{2}\right)}{\left(\frac{1}{2} - 1\right)\left(\frac{1}{2} - 0\right)\left(0 - \frac{1}{2}\right)} = 4\zeta(\zeta - 1)(2\eta - 1)$$

$$L_5 = \frac{(\xi - 0)(\xi - 0)}{\left(\frac{1}{2} - 0\right)\left(\frac{1}{2} - 0\right)} = 4z\eta$$

$$L_6 = \frac{\left(\xi - \frac{1}{2}\right)(\eta - 0)(\eta - 1)}{\left(0 - \frac{1}{2}\right)\left(\frac{1}{2} - 0\right)\left(\frac{1}{2} - 1\right)} = 4(2\zeta - 1)\,\eta(\eta - 1)$$

8.15 RECTANGULAR ELEMENT: LAGRANGE'S POLYNOMIALS

Assume a rectangular element with eight nodes (4 corner and 4 mid-side) as shown in Fig. 8.15. Assigning the natural co-ordinate located on the centroid of the element, obtain the shape function in the following manner.

The Lagrange's polynomial are obtained as:

Node i	j^{th} nodes	k^{th} nodes
1	5, 2	8, 4
2	1, 5	6, 3
3	4, 7	6, 2
4	7, 3	8, 1
5	1, 2	7
6	8	2, 3
7	4, 3	5
8	6	4, 1

There are eight shape functions that are quadratic in order. The interpolation functions obtained by Lagrange's polynomial method are listed as follows:

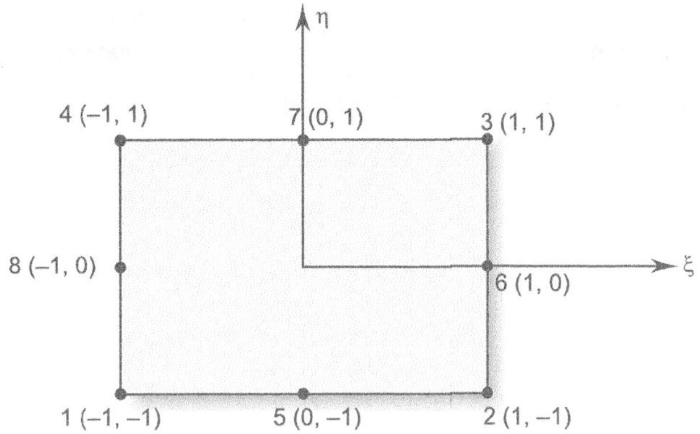

Fig. 8.15. Rectangular element.

$$L_1 = \frac{(\xi-0)(\xi-1)(\eta-0)(\eta-1)}{(-1-0)(-1-1)(-1-0)(-1-1)} = \frac{\xi\eta(\xi-1)(\eta-1)}{4}$$

$$L_2 = \frac{(\xi+1)(\xi-0)(\eta-1)(\eta-0)}{(1+1)(1-0)(-1-1)(-1-0)} = \frac{z\eta}{4}(\xi+1)(\eta-1)$$

$$L_3 = \frac{(\xi+1)(\xi-0)(\eta+1)(\eta-0)}{(1+1)(1-0)(1+1)(1-0)} = \frac{z\eta}{4}(\xi+1)(\eta+1)$$

$$L_4 = \frac{(\xi-0)(\xi-1)(\eta+1)(\eta-0)}{(-1-0)(-1-1)(1+1)(1-0)} = \frac{\eta\xi}{4}(\xi-1)(\eta+1)$$

$$L_5 = \frac{(\xi+1)(\xi-1)(\eta-1)}{(0+1)(0-1)(-1-1)} = \frac{1}{2}(\xi+1)(\xi-1)(\eta-1)$$

$$L_6 = \frac{(\xi+1)(\eta-1)(\eta+1)}{(1+1)(0-1)(0+1)} = -\frac{1}{2}(\zeta+1)(\eta-1)(\eta+1)$$

$$L_7 = \frac{(\xi+1)(\xi-1)(\eta+1)}{(0+1)(0-1)(1+1)} = -\frac{1}{2}(\xi+1)(\xi-1)(\eta+1)$$

$$L_8 = \frac{(\xi-1)(\eta+1)(\eta-1)}{(-1-1)(0+1)(0-1)} = \frac{1}{2}(\xi-1)(\eta+1)(\eta-1).$$

8.16 SIMPLEX ELEMENTS

Element Configuration	Features
1. Two noded bar: 1D element  2. Three noded triangle: 2D 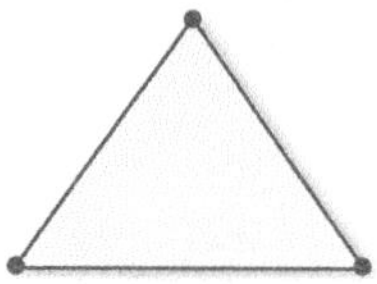 3. Four noded quadrilateral: 2D  4. 8 noded hexahedron: 3D 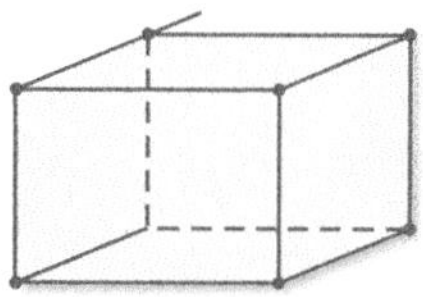	• Simplex elements are the linear elements. • The polynomials that represent the displacement functions of simplex elements are linear. • The accuracy of solution is governed by *h*-refinements. • The solutions are less accurate as compared to higher order element. • The computation of shape functions and stiffness matrix are simpler. • The computational time and efforts required as less. • Such element are not recommended in the critical mesh areas. • The examples of simplex elements are linear bar element, CST, linear quad, linear hexahedron.

8.17 COMPLEX ELEMENTS

Element Configuration	Features
1. Three noded bar: 1D  2. Six noded triangle: 2D 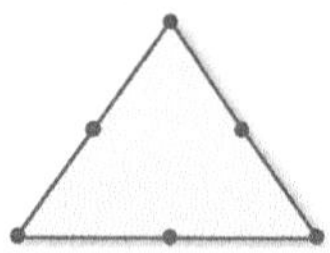 3. Eight noded quad: 2D 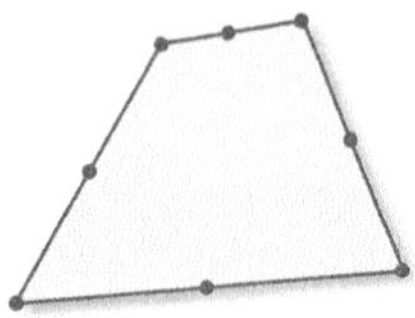 4. 20 noded brick: 3D	• Higher order elements are included in this category. • The degree of the polynomials used to form the displacement functions is 2 or more. • The accuracy of solution is governed by p-refinement. • Provides reasonable accuracy with the coarse mesh also. • Better accuracy and better convergence rate. • Computational efforts and time required are more. • Usage of such elements in critical area is reasonable. • Examples are: 3 noded bar, six noded triangle, eight noded quad, 20 noded brick.

8.18 MULTIPLEX ELEMENT

Element Configuration	Features
1. Six noded right angled triangular element 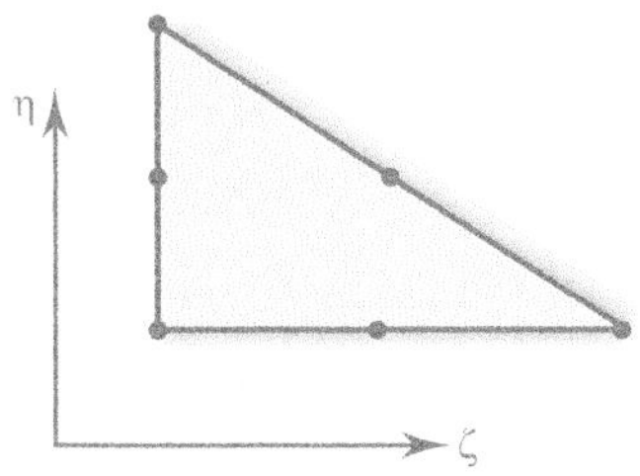	• These are higher order elements with the edges aligned with local co-ordinate axes. • Such elements favour the isoparametric formulation. • Multiplex elements can be identified by parent element. • The shape functions can be easily obtained by Lagrange's Polynomial method.
2. Nine noded square element 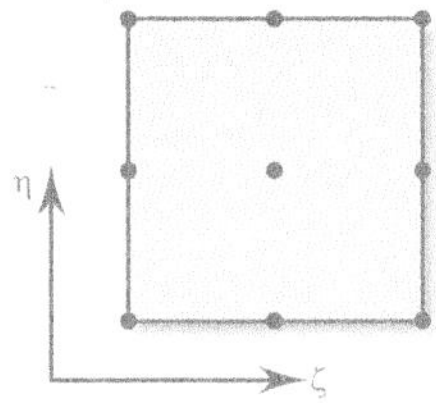	• They can be related to arbitrary shaped image element by mapping. • The solution accuracy is maintained by p-refinement and convergence is faster. • Example for multiplex elements are six noded right angle triangle, nine noded square or rectangle.

8.19 PASCAL'S TRIANGLE: 1D BAR ELEMENT

Physical Representation	Polynomial Function	Pascal's Triangle
1. Linear element 	$u = a_0 + a_1x$	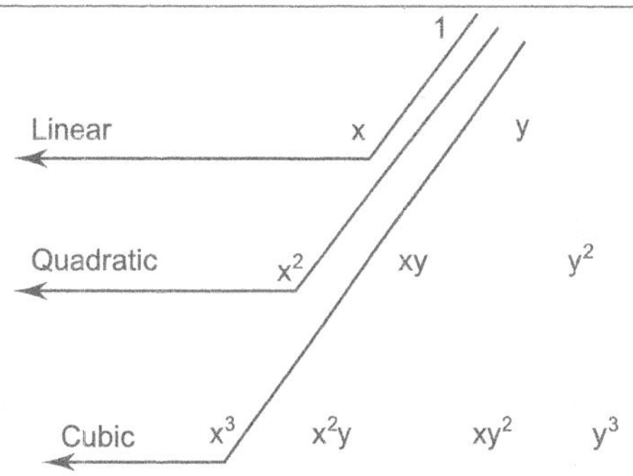
2. Quadratic element 	$u = a_0 + a_1x + a_2x^2$	
3. Cubic element 	$u = a_0 + a_1x + a_2x^2 + a_3x^3$	

The linear element is represented by the displacement function of degree one with two terms—one a constant and other a term with one degree in x. The quadratic element has the displacement polynomial with three terms with addition of a term with degree two. The cubic element has the displacement function having four terms with a term of highest degree of 3.

8.20 PASCAL'S TRIANGLE: TRIANGULAR ELEMENT (2D)

Element Type	Function	Pascal's Triangle
1. Three noded 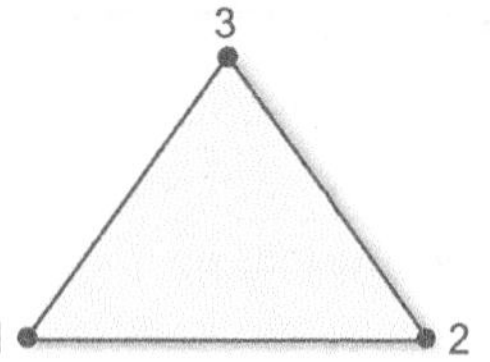 Linear element	$u = a_0(1) +$ $a_1(x) + a_2(y)$ Linear in x and y	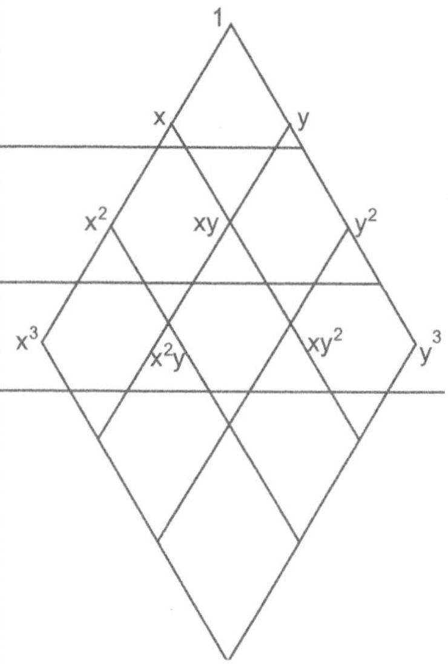
2. Six noded 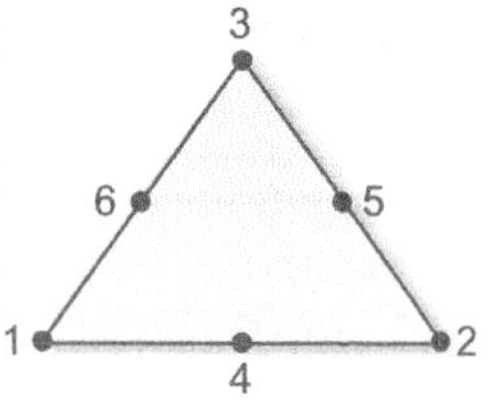 Quadratic element	$u = a_0(1) +$ $a_1(x) + a_2(y) +$ $a_3(x^2) + a_4(xy) + a_5(y^2)$	
3. Ten noded 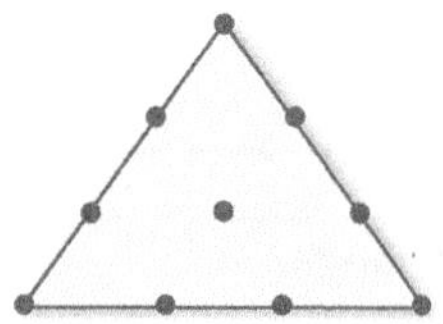 Cubic element	$u = a_0(1) +$ $a_1(x) + a_2(y) +$ $a_3(x^2) + a_4(xy) + a_5(y^2) +$ $a_6(x^3) + a_7(x^2y) + a_8(xy^2) + a_9(y^3)$	

8.21 PASCAL'S TRIANGLE: 2D QUADRILATERAL

Element Type	Polynomial Function	Pascal's Triangle
1. Linear Quad. 	$u(x) = a_0(1) + a_1(x) +$ $a_2(y) + a_3(xy)$ Polynomial contains four terms including const.	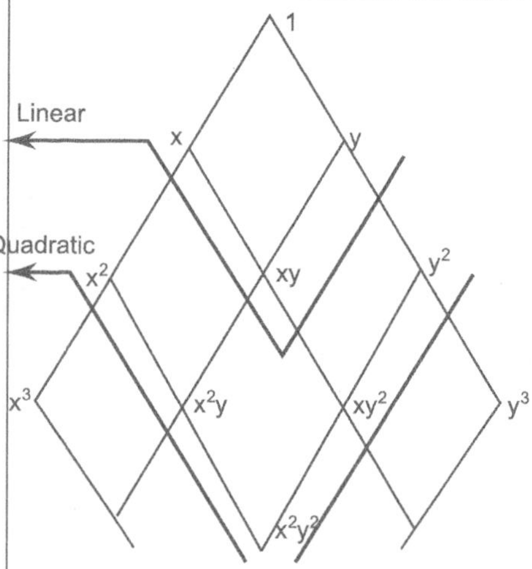
2. Nine Noded Quad. 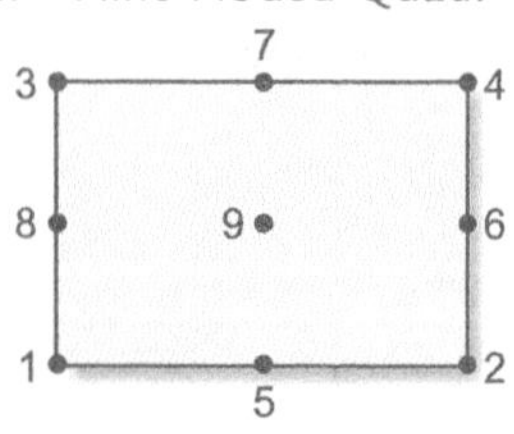	$u(x) = a_0(1) +$ $a_1(x) + a_2(y) +$ $a_3(x^2) + a_4(xy) + a_5(y^2) +$ $a_6(x^2y) + a_7(xy^2) + a_8(x^2y^2)$ The polynomial contains nine terms including the constant.	

8.22 PASCAL'S PYRAMID

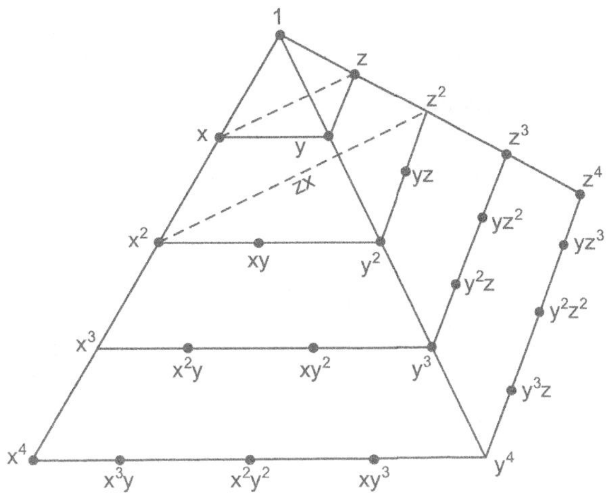

Fig. 8.16. Pascal's pyramid.

The displacement polynomial for a linear tetrahedron element is

$$u = a_0 + a_1x + a_2y + a_3z$$

The displacement polynomial for a linear brick element is

$$u = a_0 + a_1x + a_2y + a_3z + a_4xy + a_5yz + a_6zx.$$

8.23 INTRODUCTION TO CONVERGENCE CRITERIA

The chosen displacement function should ensure that the displacement solution converges to the exact answer as the mesh is refined. For the convergence of result with reduction in mesh size the assumed displacement function has to satisfy certain norms known as convergence criteria.

If FE solution results compares with acceptable accuracy to the exact solution of the mathematical models it may be decided that the solution has converged. The FE scheme exhibits convergence if the discretization approaches zero if the mesh is made infinitely fine (*i.e.*, the element size approaches zero.

* **Order of convergence:** As the order of convergence the following observations are to be made.

- Isoparametric elements always have capability to represent rigid body modes.
- The displacements are based on polynomial expansion.
- Assuming uniform meshes with characteristic dimension.
- Exact solution is assumed to vary smoothly. (Normally exact solution is not smooth).
- Mesh grading is to be employed in a non-smooth distribution.
- With regular meshes the density of solution error is constant.
- Mesh refinement.

There are two types of refinement in mesh is possible to achieve convergence.

1. ***h*-refinement:** If the refinement is based on the element size it is called *h*-method of refinement. In this the element size is reduced to arrive at the convergence of the solution.

2. ***p*-refinement:** If the convergence is achieved by using higher order polynomial to the approximation of the displacement function the refinement is called the *p*-method of refinement.

8.24 CONVERGENCE OF ENERGY

If u is the exact solution to displacement of an elastic structural problem that makes the potential energy minimum. If the corresponding stress and strain value are $\sigma(u)$ and $\in(u)$, the expression for strain energy is

$$U = \frac{1}{2}\int_V \sigma^T \in dV \qquad ...(8.47)$$

Let u_h be the FE solution with h-refinement to the displacement field, then the corresponding stress and strain being $\sigma_h(u_h)$ and $\in_h(u_n)$, the strain energy obtained is

$$U_h = \frac{1}{2}\int_V \sigma_h^T \in_h dV \qquad ...(8.48)$$

The convergence of energy is achieved if

$|U - U_h| \to 0$ as h tends to zero.

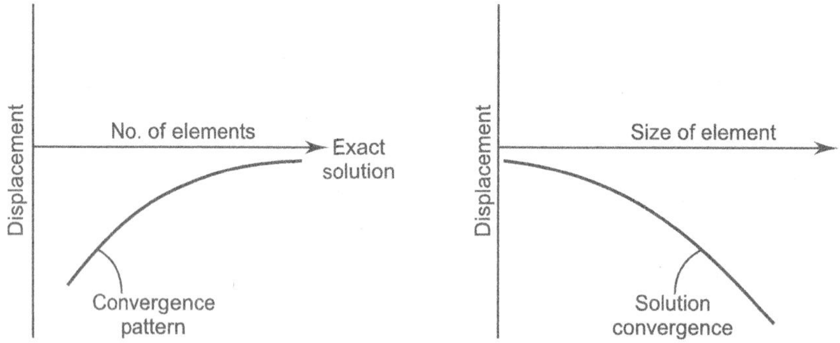

Fig. 8.17. Convergent solution.

In terms of displacements

$$|U - U_h|_{h \to 0} = \sqrt{(u - u_h)^2 + (V - V_h)^2} \qquad dV \to 0 \text{ as } h \to 0 \qquad ...(8.49)$$

If the solution is convergent the plot of the displacement against the element size (or number of elements) is as observed in Fig. 8.17.

The Fig. 8.18 shows the comparison between actual variation of strain with the computed one using FE technique. For a one-dimensional bar the strain is assumed to be constant when linear element is taken. But the actual variation is not constant but non-linear. By assuming a linear interpolation polynomial constant strain state is achieved.

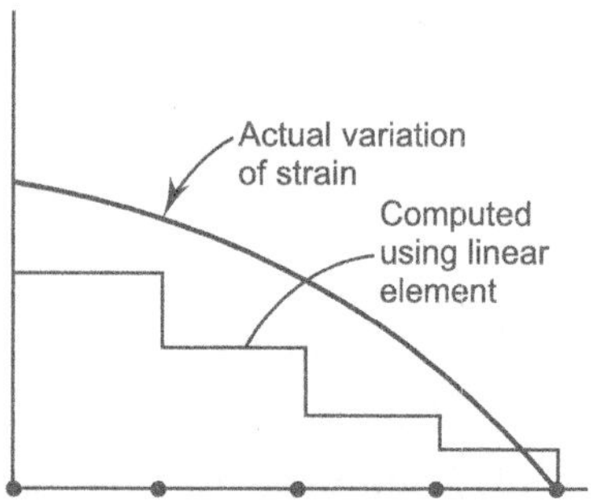

Fig. 8.18. Strain variation.

8.25 CONVERGENCE CRITERIA

- **Convergence Criterion 1:**

Continuity of displacement function in an element. This condition is met with all the polynomials because of their inherent property.

For example consider a 1D linear element. The polynomial chosen is

$$u(x) = a_0 + a_1x$$

For any value of x, there exists a value for the displacement $u(x)$. Hence for any point in the element the displacement field described by the polynomial is continuous.

- **Convergence Criterion 2:**

The representation in the displacement field should contain rigid body motion of the element.

A body is said to be in rigid body motion if there is no straining of the element and stored strain energy is zero during the motion. Hence the stress developed is zero in a rigid body motion.

For example, consider a 2D triangular element. The polynomial assumed for the displacements are

$$u(x, y) = a_0 + a_1x + a_2y$$
$$v(x, y) = b_0 + b_1x + b_2y \quad ...(8.50)$$

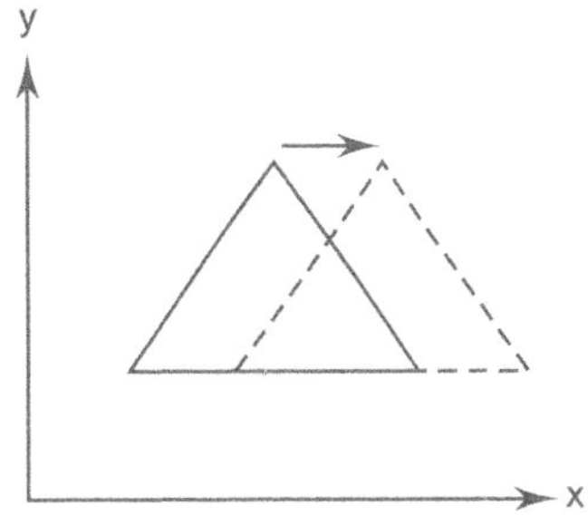

Fig. 8.19. x-translation.

Case 1:

$a_0 \neq 0$ and all other coefficients assumed zero represent the rigid body translation in the x-direction. Refer Fig. 8.19.

Case 2:

$b_0 \neq 0$ but all other co-efficients assumed zero represent the translation along the y-direction. There is zero stress and strain in the element. Refer to the Fig. 8.20.

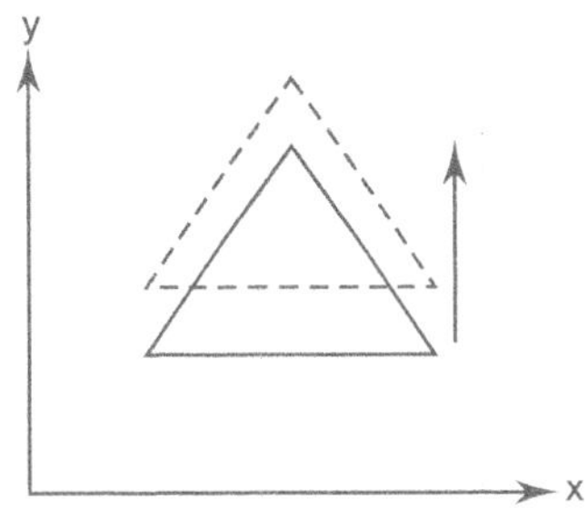

Fig. 8.20. y-translation.

Case 3:

When $a_0 = b_0 = a_1 = b_2 = 0$ but $a_2 = -b_1$ the case represents rigid body rotation about the z-axis (in the xy plane). Refer Fig. 8.21. As it is explained in a rigid body motion the components of strain are zero.

i.e.,
$$\in = \frac{du}{dx} = 0 \quad ...(8.51)$$

Integrating $\quad u(x) = a_0$

Hence the polynomial assumed to represent the displacement field, has to contain the lowest order term in the form of a constant, if it has to represent the rigid motion of the element and the criterion 2 is to be satisfied. The rigid body motion is the most elementary pattern of deformation.

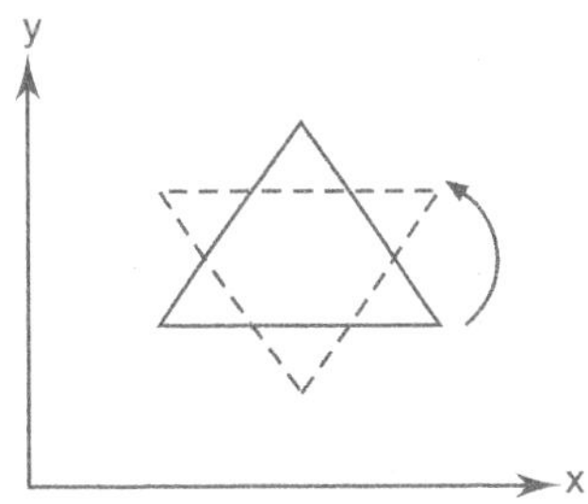

Fig. 8.21. Rotation.

In the polynomial assumed for the displacement of 1D linear bar element,

$$u(x) = a_0 + a_1x$$

there exists a lower order term a_0 and hence it represents rigid body motion.

- **Convergence Criterion 3:**

  The displacement field chosen must represent a state of constant strain in the element.

  With the mesh refinement the element size becomes smaller and the variation of strain becomes so small that is almost zero. Hence this leads to a state of constant strain. The state of constant strain is one of the requirements for convergence.

  For example the assumed linear polynomial for 1D bar element is

$$u(x) = a_0 + a_1 x \qquad \text{...(8.52)}$$

Differentiating we get the strain to be

$$\in = \frac{du}{dx} = a_1$$

Further differentiating

$$\frac{\partial \in}{\partial x} = 0$$

which signifies that for a linear elements (1*D* bar and 2*D* triangular) the strain remains constant.

$$\text{Increment in strain} = \{\text{Rate of strain}\} \times \{\text{Element size}\}$$

$$\Delta\in = \left[\frac{\delta \in}{\delta x}\right]\Delta x \qquad \text{...(8.53)}$$

$\Delta\in$ approaches zero as Δx tends to zero (*i.e.*, possible with mesh refinement). Hence the increment in strain becomes zero indicating constant state of strain providing convergence of the solution.

Completeness:

The elements satisfying criteria 2 and 3 are called complete elements. The requirement of completeness is that the displacement approximation be at least a linear polynomial of degree (k = 1). Any two noded element and higher are complete elements.

1D-formulation	2D-formulation				
1		1			
x	x		y		k = 1
x^2	x^2	xy	y^2		(Linear)
⋮					
x^n					

$$u = a_0 + a_1 x \qquad u(x, y) = a_0 + a_1 x + a_2 y$$

$$v(x, y) = b_0 + b_1 x + b_2 y$$

The two are complete polynomial of degree 1.

Differentiating the two expressions

$$\frac{du}{dx} = a_1 = \text{constant} \qquad \in_x = \frac{\partial u}{\partial x} = a_1$$

$$= \text{strain} \qquad \in_y = \frac{\partial u}{\partial y} = b_2 \qquad \text{...(8.54)}$$

$$\gamma_{xy} = \frac{\partial u}{\partial y} + \frac{\partial u}{\partial x} = a_2 + b_1$$

For convergence the element formulation chosen has to be complete. Hence for a 1D linear element.

$$u(x) = a_0 \Rightarrow \text{Rigid body mode} \\ + a_{1x} \Rightarrow \text{Constant strain state} \Bigg\} \text{Complete}$$

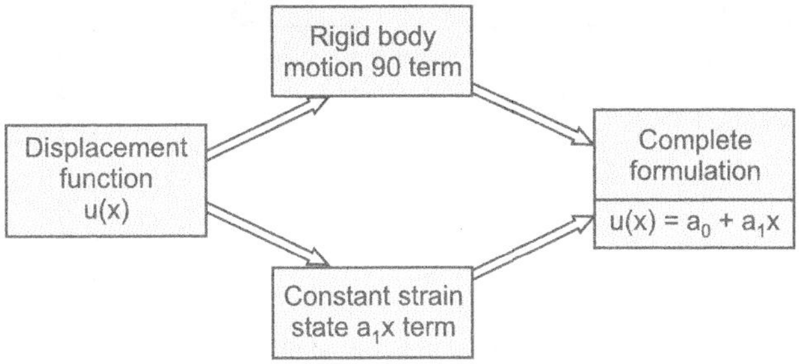

Fig. 8.22. Completeness condition.

- **Convergence Criterion 4:**

  The displacements should meet the compatibility requirements between the adjacent elements.

  The continuity must exist in the interface (boundary) between the two elements. This is called interelement compatibility requirement.

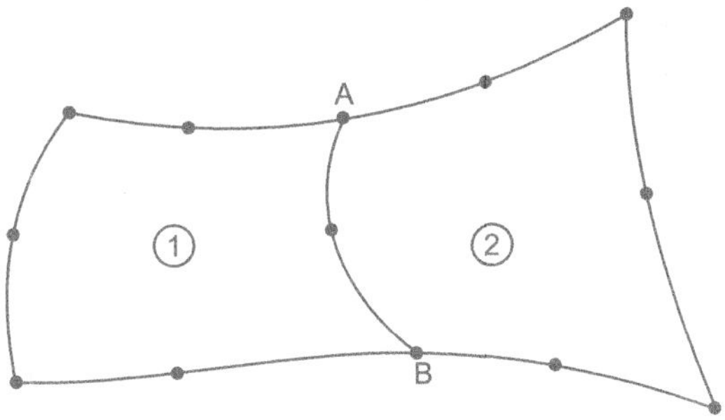

Fig. 8.23. Compatible elements assemblage.

The co-ordinates and the displacements vary parabolically along the common edge AB of both elements shown in Fig. 8.23. This is considered to be the compatible element assemblage.

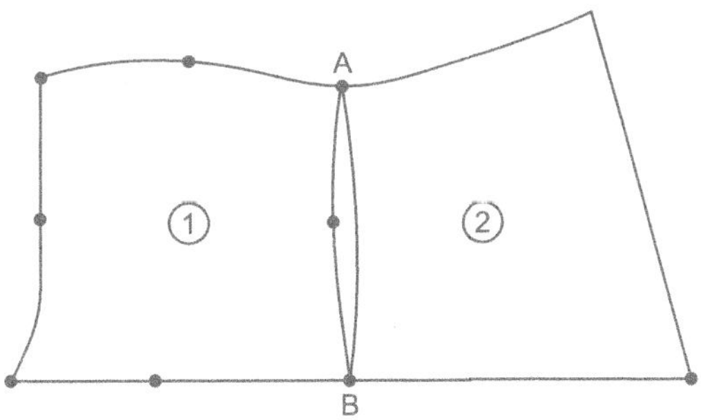

Fig. 8.24. Incompatible element assemblage.

The edge AB for the element (1) has the nodal co-ordinates to vary linearly but the displacements vary parabolically. But for element (2) along AB both vary linearly. This arrangement in Fig. 8.24 is an incompatible assemblage.

Compatible transitions:

1. 4 node to 4 node element transition

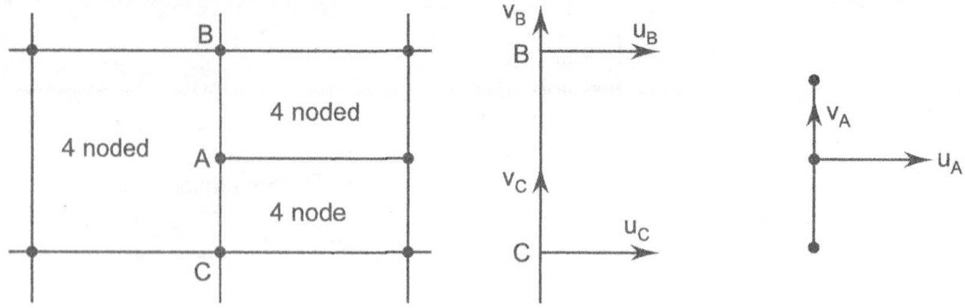

Fig. 8.25. Element transition.

$$u_A = (u_C + u_B)/2 \quad \text{...(8.55)}$$
$$v_A = (v_C + v_B)/2$$

2. 9 node – 9 node transition

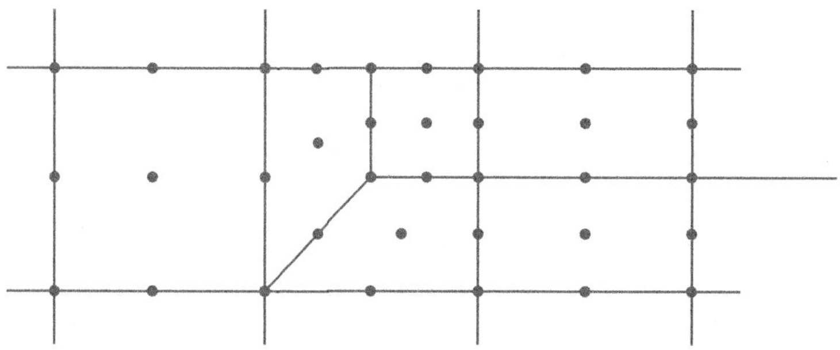

Fig. 8.26. 9 noded element.

To understand the compatibility aspect between the elements consider a pair of quadrilateral elements as shown in Fig. 8.27.

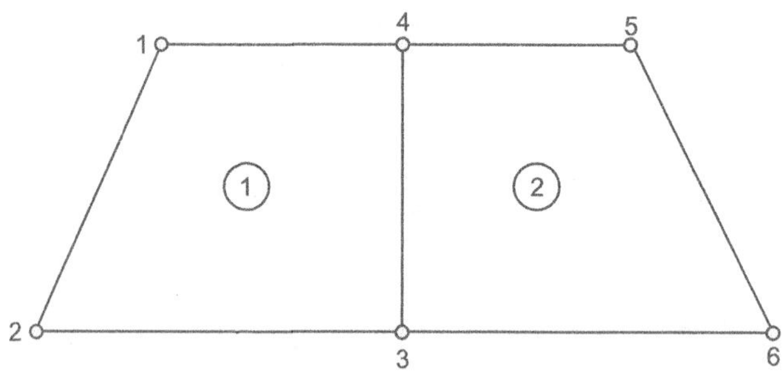

Fig. 8.27. Interelement compatibility.

The displacement approximation for the element # 1 which is a linear quadrilateral element, is written using the shape function to be

$$u = N_1u_1 + N_2u_2 + N_3u_3 + N_4u_4 \quad \text{...(8.56)}$$

where N_i for $i = 1$ to 4 are the shape function. On the element boundary 3 – 4, $N_1 = 0$ $N_2 = 0$.

Hence

$$u_{3-4}^{(1)} = N_3u_3 + N_4u_4 \quad \text{...(8.57)}$$

For the element #2, the displacement function is

$$u = N_3u_3 + N_4u_4 + N_5u_5 + N_6u_6 \quad ...(8.58)$$

Considering the displacement of the boundary 3–4 and with the incorporation of $N_5 = 0$, $N_6 = 0$ we get interface displacement to be

$$u_{3-4}^{(2)} = N_3u_3 + N_4u_4 \quad ...(8.59)$$

From the relations (8.57) and (8.59) it is evident that

$$u_{3-4}^{(1)} = u_{3-4}^{(2)} \quad ...(8.60)$$

The displacement obtained at the interface is same for the two element signifying the existence of interelement continuity. The Fig. 8.28 shows the requirement of compatibility. The criterion 1 of elemental continuity and criterion 4 of interface continuity form the requirements of compatible formulations. The elements that satisfy compatible formulations are called "conforming elements".

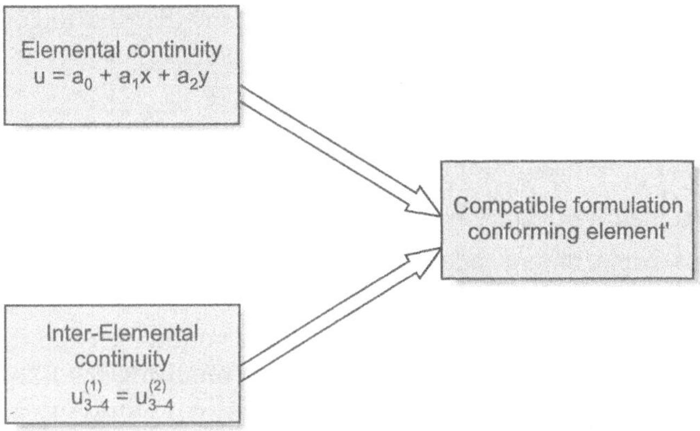

Fig. 8.28. Compatible formulation.

- **Convergence Criterion 5:**

The number of generalized co-efficients used in the displacement functions must be equal to the degree of freedom in the element.

Fig. 8.29. 1D linear element. Bar

For example consider 1D linear bar element shown in Fig. 8.29 which has two nodes with one DOF at each node. The displacement approximate function used is

$$u(x) = [1,\ x]\begin{Bmatrix} a_0 \\ a_1 \end{Bmatrix} \quad ...(8.61)$$

The generalized co-ordinates $[a_0\ a_1]^T$ are two in the function and the degrees of freedom are also two. To solve for a_i for $i = 1, 2$, it requires two displacement boundary conditions called the Dirichlet boundary conditions, *i.e.*, $u(0) = u_1$ and $u(l) = u_2$. Hence this is called the C_0 continuity problem.

Now consider a beam element with two nodes having a translational and a rotational degree of freedom at each node. Totally the Euler-Bernoulli's beam has four degrees of freedom, as shown in the Fig. 8.30.

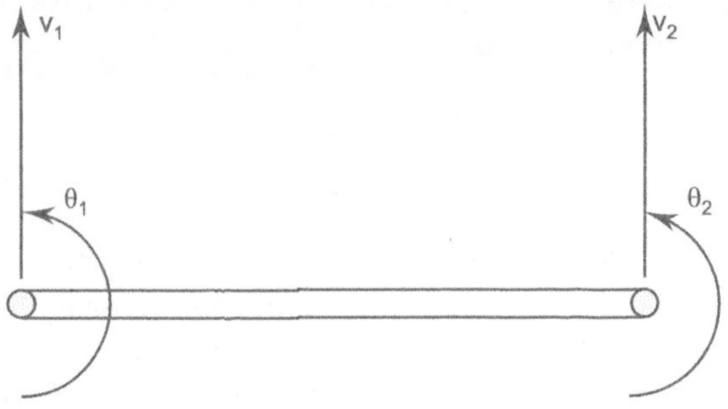

Fig. 8.30. Euler-Bernoulli's beam.

The displacement polynomial used to approximate the distribution of deformation in the element is written as

$$u(x) = a_0 + a_1x + a_2x^2 + a_3x^3 \quad \text{...(8.62)}$$

The four constants, a_i for $i = 1, 2, 3$, and 4 known as generalized co-ordinates are determined using

1. The displacement boundary conditions.
2. The slope of displacement called Neumann boundary condition. The slope is obtained by the first derivative of displacement function.

The problem requiring the displacement and slope boundary condition to solve is called the C_1 continuity problem. In general the problem requiring the nth degree derivative to obtain its generalized co-ordinates is called the 'C_n'– continuity problem.

- **Convergence Criterion 6:**

The displacement function should exhibit geometric isotropy or invarience for the convergence of FE solutions.

It the displacements are independent of the local co-ordinates system, this is known a geometric isotropy or geometric invarience. The interchange of co-ordinate axes produces no effect on the displacement patterns. This statement is irrelevant to one dimensional problems. But for two and three dimensional elements, the displacement polynomials should approximate the displacement pattern independent of local co-ordinate.

For example consider a linear triangular element with three nodes. This CST element is 2-dimensional. The displacement functions in x and y directions are

$$u(x, y) = a_0 + a_1x + a_2y \quad \text{...(8.63)}$$
$$v(x, y) = b_0 + b_1x + b_2y$$

The polynomials for u and v can be interchanged without any difference. The generalized co-ordinates depend on the boundary condition. Hence the polynomial functions exhibit geometric isotropy.

Consider for example for the quadrilateral element, the assumed functions are

$$u(x, y) = a_0 + a_1x + a_2y + a_3xy$$
$$u(x, y) = b_0 + b_1x + b_2y + b_3y^2 \quad \text{...(8.64)}$$

The polynomials for u and v functions cannot be interchanged as they are not geometrically isotropic.

8.26 ISO-, SUB-, SUPER PARAMETRIC FORMULATION

ISO-Parametric model	Sub-parametric model	Super parametric model
• Displacement order 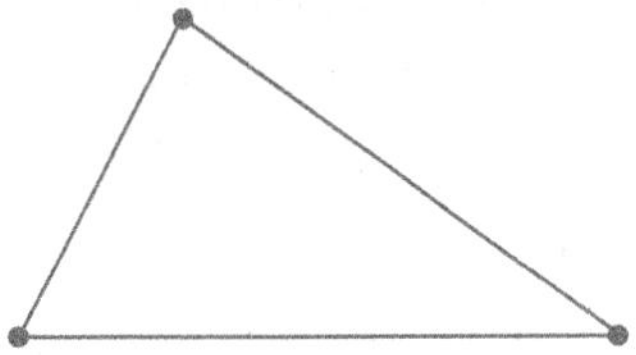 • Geometric order 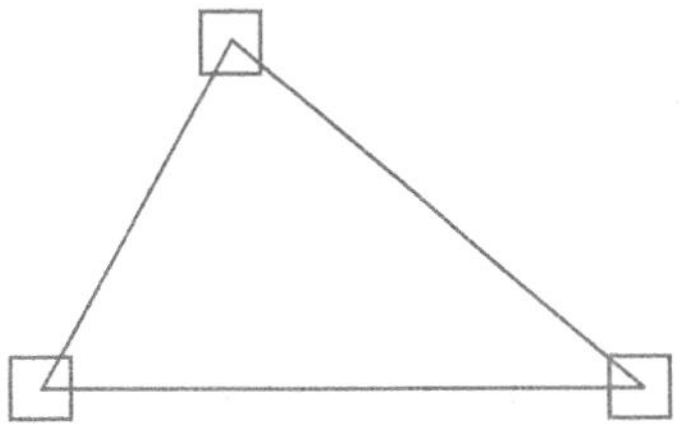	• Displacement model 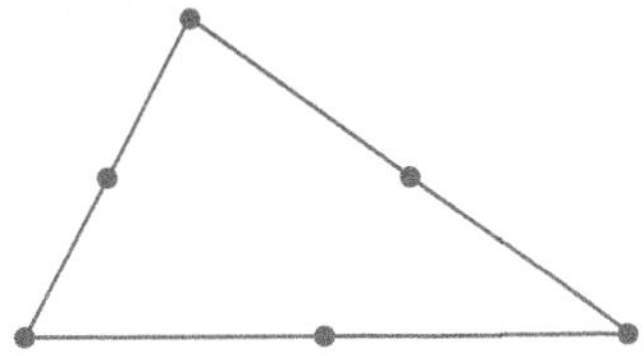 • Geometry model 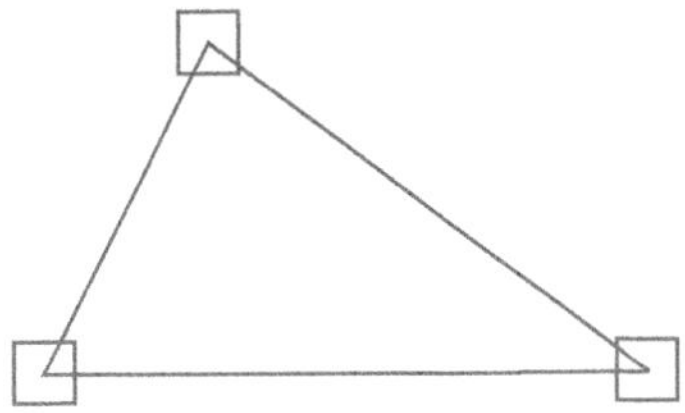	• Displacement model 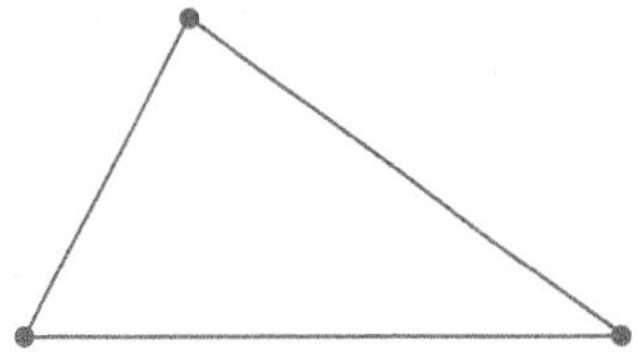 • Geometry model 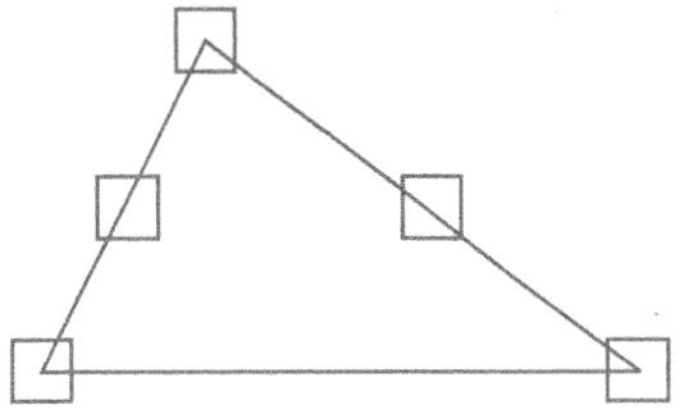
• The displacements and geometry of the element are described by the same parameters and are of the same order. • The element formulated satisfies completeness, compatibility and are isotropic in geometry. • Facilitates formulation of curved elements.	• Displacement model is of higher order. • Geometry model is of lower order. • The order of the shape functions for the displace-ment function is higher than that of the shape functions of geometry model. $u = (N)_D\{u_i\}$ $x = (N)_G\{x_i\}$ $(N)_D > (N)_G$ - order.	• Displacement model is of lower order. • Geometry model is of higher order than displacement model. • The order of the shape functions representing displace-ment is lower than that of shape functions representing geometry. order $(N)_D <$ order $(N)_G$.

8.27 PATCH TEST

This test for convergence is explained for its definition, for its purpose, for its requirements and for its procedure by answering following set of questions which are needed to understand the concept.

- **What is patch test?**

  Patch test is a simple numerical test that is performed on an element assemblage so as to check the completeness and compatibility requirements. The patch test may be regarded as the test that serves the necessary and sufficient condition for the convergence of a finite element formulation. To ensure the confirmance of all criteria of convergence the element should pass through the patch test. The elements passing the patch test ensure valid usage in the modelled structure. With the mesh refinement the approximate solution converges to an exact solution.

  "A group of similar elements used to discretize a region is called patch".

- **Why the patch test is to be performed?**
  - The displacement formulation provides upper bound solution to the true stiffness of structure. Hence a situation is reached where under the load the FE simulated structure deforms less than the actual structure.
  - But if the elements used are complete and compatible, with the mesh refinement the stresses would converge to an exact solution. This is called monotonic and upper bound convergence. The strain energy with the model having finite number of elements is greater than true system with infinite number of element. Hence it is deemed to be the upper bound solution.
  - In actual practice the interelement continuity may not exist in all the nodal point layouts. In some cases elements may be inherently incompatible like in the C_1 continuity problems.
  - Even then the solution may lead to convergence. The strain energy may be not necessarily an upper bound to the exact solution. Hence the convergence cannot be monotonic in nature.
  - The assemblage of incompatible elements in total can represent a constant strain state. The requirement of constant strain condition is not a criteria for the individual element but a group of elements. Hence completeness condition on an assemblage of elements can lead to non-monotonic convergence.

  "Hence the purpose of patch test is to investigate the requirement of completeness condition of an assemblage to produce the convergent solution. This provides a test of acceptability of non-conforming elements under the violation of the condition of elasticity".

- **What are the requirements of Patch Test?**
  1. There must be at least one internal node in the patch under consideration.
  2. The load produces constant stress in the patch.
  3. The patch should be subjected to uniform displacement to ensure constant strain in the system.
  4. The internal node or nodes should not be loaded.
  5. The internal node/s should not have restrained degree of freedom.
  6. The rigid body motion of the patch should be prevented by suitable support boundary conditions.
  7. Uniformly distributed loads should be converted to consistent loads or work equivalent loads.

Acceptable Patches

Two internal nodes

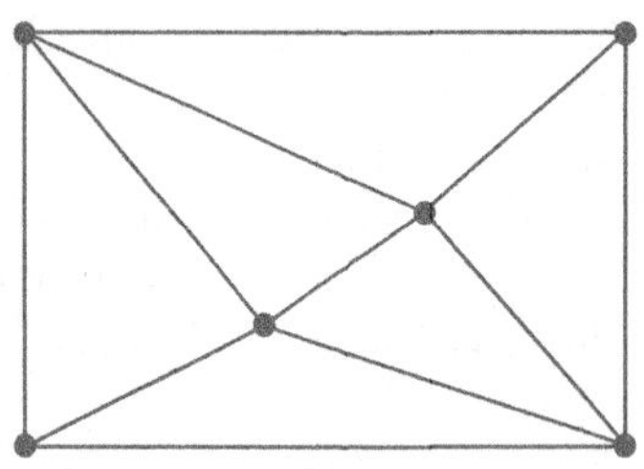

Unacceptable Patches

No internal nodes

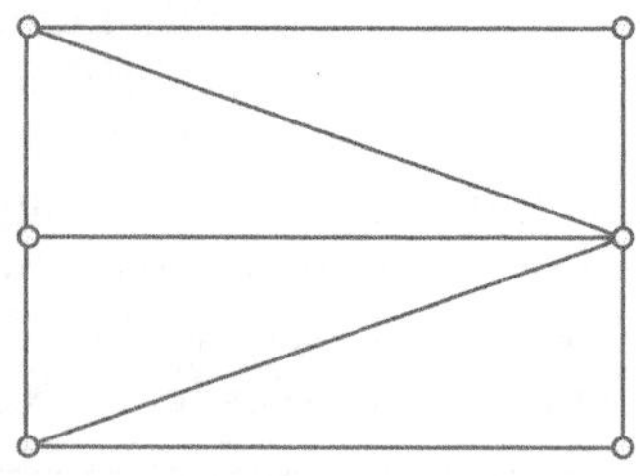

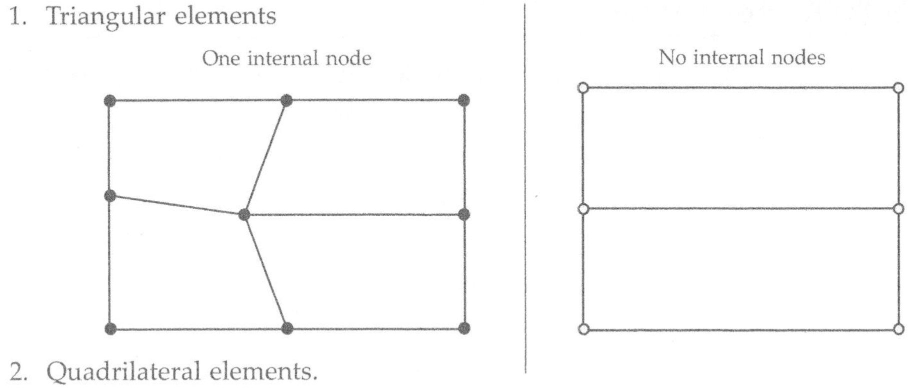

Fig. 8.31. Acceptable and unacceptable patches.

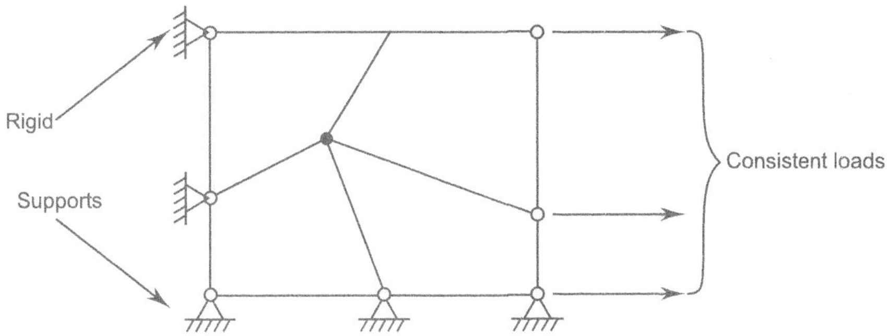

Fig. 8.32. Loads and supports.

- **How is the patch test done?**

1. The modelled patch is subjected to standard finite element analysis. It is loaded with the consistent load. The computed stress and strains are compared with the exact solution, neglecting round of error occuring in the digital computer.
2. The patch test is conducted with the model with finer element. Passing of patch test is required for the mesh refinement in the model to ensure convergence.
3. The patch test should conform to more than one patch configuration, geometry and mesh pattern.
4. Due to non-monotonic convergence patch test does not reveal the rate of convergence. It gives only the degree of convergence. Hence the patch test only gives sufficient condition of convergence.
5. The patch test checks requirements following criteria of convergence.
    - Rigid body modes
    - Constant strain state
    - Interelement compatibility.

Rigid body mode is the special case of constant strain where the value of strain is zero. With fine meshes the constant strain condition and interelement conditions are preserved. The completeness criteria is satisfied by patch test when the assemblage contains large sized elements.

POINTS TO BE REMEMBER

- The three noded bar element uses quadratic polynomial for the displacement function.
- The stress and strain in an quadratic element vary linearly.
- The quadratic bar element has three shape functions that have the degree 2.
- Four noded bar element assumes the cubic displacement function.
- The stress and strain in a cubic elements vary quadratically.
- A cubic bar element has four shape functions.
- The quadratic and cubic elements are called higher order elements.
- The higher order elements give more accurate result than linear elements.
- The factors that contribute to stiffness of linear and higher order elements are same *i.e.*, Young's modulus, area of cross-sections and the length.
- The stiffness matrix (elemental) of quadratic bar element is a 3 × 3 symmetric square matrix.
- Lagrange's Polynomial directly provides a way for obtaining the shape function for linear and higher order elements in both Cartesian and natural co-ordinates.
- The linear 1D, 2D and 3D elements are called simplex elements.
- Higher order 1D, 2D and 3D elements are called complex elements.
- The higher order elements with the edges parallel to local co-ordinate axes are called the multiplex elements.
- Pascal's triangle acts as a tool for assuming displacement polynomial for various elements in FEM (1D and 2D elements).
- The Pascal's Pyramid is a tool for suggesting displacement polynomial for 3D elements.
- The mesh refinement based on the size of the element is called *h*-refinement.
- The convergence achieved through higher order displacement polynomial is called *p*-refinement.
- For the convergence of solution to FEM problem requires the following criteria.
  1. continuity of displacement function.
  2. representation of rigid body mode.
  3. state of constant strain.
  4. inter-element compatibility.
  5. The generalized co-efficients equal to degree of freedom of the element.
  6. Geometric isotropy of displacement function.
- Patch test is a numerical test used to check the completeness and compatibility of mesh assembly.
- This provides a test of acceptability of non-conforming elements.
- The requirements of patch test are 1. rigid body modes 2. constant strain state 3. Interelement compatibility.

EXERCISE

1. Derive the shape functions in cartesian system for three noded 1D bar element.
2. Starting with the shape functions derived in (8.1) obtain the [B] matrix and the elemental stiffness matrix.
3. Obtain for the three noded bar element (1D) the shape functions in natural co-ordinate system.
4. With the shape functions obtained in (8.3) derive the strain-displacement [B] matrix and the elemental stiffness matrix.
5. Write using Lagrange's polynomial, the shape functions in cartesian system for:
   (*a*) Two noded 1D bar element
   (*b*) Three noded 1D bar element
   (*c*) Four noded 1D bar element
6. With the help of Lagrange's polynomial obtain the expressions for shape functions in natural co-ordinate system for the following:
   (*a*) 1D quadratic bar element
   (*b*) 1D cubic bar element
   (*c*) 3 noded right angled triangular element
   (*d*) Four noded quadrilateral element
   (*e*) Six-noded triangular (right) element
   (*f*) 8-noded rectangular element.
7. What is a simplex element? What are the features of simplex element?
8. What is a complex element? Give examples.
9. What are the features of complex element?
10. What is a multiplex element? Give examples.
11. What are the features of multiplex elements?
12. Write displacement function using Pascal triangle for 2D linear, quadratic and cubic triangular elements.
13. Using Pascal's triangle write the expressions for the displacement functions for linear and quadratic quad element.
14. What is Pascals pyramid? Give examples for its use.
15. What are the observations to be made so as to decide the order of convergence in FEA solutions.
16. What is mesh refinement? Explain types of refinement methods.
17. Explain the convergence with respect to strain energy.
18. State the convergence criteria in FEM.
19. Explain the importance of continuity of displacement function in convergence.
20. Explain the importance of rigid body modes of displacement in the convergence concept.
21. Explain the importance of constant strain state in displacement function for convergence.
22. What is completeness? Explain with example.

23. Explain the importance of compatibility requirement in solution convergence.
24. What is C° continuity? Explain with an example.
25. What is geometric isotropy? Explain with an example.
26. What is patch test? Why the patch test is to be performed?
27. What are the requirements of patch test?
28. Give examples for acceptable and unacceptable patches.
29. How is the patch test done in FEM?
30. Explain the concept of Iso-, Sub-, and super parametric element formulation in FEM.

Chapter 9

HEAT TRANSFER AND FINITE ELEMENT METHOD

अन्नदानं परं दानं।
विद्यादानं ततः परम्॥
अन्नेन क्षणिका तृप्तिः।
थावज्जीवंच विद्यया॥

[Offering free food to others is great. Giving freely the education is greater than that, because the food gives satisfaction for a moment. But education satisfies for the life time.]

9.1 INTRODUCTION

Heat transfer is one of the important processes studied by mechanical engineers. The principles of finite element method can also be applied to heat transfer and fluid flow problems effectively. Since the governing equation describing the heat flow is differential equation the variational principles and weighted residual methods can be applied to obtain the temperature distribution. FEM can be successfully employed to obtain the temperature distribution in composite wall, extended surfaces like fins. Heat transfer through composite furnace wall, reactor wall, steam pipes are some of the examples which can be analysed using the procedures of FEM.

This chapter deals with the following issues in brief.

- The introduction to modes of heat transfer
- Fourier's Law of conduction
- 1 D steady state heat conduction

- Solution of composite wall problem using FE methodology
- Formulation of FE equation through fins.
- Solution of 1 D fin problem using FE procedures.

Detailed study on heat transfer is not within the scope of this book.

9.2 MODES OF HEAT TRANSFER

Heat energy moves from a hotter body to a colder object by the process of heat transfer. There are three modes of heat transfer based on the media through which it occurs. The three modes of heat transfer are:

- *Conduction*–When the two solid bodies at different temperature are in contact the heat transfer takes place by conduction.
- *Convection*–Heat is carried by moving gas or liquid.
- *Radiation*–If the flow of heat does not require the intervening medium the mode is called radiation.

9.2.1 Conduction Heat Transfer

Conduction heat transfer is the process whereby the heat is transferred directly through a material with bulk motion of the material playing no role in the heat transfer. The materials that conduct heat well are called the thermal conductors, while those that conduct heat poorly are known as thermal insulators.

Most metals are good thermal conductors while wood, glass and most plastics are thermal insulators. The free electrons in the metals are responsible for excellent thermal conductivity for the metals.

9.2.2 Convection

Convection is the process in which heat is carried from place to place by the bulk movement of a fluid. When a pan of water is heated convection currents are setup. The convection is classified into free convection and forced convection.

9.2.3 Radiation

Radiation is the process in which the energy is transferred by means of electromagnetic waves. Heat transfer by radiation can take place through vacuum. Electromagnetic waves travel without medium, but can travel in an empty space some of the examples for radiation are:

1. Energy carried by electromagnetic waves.
2. Light, micro waves, radiowaves, X-rays. In radiation the wavelength is related to electromagnetic frequency of oscillation.

9.3 MORE ON HEAT CONDUCTION

Consider a metallic rod of uniform cross-sectional area A and length equal to L as shown in Fig. 9.1 with one end at a temperature T_1 and other end T_2.

Assuming the rod to be isotropic the thermal conductivity is K. The amount of heat transferred is given by

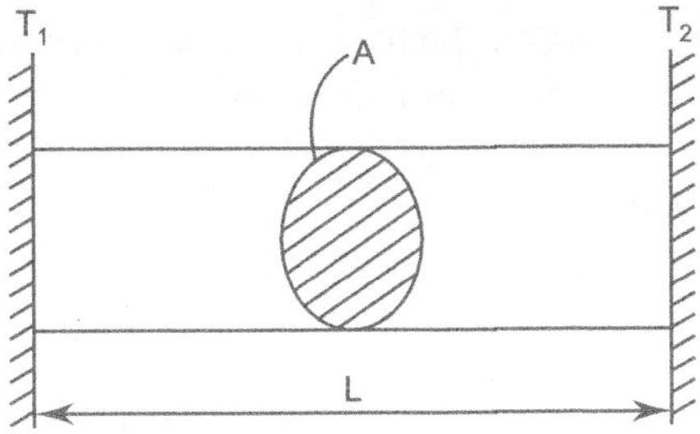

Fig. 9.1. Conduction.

$$Q = kA\left(\frac{\Delta T}{L}\right)t \quad ...(9.1)$$

The equation (9.1) composes of

k = Thermal conductivity

Q = Heat transfer

A is the cross-sectional area

L is the length between the ends

t is the duration of heat transfer.

ΔT is the difference in temperature between two ends.

9.4 FOURIER LAW OF HEAT CONDUCTION

Fourier law states that the heat transfer rate is directly proportional to temperature gradient and the constant of proportionality is called the thermal conductivity

$$\{\text{Heat Transferred}\} \propto \frac{(\text{Area})\{\text{Temperature difference}\}}{\{\text{Distance}\}} \quad ...(9.2)$$

The rate of heat transfer can be written using as (9.2)

$$\mathring{Q} = -kA\frac{dT}{dx} \quad ...(9.3)$$

The minus (–) sign in equation (9.3) denotes the direction of heat flow. Heat transfer rate is a vector parameter directed from high temperature to low temperature.

i.e.,

$$\mathring{Q} = -kA\,\nabla T = -k\left(\frac{\partial T}{\partial x}i + \frac{\partial T}{\partial y}j + \frac{\partial T}{\partial \zeta}k\right) \quad ...(9.4)$$

The direction of heat transfer is perpendicular to the isothermal surfaces. In general thermal conductivity depends on the direction. If the material is isotropic the thermal conductivity is independent of direction and is a constant, otherwise it is second order tensor k_{ij}.

9.4.1 Heat Conduction Equation

In general the heat conduction equation in Cartesian system can be written as

$$\rho C_p\frac{\partial T}{\partial t} = \frac{\partial}{\partial x}\left(k\frac{\partial T}{\partial x}\right) + \frac{\partial}{\partial y}\left(k\frac{\partial T}{\partial y}\right) + \frac{\partial}{\partial \zeta}\left(k\frac{\partial T}{\partial \zeta}\right) \quad ...(9.5)$$

The equation (9.5) describes a transient heat conduction process. The steady state conduction can be obtained by equating LHS of equation (9.5) to zero.

9.4.2 Initial and Boundary Condition

The initial condition describes the temperature distribution when the time, $t = 0$.

i.e., $$T(x, y, \zeta, t = 0) = T_0 (x, y, \zeta) \quad ...(9.6)$$

The boundary conditions are assigned from physical observation of the domain boundary.

- *BC of the first kind:*

  This denotes the known temperature of the boundary of the system. *e.g.,* phase change at interface, temperature controlled surface.

  $$T(x = 0, t) = T_0 \quad ...(9.7)$$

- *BC of the second kind:*

  The known heat flux at the boundary of the system *e.g.,* insulated surface.

  For an insulated surface,

  $$\left.\frac{\partial T}{\partial x}\right|_{x=0} = 0 \quad \text{implies insulated boundary} \quad ...(9.8)$$

9.5 1D STEADY HEAT CONDUCTION

In equation (9.5) for a steady case heat process

$$\frac{\partial}{\partial t} = 0 \quad ...(9.9)$$

1D conduction implies

$$\frac{\partial}{\partial y} = 0 \qquad \frac{\partial}{\partial \zeta} = 0 \quad ...(9.10)$$

Substituting equations (9.9) and (9.10) in equation (9.5) the governing equation is

$$\frac{d}{dx}\left(k \frac{dT}{dx}\right) = 0 \quad ...(9.11)$$

For an isotropic material thermal conductivity is constant, hence the equation (9.11) reduces to

$$\frac{d^2 T}{dx^2} = 0 \quad ...(9.12)$$

9.5.1 FEM Formulation : Heat Conduction

1D steady conduction through isotropic material is depicted in Fig. (9.2)

On integration of equation (9.12)

$$T(x) = C_1 x + C_2 \quad ...(9.13)$$

The Boundary conditions are

at $x = 0$ $\quad T(x) = T_1$

at $x = L$ $\quad T(x) = T_2$

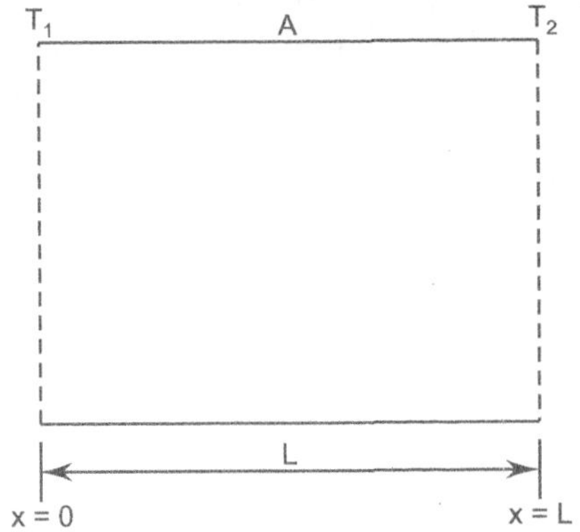

Fig. 9.2. Plane wall.

Substituting these BC's in equation (9.13)

$$C_2 = T_1 \qquad C_1 = \frac{(T_2 - T_1)}{L}$$

Hence

$$T(x) = \left(\frac{T_2 - T_1}{L}\right) x + T_1 \qquad \text{...(9.14)}$$

Manipulating equation (9.14)

$$T(x) = \left(\frac{1-x}{L}\right) T_1 + \frac{x}{L} T_2 \qquad \text{...(9.15)}$$

In the matrix form of the equation (9.15)

$$T(x) = \left[1 - \frac{x}{L} \quad \frac{x}{L}\right] \begin{Bmatrix} T_1 \\ T_2 \end{Bmatrix} \qquad \text{...(9.16)}$$

$$= [N_i] \{T\}$$

where $N_1 = 1 - x/L$ and $N_2 = x/L$ are shape functions.

The distribution of shape functions are linear.

From the Fourier's law of heat conduction

$$q = -kA \frac{dT}{dx} \qquad \text{...(9.17)}$$

Form equations (9.16) and (9.17)

$$q = -kA \frac{d}{dx} \lfloor N_i \rfloor \{T\}$$

where

$$\frac{d}{dx}[N_i] = [B] = [-1/L, 1/L] \qquad \text{...(9.18)}$$

For a linear element

$$[K]^e = \int_0^L [B]^T kA [B] \, dx \qquad \text{...(9.19)}$$

$$= \int \frac{1}{L} \begin{bmatrix} -1 \\ 1 \end{bmatrix} \frac{kA}{L} [-1 \quad 1] \, dx$$

$$= \frac{kA}{L}\begin{bmatrix} 1 & -1 \\ -1 & 1 \end{bmatrix} \quad ...(9.20)$$

Hence $$[K]^e \{\phi\} = Q \quad ...(9.21)$$

where $[K]^e$ is analogus to stiffness matrix for the linear bar element.

Also from equation (9.14) and (9.17), the rate of heat transfer

$$q = \frac{-kA\,(T_2 - T_1)}{L}$$

or $$= \frac{kA\,(T_1 - T_2)}{L} = \frac{(T_1 - T_2)}{L/kA} \quad ...(9.22)$$

L/k A is the thermal resistance, by electrical analogy that $i = \dfrac{V_1 - V_2}{R}$.

9.5.2 Heat Transfer Across Composite Wall

Heat transfer through three layered composite wall as shown in Fig. 9.3 is governed by the equation

$$\underbrace{kA\frac{d^2\phi}{dx^2}}_{\text{Heat Conduction}} + \underbrace{q = 0}_{\text{Heat Supply}} \quad ...(9.23)$$

The following steps to be followed to solve the equation (9.23)

1. Governing equation

$$D\frac{d^2\phi}{dx^2} + Q = 0 \quad ...(9.24)$$

$$BC's\ \phi(0) = \phi_0,\ \phi(H) = \phi_H$$

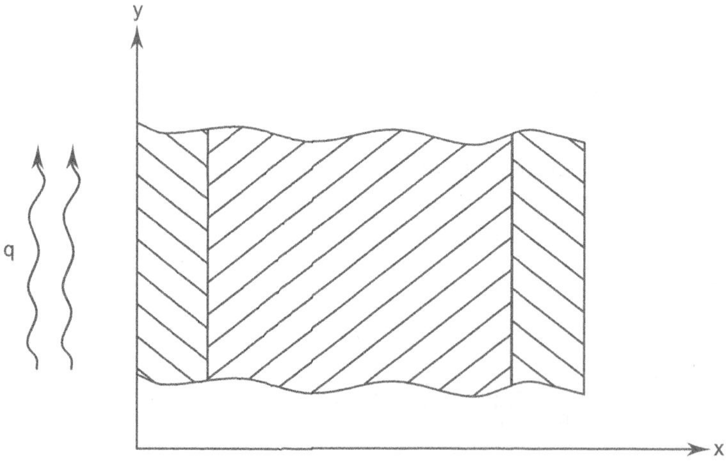

Fig. 9.3. Composite wall.

2. Define elements and nodes

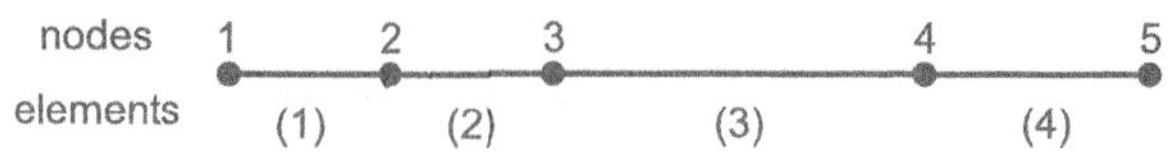

3. Weight and integrate system of equations

$$\int_0^H W(x)\left[D\frac{d^2\phi}{dx^2}+Q\right]dx = 0 \qquad ...(9.25)$$

4. Solve for nodal values $\Rightarrow \phi_2, \phi_3, \phi_4$.

The conductance matrix for a wall in touch with hot gases on one side is given by

$$[K]^e = \frac{kA}{L}\begin{bmatrix} 1 & -1 \\ -1 & 1 \end{bmatrix} + [K]_h \qquad ...(9.26)$$

Example 9.1. *Calculate the temperature distribution across the two layered wall shown in Fig. 9.4 using finite element formulation.*

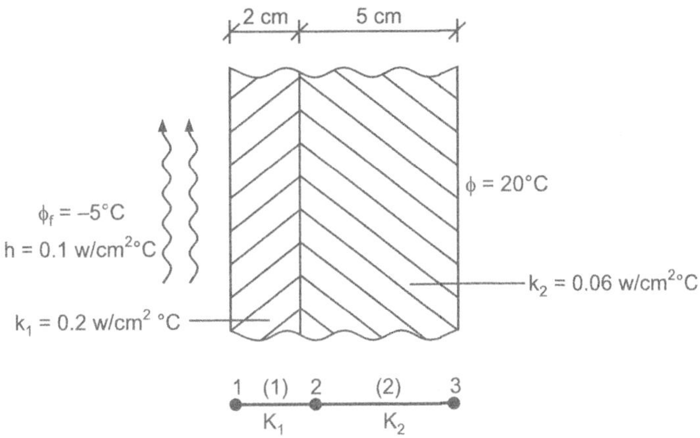

Fig. 9.4.

Solution. The composite wall with two layers of different materials is modelled with two linear elements with three nodes as shown in Fig. 9.3. Assuming unit area (1 cm^2) perpendicular to the direction of heat flow the solution is obtained as follows.

For element (1)

$$\frac{K_1A}{L_1} = \frac{0.2(1)}{2} = 0.1 \text{ W/°C}$$

$$hA = 0.1(1) = 0.1 \text{ W/°C}$$

$$hAQ_f = 0.1(1)(-5) = -0.5 \text{ W}$$

Elemental matrix

$$[K]_1^e = \frac{k_1A}{L_1}\begin{bmatrix} 1 & -1 \\ -1 & 1 \end{bmatrix} + \begin{bmatrix} hA & 0 \\ 0 & 0 \end{bmatrix}$$

$$= \begin{bmatrix} 0.2 & -0.1 \\ -0.1 & 0.1 \end{bmatrix}$$

$$[f]_1^e = \begin{Bmatrix} hAQ_f \\ 0 \end{Bmatrix} = \begin{Bmatrix} -0.5 \\ 0 \end{Bmatrix}$$

For element (2)

$$\frac{k_2 A}{L_2} = \frac{0.06(1)}{5} = 0.012 \text{ W/°C}$$

$$[K]_2^e = \frac{k_2 A}{L_2}\begin{bmatrix} 1 & -1 \\ -1 & 1 \end{bmatrix} = \begin{bmatrix} 0.012 & -0.012 \\ -0.012 & 0.012 \end{bmatrix}$$

Upon assembly of $[K]^e$ matrices for the two elements

$$\begin{bmatrix} 0.2 & -0.1 & 0 \\ -0.1 & 0.112 & -0.012 \\ 0 & -0.012 & 0.012 \end{bmatrix}\begin{Bmatrix} \phi_1 \\ \phi_2 \\ \phi_3 \end{Bmatrix} = \begin{Bmatrix} -0.5 \\ 0 \\ 0 \end{Bmatrix} + \begin{Bmatrix} 0 \\ 0 \\ Q^* \end{Bmatrix}$$

The data that is provided is $\phi_3 = 20°C$

The equation gets reduced to

$$\begin{bmatrix} 0.2 & -0.1 \\ -0.1 & 0.112 \end{bmatrix}\begin{Bmatrix} \phi_1 \\ \phi_2 \end{Bmatrix} = \begin{Bmatrix} -0.5 \\ 0.24 \end{Bmatrix} \longrightarrow (= 20 \times 0.012)$$

Solving the system of equation for the variables ϕ_1 and ϕ_2.

$$\begin{Bmatrix} \phi_1 \\ \phi_2 \\ \phi_3 \end{Bmatrix} = \begin{Bmatrix} -2.5806 \\ -0.1613 \\ 20 \end{Bmatrix}$$

Example 9.2. *Three thin film layers as shown in Fig. 9.5 heat the plasma to 150°C. Calculate the temperature distribution at the interface of the layers.*

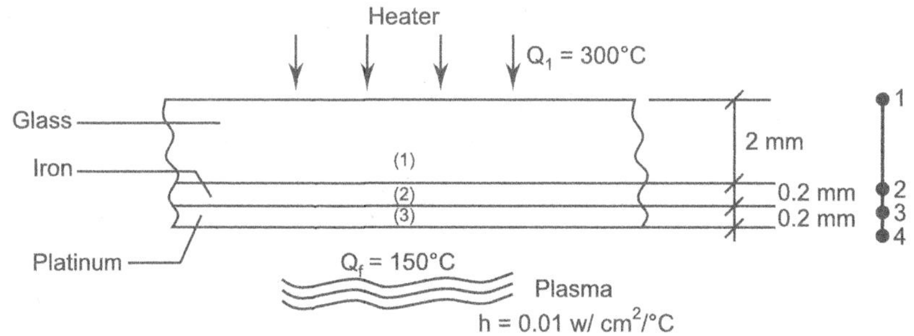

Fig. 9.5. Thin film layers.

Solution. Thermal conductivities

$$K_1 = 0.1 \text{ w/cm}^2\text{°C}$$
$$K_2 = 0.5 \text{ w/cm}^2\text{°C}$$
$$K_3 = 0.4 \text{ w/cm}^2\text{°C}$$

The layers are modelled with three linear elements having four nodes. The temperature of the first node adjucent to the heater is 300°C and the plasma is to be 150°C in temperature. Unit area is assumed for heat transfer.

For element (1)

$$\frac{K_1A}{L_1} = \frac{0.1(1)}{0.2} = 0.5 \text{ W/°C}$$

$$[K]_1^e = \frac{K_1A}{L_1}\begin{bmatrix} 1 & -1 \\ -1 & 1 \end{bmatrix} = \begin{bmatrix} 0.5 & -0.5 \\ -0.5 & 0.5 \end{bmatrix}$$

For element (2)

$$\frac{K_2A}{L_2} = \frac{0.5(1)}{0.02} = 25 \text{ W/°C}$$

$$[K]_2^e = \frac{k_2A}{L_2}\begin{bmatrix} 1 & -1 \\ -1 & 1 \end{bmatrix} = \begin{bmatrix} 25 & -25 \\ -25 & 25 \end{bmatrix}$$

For element (3)

$$\frac{k_3A}{L_3} = \frac{0.4(1)}{0.02} = 20 \text{ W/°C}$$

$$h_fA = 0.01(1) = 0.01 \text{ W/°C}$$

$$h_fAQ_f = 0.01(1)\,(150) = 1.5 \text{ W}$$

$$[K]_3^e = \frac{K_3A}{L_3}\begin{bmatrix} 1 & -1 \\ -1 & 1 \end{bmatrix} + \begin{bmatrix} 0 & 0 \\ 0 & hA \end{bmatrix}$$

$$= \begin{bmatrix} 20 & -20 \\ -20 & 20.01 \end{bmatrix}$$

On assembly of element matrices for three layers

$$\begin{bmatrix} 0.5 & -0.5 & 0 & 0 \\ -0.5 & 25.5 & -25 & 0 \\ 0 & -25 & 45 & -20 \\ 0 & 0 & -20 & 20.01 \end{bmatrix}\begin{Bmatrix} \phi_1 \\ \phi_2 \\ \phi_3 \\ \phi_4 \end{Bmatrix} = \begin{Bmatrix} Q^* \\ 0 \\ 0 \\ 1.5 \end{Bmatrix}$$

$\phi_1 = 300°\text{C}$ (Heater temperature)

$$\begin{bmatrix} 25.5 & -25 & 0 \\ -25 & 45 & -20 \\ 0 & -20 & 20.01 \end{bmatrix}\begin{Bmatrix} \phi_2 \\ \phi_3 \\ \phi_4 \end{Bmatrix} = \begin{Bmatrix} 0.5 \times 300 \\ 0 \\ 1.5 \end{Bmatrix}$$

Hence,

$$\begin{Bmatrix} \phi_1 \\ \phi_2 \\ \phi_3 \end{Bmatrix} = \begin{Bmatrix} 297.1 \\ 297 \\ 296.9 \end{Bmatrix}$$

$$Q^* = 0.5(300) - 0.5(297.1) = 1.45 \text{ w/cm}^2$$

9.6 FIN ANALYSIS

Fin is the extended surface on a hot body, that transfers heat to effect the cooling in less time. Because the extended surfaces increase the area of heat transfer the cooling is faster. In this section the issues of main concern are establishing the governing equations and solving them for the temperature distribution by using FEM procedures. The scope of study is limited to 1D steady state problems only.

9.6.1 Governing Equation

The assumptions are

- 1D linear problem
- Steady state analysis
- The cross-sectional area is constant.

The heat balance equation is

$$q_x + dq_{conv} = q_{x + dx} \qquad \text{...(9.27)}$$

where

$$q_x = -kA\frac{dT}{dx} \qquad \text{...(9.28)}$$

$$dq_{conv} = hp_{dx}(T - T_\infty) \qquad \text{...(9.29)}$$

p is the perimeter

$$q_{x+dx} = q_x + \frac{dq_x}{dx}\partial x \qquad \text{...(9.30)}$$

The steady (time independent) equation is

$$\frac{d^2T}{dx^2} - \frac{hP}{KA_s}(T - T_\infty) = 0 \qquad \text{...(9.31)}$$

The equation (9.31) is obtained by manipulating the equations from (9.27) to (9.30).

9.6.2 Fin Performance

Fin resistance:

$$R_{tf} = \frac{T_b - T_\infty}{q_f} \qquad \text{...(9.32)}$$

Fin effectiveness:

$$\in_f = \frac{q_f\ (\text{with fin})}{q_f\ (\text{without fin})}$$

$$= \frac{q_f}{hA_{sb}(T_b - T_\infty)} \qquad \text{...(9.33)}$$

Fin efficiency:

$$\eta_f = \frac{q_f}{q_{max}}$$

$$= \frac{q_f}{hA_f(T_b - T_\infty)} \qquad \text{...(9.34)}$$

9.7 1-D FIN: FEM FORMULATION

The governing heat transfer equation for 1D fin in steady state is

$$\underbrace{KA\frac{d^2\phi}{dx^2}}_{\text{Heat Conduction}} \underbrace{- hP\phi + hPQ_f}_{\text{Heat Convection}} + \underbrace{q}_{\text{Heat Source}} = 0 \qquad ...(9.35)$$

where
- k = thermal conductivity
- h = convection coefficient
- A = cross-sectional Area of fin
- P = perimeter of the fin
- ϕ = temperature distribution
- ϕ_f = ambient temperature

Specified BC's are

$$\phi(0) = \phi_0$$

Convective heat loss at the free end

$$-kA\frac{d\phi}{dx} = hA(Q_b - Q_f) \qquad ...(9.36)$$

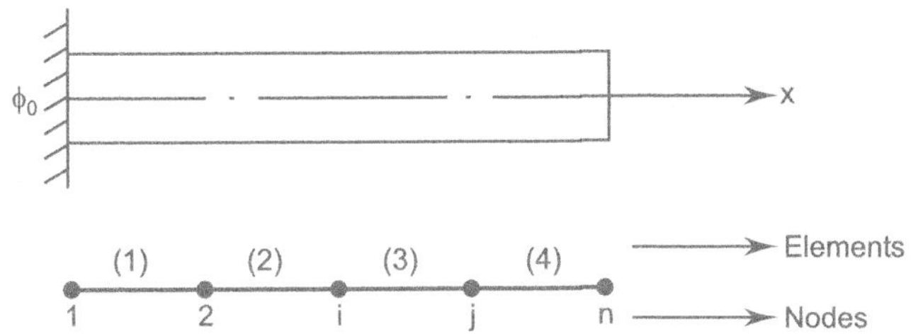

Fig. 9.6. One dimensional fin.

With the application of Galerkin's approach to the governing equation (9.35) the residual obtained is

$$R^e = -\int_{x_i}^{x_j} N^T\left(D\frac{d^2\phi}{dx^2} - g\phi + Q\right)dx \qquad ...(9.37)$$

$$= -\int_{x_i}^{x_j} N^T\left(D\frac{d^2\phi}{dx^2} + Q\right) + \int_{x_i}^{x_j} gN^T\phi\, dx \qquad ...(9.38)$$

In equation (9.37) the following notations may be assumed

$$D = kA; \quad g = hP; \quad Q = q + hPQ_f$$

Intigration by parts applied to the first term on RHS of the equation (9.38) yields the following

$$R^e = -N^T\frac{d\phi}{dx}\Big[_{x_i}^{x_j} + \int_{x_i}^{x_j}\frac{dN^T}{dx}.D\frac{d^2\phi}{dx^2}dx - \int_{x_i}^{x_j} QN^T dx$$

$$+ \int_{x_i}^{x_j} gN^T\phi\, dx \qquad ...(9.39)$$

$$\phi(x) = N(x)\Phi^{(e)}$$

Hence

$$R^e = -N^T D \frac{d\phi}{dx}\Big|_{x_i}^{x_j} \longrightarrow b^{(e)}$$

$$+ \left(\int_{x_i}^{x_j} \frac{dN^T}{dx} D \frac{dN}{dx}\, dx \right) \Phi^e \longrightarrow K_D^{(e)}$$

$$- \int_{x_i}^{x_j} QN^T dx \longrightarrow f_Q^{(e)}$$

$$+ \int_{x_i}^{x_j} (gN^T N\, dx)\Phi^e \longrightarrow kg^{(e)} \qquad \text{...(9.40)}$$

The equation (9.40) can be written in short as

$$R^{(e)} = b^{(e)} + [K_D^{(e)} + K_g^{(e)}]\, \Phi^e - f_Q^{(e)} \qquad (9.41)$$

The terms in equation (9.40) can be described to be

$$[K_D^e] = \int_{x_i}^{x_j} \frac{dN^T}{dx}, D \frac{dN}{dx}\, dx = \int$$

$$= \int_{x_i}^{x_j} B^T DB\, dx = \text{Thermal resistance}$$

$$[K_g^e] = \int_{x_i}^{x_j} gN^T N\, dx = \text{Thermal convection}$$

$$\{f_Q^e\} = \int_{x_i}^{x_j} QN^T dx = \text{External heat supplied}$$

$$b^e = -N^T D \frac{d\phi}{dx} = \text{Temperature gradient at the two ends of the element}$$

9.7.1 Thermal Resistance

Considering the linear element, the interpolation functions are

$$N(x) = [N_i, N_j] = \left[\frac{x_j - x}{l}, \frac{x - x_i}{l} \right] \qquad \text{...(9.42)}$$

$$B = \frac{dN}{dx} = \left[-\frac{1}{l}, \frac{1}{l}\right] \qquad \text{...(9.43)}$$

$$[K_D^e] = \int B^T D B \, dx$$

$$= \int_{x_i}^{x_j} \begin{bmatrix} -1/l \\ 1/l \end{bmatrix} D [-1/l \quad 1/l] \, dx$$

$$= \frac{kA}{l} \begin{bmatrix} 1 & -1 \\ -1 & 1 \end{bmatrix} \qquad \text{...(9.44)}$$

9.7.2 Thermal Convection

$$K_g^e = \int_{x_i}^{x_j} g \begin{bmatrix} (x_j - x)/l \\ (x - x_i)/l \end{bmatrix} \left[\frac{x_j - x}{l}, \frac{x - x_i}{l}\right] dx$$

$$= \frac{hpl}{6} \begin{bmatrix} 2 & 1 \\ 1 & 2 \end{bmatrix} \qquad \text{...(9.45)}$$

9.7.3 External Heat Supplied

$$f_Q^e = \int_{x_i}^{x_j} Q \begin{bmatrix} (x_j - x)/l \\ (x - x_i)/l \end{bmatrix} dx$$

$$= \frac{Ql}{2} \begin{bmatrix} 1 \\ 1 \end{bmatrix} = (q + hpQ_f)\, l/2 \begin{Bmatrix} 1 \\ 1 \end{Bmatrix} \qquad \text{...(9.46)}$$

9.7.4 Temperature Gradient

$$b^e = -N^T D \frac{d\phi}{dx} \Big|_{x_i}^{x_j}$$

$$= \begin{Bmatrix} kA \dfrac{d\phi}{dx}\Big|_{x = xi} \\ -kA \dfrac{d\phi}{dx}\Big|_{x = j} \end{Bmatrix}$$

$$= \underbrace{\begin{Bmatrix} kA \dfrac{d\phi}{dx}\Big|_{x = x_i} \\ 0 \end{Bmatrix}}_{b_R} + \underbrace{\begin{Bmatrix} 0 \\ -kA \dfrac{d\phi}{dx}\Big|_{x = xj} \end{Bmatrix}}_{b_L} \qquad \text{...(9.47)}$$

$$\{b^e\} = \{b_R^e\} + \{b_L^e\}$$

When there is heat convection at the boundary

$$\{b_R^{(e)}\} = \begin{Bmatrix} 0 \\ hA(\phi_b - \phi_f) \end{Bmatrix} = \begin{Bmatrix} 0 \\ hA\phi_j \end{Bmatrix} = -\begin{Bmatrix} 0 \\ hA\phi_f \end{Bmatrix}$$

ϕ_b is the temperature of the boundary point, *i.e.*, $\phi_b = \phi_j$

$$\text{Therefore} \quad \{b_R^e\} = \underbrace{\begin{bmatrix} 0 & 0 \\ 0 & hA \end{bmatrix}}_{K_m^{(e)}} \begin{Bmatrix} \phi_i \\ \phi_j \end{Bmatrix} - \underbrace{\begin{Bmatrix} 0 \\ hA\phi_f \end{Bmatrix}}_{\{f_s^{(e)}\}} \quad \text{...(9.48)}$$

$$\text{Hence} \quad \{b_R^{(e)}\} = \left[K_m^{(e)}\right]\Phi^e - \left\{f_s^{(e)}\right\} \quad \text{...(9.49)}$$

The residual is given by

$$R^e = \left[K_D^e + K_g^e + K_m^e\right]\Phi^e - \{f_Q^e + f_s^e\}$$

$$= [K^{(e)}\}\,\{\phi^{(e)}] - f^{(e)} \quad \text{...(9.50)}$$

Residuals are assembled for all the elements and equated to zero.

$$[K]\{\phi^e\} = F \quad \text{...(9.51)}$$

This resembles the force displacement relation similar to static analysis in mechanics. After assembling the elemental matrices the equation (9.51) can be expanded to the matrix form in the following way

$$\begin{bmatrix} K_{11}^{(1)} & K_{12}^{(1)} & 0 \\ K_{21}^{(1)} & K_{22}^{(1)} + K_{11}^{(2)} & K_{12}^{(2)} \\ 0 & K_{21}^{(2)} & K_{22}^{(2)} \end{bmatrix} \begin{Bmatrix} \phi_1 \\ \phi_2 \\ \phi_3 \end{Bmatrix} = \begin{Bmatrix} f_1^{(1)} \\ f_2^{(1)} + f_1^{(2)} \\ f_2^{(2)} \end{Bmatrix} \quad \text{...(9.52)}$$

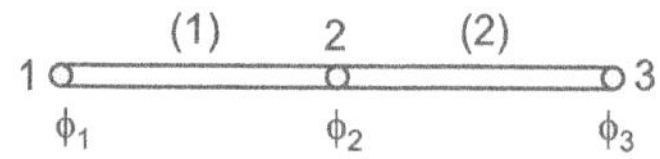

Fig. 9.7. Two linear elements.

Example 9.3. *Calculate the temperature distribution at discrete points (equally spaced) in the 1-D fin as shown by the Fig. 9.8 using FE methods.*

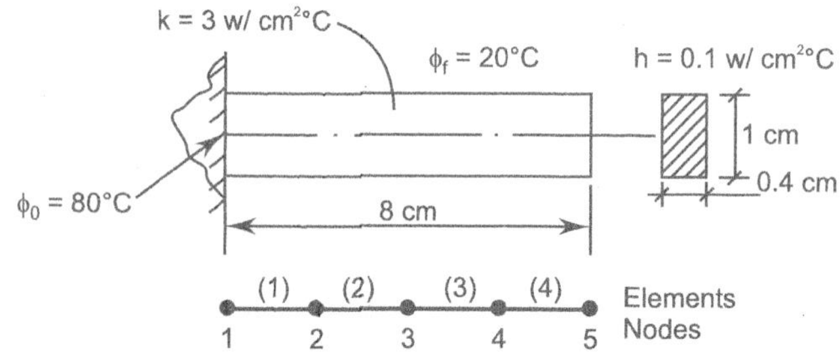

Fig 9.8. Problem on 1-D Fin.

Solution. The fin is discretized into four linear elements with five nodes. The node 1 is at $\phi_1 = 80°C$. The ambient temperature $\phi_f = 20°C$.

To obtain various factors forming elemental matrices

$$\frac{KA}{L} = \frac{3(0.4)}{2} = 0.56 \text{ W/°C}$$

$$hA = 0.1(0.4) = 0.04 \text{ W/°C}$$

$$\frac{hPL}{6} = \frac{0.1(2.8)(2)}{6} = 0.093 \text{ W/°C}$$

$$\left(\frac{hPL}{2}\right)\phi_f = \frac{0.1(2.8)(2)(20)}{2} = 5.6 \text{ W}$$

$$hA\phi_f = 0.1(0.4)(20) = 0.8 \text{ W}$$

For elements 1, 2, 3

$$[K^{(e)}] = [k_D^{(e)}]+[K_g^{(e)}] \qquad f^{(e)} = f_Q^{(e)}$$

$$[K^e] = \frac{kA}{L}\begin{bmatrix} 1 & -1 \\ -1 & 1 \end{bmatrix} + \frac{hPL}{6}\begin{bmatrix} 2 & 1 \\ 1 & 2 \end{bmatrix}$$

$$[K^{1,2,3}] = \begin{bmatrix} 0.786 & -0.507 \\ -0.507 & 0.786 \end{bmatrix}$$

$$f^{(e)} = \frac{hPL\phi_f}{2}\begin{Bmatrix} 1 \\ 1 \end{Bmatrix} \Rightarrow \{f^{1,2,3}\} = \begin{Bmatrix} 5.6 \\ 5.6 \end{Bmatrix}$$

For elements 4

$$[K^e] = [K_D^e]+[K_g^e]+[K_m^e] \qquad \{f^e\} = f_Q^e + f_S^e$$

$$[K^e] = \frac{kA}{L}\begin{bmatrix} 1 & -1 \\ -1 & 1 \end{bmatrix} + \frac{hPL}{6}\begin{bmatrix} 2 & 1 \\ 1 & 2 \end{bmatrix} + \begin{bmatrix} 0 & 0 \\ 0 & hA \end{bmatrix}$$

$$[K^4] = \begin{bmatrix} 0.786 & -0.507 \\ -0.507 & 0.826 \end{bmatrix}$$

$$[f^4] = \frac{hPL\phi_f}{2}\begin{Bmatrix} 1 \\ 1 \end{Bmatrix} + \begin{Bmatrix} 0 \\ hA\phi_f \end{Bmatrix} = \begin{Bmatrix} 5.6 \\ 6.4 \end{Bmatrix}$$

Now the assembled equation is

$$\begin{bmatrix} 0.786 & -0.507 & 0 & 0 & 0 \\ -0.507 & 1.507 & -0.507 & 0 & 0 \\ 0 & -0.507 & 1.572 & -0.507 & 0 \\ 0 & 0 & -0.507 & 1.572 & -0.507 \\ 0 & 0 & 0 & -0.507 & 0.826 \end{bmatrix}\begin{Bmatrix} \phi_1 \\ \phi_2 \\ \phi_3 \\ \phi_4 \\ \phi_5 \end{Bmatrix} = \begin{Bmatrix} Q^* \\ 0 \\ 0 \\ 0 \\ 0 \end{Bmatrix} + \begin{Bmatrix} 5.6 \\ 11.2 \\ 11.2 \\ 11.2 \\ 6.4 \end{Bmatrix}$$

ϕ_1 = 80°C, the other four temperatures at nodes 2, 3, 4 and 5 are unknowns (ϕ_2, ϕ_3, ϕ_4, ϕ_5). After we eliminate Q* in the system of matrix equation

$$\begin{bmatrix} 0.786 & -0.507 & 0 & 0 & 0 \\ -0.507 & 1.572 & -0.507 & 0 & 0 \\ 0 & -0.507 & 1.572 & -0.507 & 0 \\ 0 & 0 & -0.507 & 1.572 & -0.507 \\ 0 & 0 & 0 & -0.507 & 0.826 \end{bmatrix}\begin{Bmatrix} \phi_1 = 80 \\ \phi_2 \\ \phi_3 \\ \phi_4 \\ \phi_5 \end{Bmatrix} = \begin{Bmatrix} 5.6 + Q^* \\ 11.2 + 80(0.507) \\ 11.2 \\ 11.2 \\ 6.4 \end{Bmatrix}$$

Solving the system of matrix equation

$$\{\phi^T\} = \{80.0, 42.0, 28.2, 23.3, 22.1\}$$

$$Q^* = 0.786\,(80) - 0.507(42) - 5.6$$

$$= 36 \text{ W}$$

POINTS TO BE REMEMBER

- The mechanism by which the heat is transferred from hotter body to cold body is called the mode of heat transfer. There are three modes of heat transfer.
- Conduction—Heat transfer between two bodies in contact but a different temperature is through conduction.
- Convection—Heat transfer taking place through the moving fluid (gas or liquid).
- Radiation—Flow of heat through vacuum, *i.e.*, no medium is by radiation.
- Fourier Law of conduction—Heat transfer rate is directly proportional to the temperature gradient and constant of proportionality is the thermal conductivity.
- Initial condition — Distribution of temperature at the time equal to zero.
- For an insulated surface the boundary condition specified is that temperature gradient is zero.
- In 1D conduction the temperature gradient exists in x-direction only.
- For a steady state condition the temperature does not change with time.
- The thermal conductivity is same in all direction for an isotropic material.
- For a linear thermal element the temperature distribution is obeying the linear function.
- The stiffness in structural mechanics is equivalent to the reciprocal of thermal resistance.

EXERCISE

1. Explain the process of heat conduction.
2. State and explain Fourier Law of heat conduction.
3. Explain initial condition and boundary condition with respect to heat conduction equation.
4. Derive the equation for temperature distribution for 1D steady heat conduction process.
5. Obtain the interpolation functions for a 1D linear heat conduction element.
6. What is the thermal resistance ? Obtain thermal resistance for a plane wall.
7. Explain the steps involved in the heat transfer analysis of a composite wall using FEM.
8. Derive the heat conductance matrix for 1D linear element.
9. State the assumptions made in 1D Fin analysis.
10. Explain the Galerkins method in FEM as applied 1D Fin.
11. Derive the matrix expression for thermal resistance of a linear thermal element.
12. Obtain the matrix describing the thermal convection in 1D Fin.
13. Obtain the matrix expression for the thermal gradient in a 1D Fin.
14. Explain the FE method for heat convection at boundary.

VIVA VOCE

1. What are the steps in FEM?
2. What is discretization?
3. State the force displacement relation.
4. What is node? What are its properties?
5. What is an element? What are its properties?
6. What are the different types of ID elements?
7. What is the difference between bar and beam elements?
8. What is the truss element?
9. What are the different types of 2D elements?
10. What are the degrees of freedom for a
    (a) ID two noded bar
    (b) ID two noded beam
11. What is the difference between Cartesian co-ordinate system and natural co-ordinate system?
12. What is the advantage of natural co ordinate system?
13. State the principle of minimum potential energy.
14. What is a functional?
15. What is represented by the area under the stress-strain curve?
16. What are the factors involved in the expression for stiffness of a bar?
17. What is a shape function?
18. What are the properties of shape function?
19. What are the properties of stiffness matrix?
20. What represents the stable state of conservative system?
21. What are Ritz constants? How many Ritz constant are there in a linear bar?
22. What are the possible sources of error in FEA?
23. What is truncation error? How is it minimized?
24. What is h-refinement?
25. What is p-refinement?
26. What is the advantage of p-refinement? What is the disadvantage?
27. What is convergence?
28. What is a complete element?
29. What is a compatible element?

30. Which term in the displacement function represent the rigid body mode?
31. What is the order of CST element?
32. What is higher order element?
33. How does the strain vary in a quadratic element?
34. How does the stress vary in a linear element?
35. What is a simplex element? Give example.
36. What is a complex element? Give example.
37. What is geometric invariance?
38. What is boundary condition?
39. What are different types of boundary condition?
40. What is meant by symmetric boundary condition?
41. What are different types of supports?
42. What are different types of loads?
43. What are types of force vectors?
44. Distinguish between Dirichlet and Neumann Boundary Conditions.
45. What is Co-continuity? Give example.
46. What is C_1 continuity? Give example.
47. What is isoparametric element?
48. Differentiate between sub and superparametric element.
49. What is a parent element?
50. What is an image element?
51. What is disadvantage of using a CST element?
52. How many nodes are there in a quadratic triangular element?
53. What is the use of Pascal's triangle?
54. How many shape function are there for a ID two nodded beam element?
55. How many shape functions are there for a quadratic bar element?
56. What is Poisson's ratio? What is the Poisson's ratio for a ID element?
57. What is the boundary condition at the inlet of pipe flow?
58. What is the pipe flow outlet boundary condition?
59. What is Kronecker Delta property? Where is it applicable in FEM?
60. What are modes of heat transfer?
61. What is the property that represents conduction?
62. What is the property required for convection?
63. What is the function of a fin? Give example.
64. What is the nodal variable in heat transfer?
65. What is the equivalent of force in heat transfer?
66. What is the difference between truss and frame?
67. Which is popular and efficient method of numerical integration?

68. The displacement function is related to nodal variable through which functions?
69. What is the residual function?
70. What are the methods of weighted residual function?
71. What is the order of six noded triangular element?
72. What is the difference between element stiffness matrix and global stiffness matrix?
73. How is the global stiffness matrix obtained?
74. What is Lagrange's Polynomial and how is it useful?
75. What are the types of element used in isoparametric formulation as parent element?
76. What is mapping? How is it useful?
77. What is jacobian? What is its function?
78. How many nodes are there in a linear tetrahedron element?
79. How many nodes are there in a linear hexahedron element?
80. How is curved geometric boundary taken care in a FE formulation?
81. What is the penalty method of solution?
82. What is the difference between static analysis and dynamic analysis?
83. What is the difference between steady state and transient analysis?
84. What is round-off error?
85. What is human contributed error?
86. What are three main modules in a FE package?
87. What is bandwidth in FEM?
88. What is material non-linearity?
89. Boundary conditions are specified in which module?
90. What is mode shape? What is its significance?
91. State the vectors in free damped vibration.
92. What type of method is Golerkin's method?
93. How is the boundary condition specified for simply supported beam?
94. What are sample points and sample weights?
95. What is the difference between free and map meshing?
96. What is the aspect ratio?
97. What is the disadvantage of using CST element?
98. What is meant by generalized co-ordinates?
99. To which element Hermites polynomial is applicable? And what is the degree of the polynomial?
100. What are the mechanical properties required to find the stiffness matrix of 2D finite elements?
101. What are the geometric properties needed to determine the elements of the stiffness matrix of the 1D beam elements?
102. Why Poisson's ratio is not specified for a 1D bar and beam element?
103. How is the modules of elasticity defined and estimated?

104. How is flexibility defined? Where is it useful?
105. What is automesh?
106. Why and where is the density of mesh high?
107. What is the admissible function? What is its use?
108. How is the strain obtained inside any point on the element given the strain at the nodes?
109. What are natural and essential boundary conditions?
110. What is the thermal load and thermal stress?

QUESTIONS FROM EXAM PAPER

CHAPTER 1

1.1 Explain in brief advantages and disadvantages of finite element method.
(VTU-BE, Dec. 2007)

1.2 Write short note on Principle of minimum potential energy. *(VTU-BE, May 2008)*

1.3 Define the following

(a) FEM (b) FEA (c) Cotinuum body (d) DOF

1.4 State and prove the principle of minimum potential energy method. *(VTU-BE, Dec. 2008)*

1.5 Define the following terms:

(a) Normal Stress (b) Shear Stress

(c) Longitudinal Strain (d) Shear Strain *(VTU-BE, Dec. 2007)*

1.6 Explain the concepts of plane stress and plane strain with suitable examples
(VTU-M.Tech., June 2007)

1.7 Explain the principle of minimum potential energy and derive total potential energy π of an elastic body subjected to body forces, surface traction, and concentrated point load.
(VTU-M.Tech., June 2007)

1.8 Write stress-strain and strain-stress relations for plane stress problem (2D elastic body).
(VTU-BE, Dec. 2008)

1.9 Derive strain-displacement relation in 2-dimensional cartesian co-ordinate system.
(VTU-BE, Dec. 2007)

1.10 Explain with examples

(a) Boundary Value Problem (b) Initial Value Problem
(VTU-M.Tech., Dec. 2008)

1.11 Bring out the difference between continuum method and FEM. *(VTU-BE, May 2008)*

1.12 Derive the equilibrium equation for 3D elastic body. *(VTU-BE, Dec. 2008)*

1.13 What are the advantages of variational approach. *(VTU-M.Tech., Dec. 2007)*

1.14 What is finite element method? Explain the basic steps involved in FEM for solving structural problem. *(VTU-M.Tech., Dec. 2008)*

1.15 Give the elastic matrix relating stress and strain $\sigma = DE$ for (1) plane strain, (2) plane stress. *(VTU-M.Tech., Dec. 2008)*

1.16 Solve the following equations using Gaussian elimination method.

$5x_1 - 4x_2 + x_3 = 0$

$-4x_1 + 6x_2 - 4x_3 + x_4 = 1$

$x_1 - 4x_2 + 6x_3 - 4x_4 = 0$

$x_2 - 4x_3 + 5x_4 = 0$ (VTU-BE, Dec. 2007)

1.17 Evaluate the following integral using 3 point Gauss quadrature

$$\int_1^3 \frac{dx}{x}$$ (VTU-BE, Dec. 2007)

1.18 Solve the simultaneous equations given below using Gauss elimination method

$$4x_1 = 2x_2 + 3x_3 = 10$$
$$2x_1 + 5x_2 + 2x_2 = 5$$
$$3x_1 + 4x_2 + 5x_3 = 15$$ (VTU-BE, May 2008)

1.19 Derive the value of w_1, w_2 and z_1, z_2 used in Gauss quadrature two point method.

1.20 Solve the following system of simultaneous equations

$$x_1 + 2x_2 + x_3 = 8$$
$$x_1 + x_2 + 2x_3 = 9$$
$$2x_1 - x_2 - 2x_3 = 10$$ (VTU-M. Tech., Dec. 2009)

1.21 Explain the terms Pre and Post processor and solver in FEM software package.
(VTU-M.Tech., Dec. 2008)

1.22 Show that the stationary function $\int_a^b f(x, y, y'(x)\,dx)$ must satisfy the Euler-Lagrangian equation of the form

$$\frac{\partial f}{\partial y} - \frac{d}{dx}\left(\frac{\partial f}{\partial y'}\right) = 0$$ (VTU-M.Tech., Dec. 2008)

1.23 Derive the strain-displacement relations in 2-dimensional Cartesian co-ordinate system.
(VTU-BE, Dec. 2007)

1.24 Compute the value of central deflection for the beam shown figure, considering trigonometric functions and using Rayleigh-Ritz method. (VTU-BE, Dec. 2007)

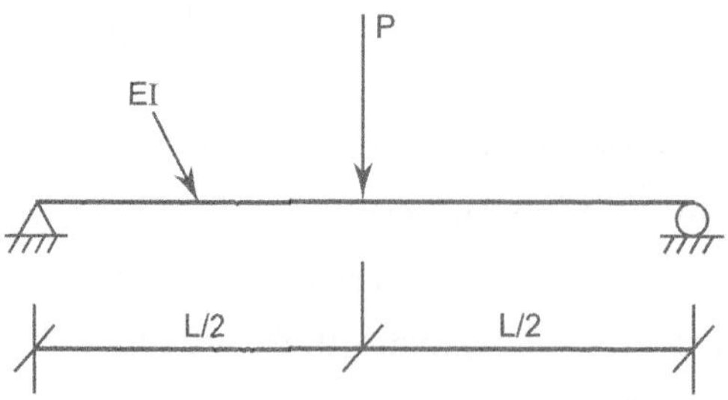

1.25 Use Rayleigh-Ritz method to find the displacement of the mid-point of the rod shown in figure. Body force per unit volume = 1. Use second degree polynomial approximation for displacement. *(VTU-BE, Dec. 2007)*

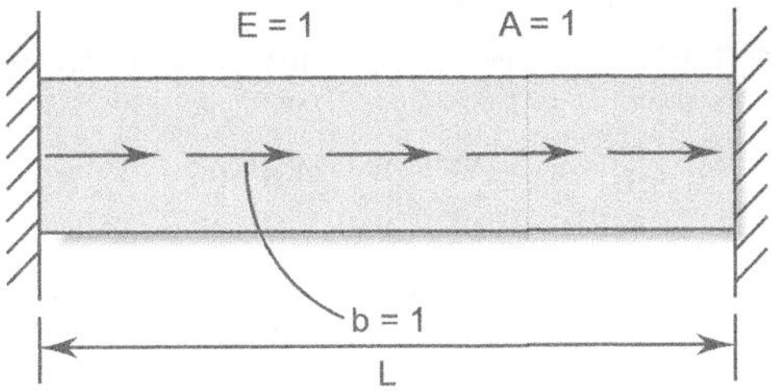

1.26 Solve for deflection at the centre of a simply supported beam carrying a point load P at its centre by Rayleigh-Ritz method. *(VTU-BE, May 2008)*

1.27 Determine the displacement and stress in the rod by using Rayleigh-Ritz method. *(VTU-BE, Dec. 2008)*

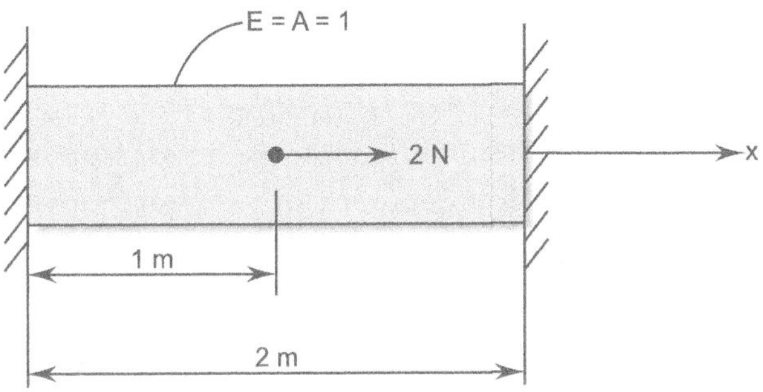

1.28 Derive Euler-Lagrange equation. *(VTU- M.Tech., Dec. 2007)*

1.29 Explain Galerkin's weighted residual method. *(VTU-M.Tech., Dec. 2007)*

1.30 Explain Rayleigh-Ritz method for solving a 1-D elasticity problem. What are the limitations of this method. *(VTU-M.Tech., Dec. 2008)*

CHAPTER 2

2.1 Find nodal displacements, stress in thickest section and left support reaction for the structure shown.

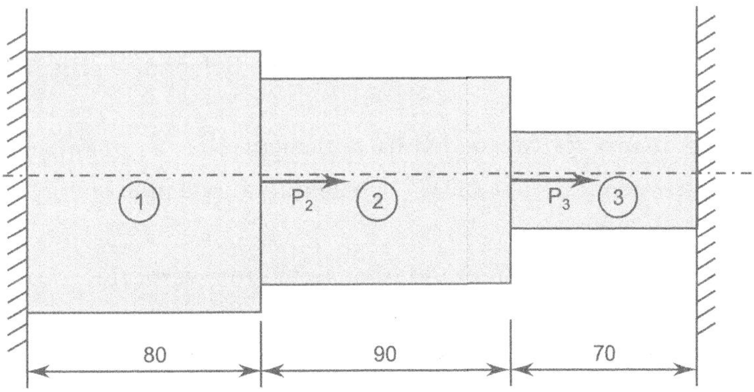

$A_1 = 900$ mm^2 $E_1 = 70$ GPa $P_2 = 20$ kN
$A_2 = 400$ mm^2 $E_2 = 105$ GPa $P_3 = 10$ kN
$A_3 = 200$ mm^2 $E_3 = 200$ GPa *(VTU-BE, Dec. 2007)*

2.2 An axial load $P = 200$ kN is applied as shown to a stepped bar. Using elimination method find

(i) The nodal displacement (ii) The stress in each member

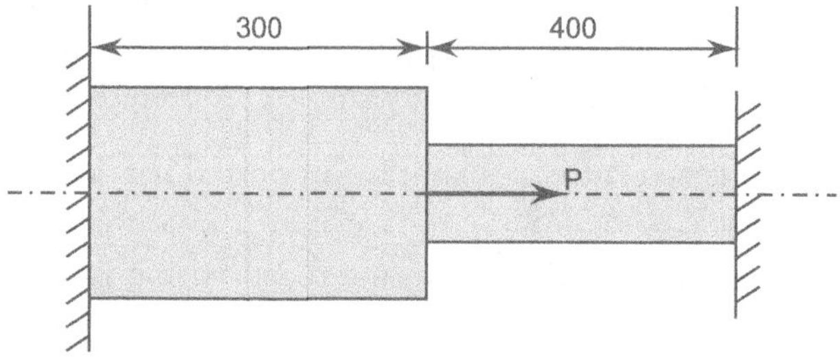

$A_1 = 2400$ mm^2 $A_2 = 600$ GPa
$E_1 = 70$ GPa $E_2 = 200$ GPa *(VTU-ME, Dec. 2008)*

2.3 Derive the expression for stiffness matrix of a two noded truss element. *(VTU-BE, May 2008)*

2.4 Derive the strain displacement matrix for 1-D linear element and also show that

$$\sigma = EBq$$ *(VTU-BE, Dec. 2008)*

2.5 Determine the stresses in stepped rod as shown in figure and find

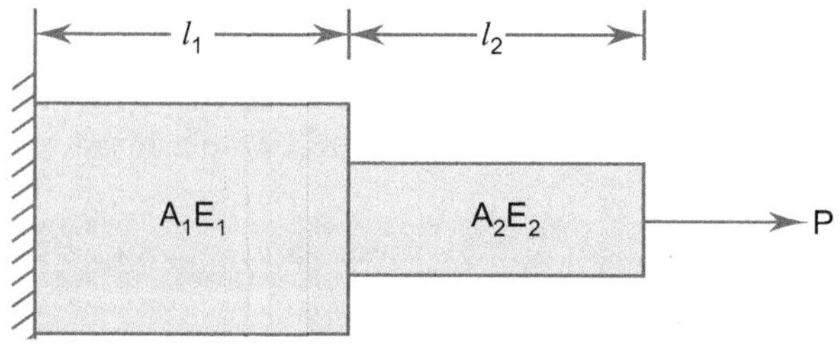

(i) Elemental stiffness matrix
(ii) Global stiffness matrix relation $E_1 = E_2 = E$
(iii) Strain and stress. *(VTU-BE, Dec. 2008)*

2.6 Derive the expression for K^e, f^e and T^e by using total potential energy method for 1D two noded element. *(VTU-BE, Dec. 2008)*

2.7 Derive stiffness matrix for two noded truss element. *(VTU-M.Tech., Dec. 2007)*

2.8 Derive the expression for K^e, f^e and T^e by using total potential energy method. *(VTU-BE, Dec. 2008)*

2.9 A three bar truss shown in Fig. Q. 2.9 is subjected to a vertical load of 50 kN. By using the principle of virtual displacement, find forces in the three members.
Area of each bar = Am^2, Modulus of elasticity = EM *(VTU-M.Tech., June 2007)*

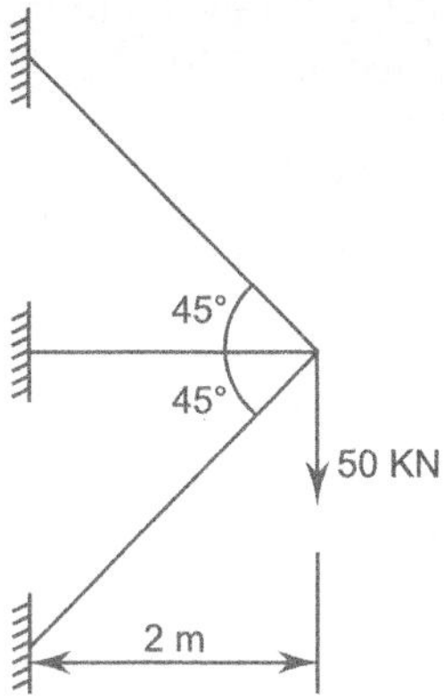

2.10 Determine the nodal displacements and support reactions using quadratic shape function and model the system with two elements

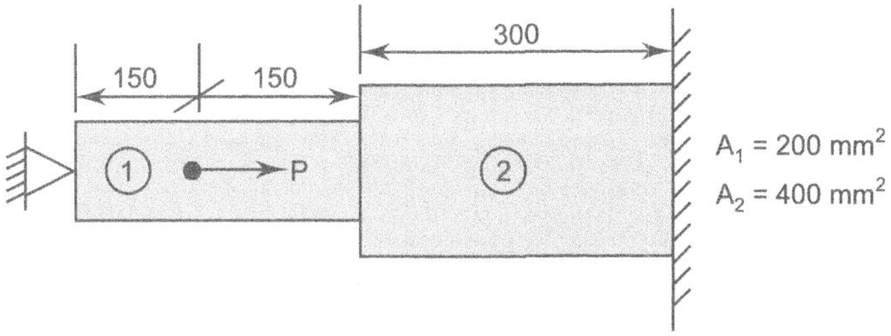

$P = 300$ kN $\qquad E = 200$ GPa

2.11 For three bar truss shown in Fig. (Q. 2.11) det. the nodal displacements and the stress in each member. Take modulus of elasticity 200 GPa. *(VTU-M.Tech., June 2007)*

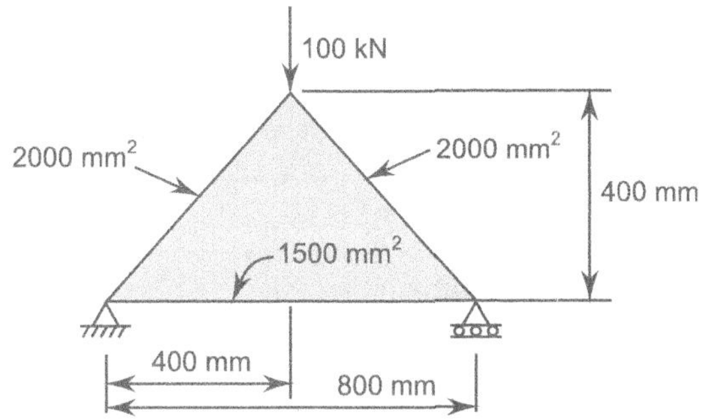

2.12 For a two bar truss determine the displacement at node 1 and stress in the element. *(VTU-M.Tech., Dec. 2008)*

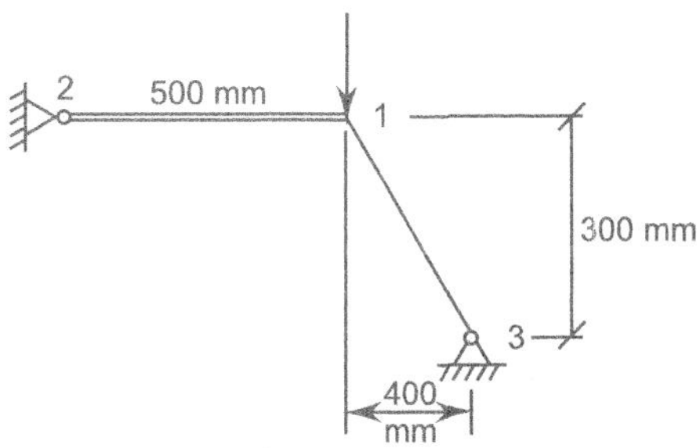

2.13 For an element shown in Fig. (Q. 2.13) evaluate the stiffness matrix with respect to XY co-ordinate system $A = 4 \times 10^{-4}$ m^2, $E = 210$ GPa, $L = 2$m, $\theta = 60^{\circ}$.

(VTU-M.Tech., Dec. 2008)

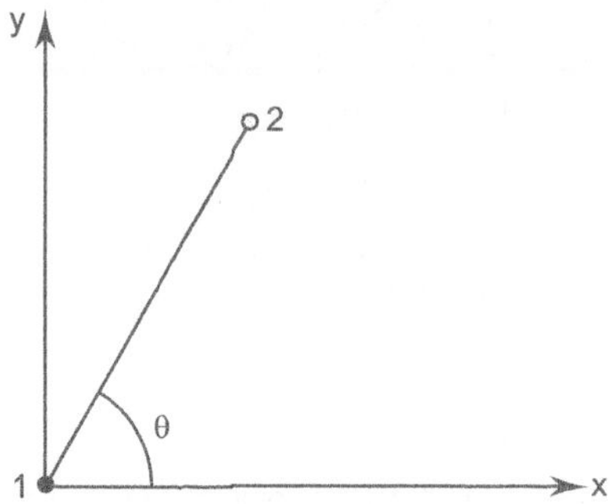

2.14 Derive the stiffness matrix for 2 noded truss element. *(VTU-M-Tech., Dec. 2007)*

2.15 For two bar truss system shown in Fig. (P2.15) determine

(i) Overall stiffness matrix

(ii) Nodal displacement

(iii) Stress in each element

(iv) Support reactions

Take cross-sectional area of each member as 100 mm^2 and E = 200 GPa.

(VTU-M.Tech., Dec. 2007)

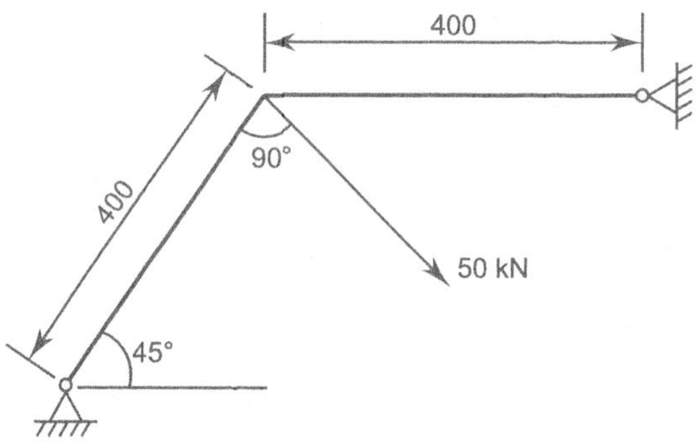

2.16 Derive an expression for interpolation function for a unit length of element with end node displacement u_1 and u_2. *(VTU-M.Tech., Dec. 2008)*

2.17 Define shape functions and explain their properties. *(VTU-M.Tech., Dec. 2008)*

2.18 Derive the general equation for determining the stiffness of an 1-D element with usual notations in the form $[K]_e = \int B^T DB\, dV$. *(VTU-M.Tech., Dec. 2008)*

2.19 Solve for deflection and slopes at points 2, and 3 using beam elements for the structure shown in Fig. P2.19. Also determine the deflection at the centre of the portion of the beam carrying UDL.

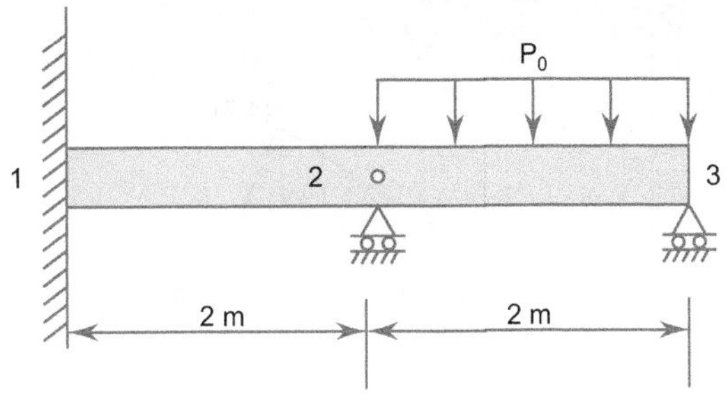

$P_0 = 12$ kN/m $E = 210$ GPa $I = 4 \times 109$ mm4   *(VTU-BE, May 2008)*

2.20 Derive Hermite shape functions for a beam element in local co-ordinate system.

(VTU-M.Tech., Dec. 2007)

2.21 A cantelever beam of length 1m is subjected to concentrated free end load of 2 kN and a uniformly distributed load of 6 kN/m acting over the whole beam as shown in figure. The moment of inertia of the beam is 4×10^{-6} m^4 and the modulus of elasticity of the beam material is 200 GPa. Determine

(i) Equivalent nodal forces and

(ii) Free end displacements

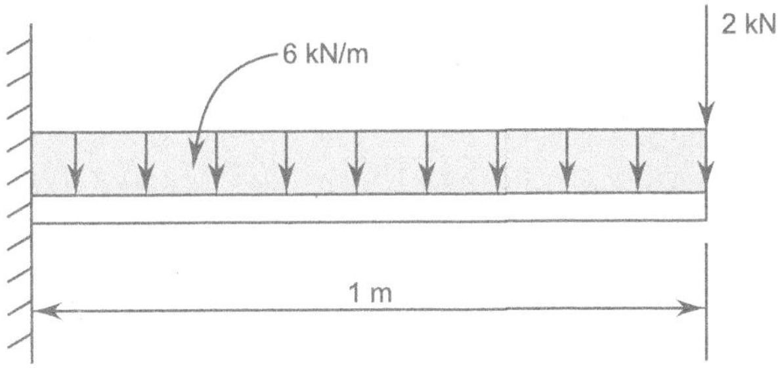

2.22 For the beam and loading shown in figure, determine the slopes at 2 and 3 and the vertical deflection at the mid-point of the distributed load $E = 200$ GPa, $I = 4 \times 10^6$ mm^4.

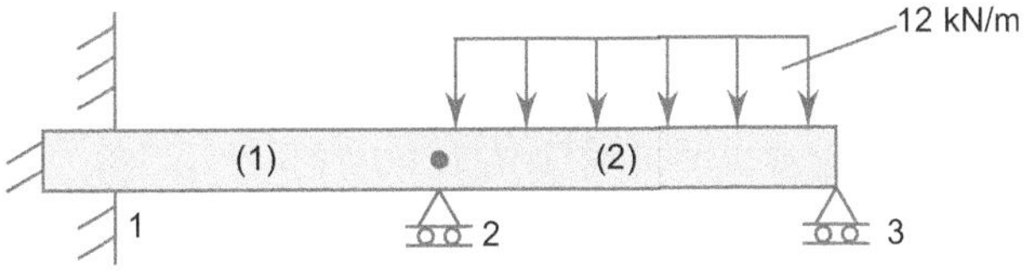

CHAPTER 3

3.1 Derive Jacobian matrix for constant strain triangle (CST) element. *(VTU-BE, Dec. 2008)*

3.2 Determine the Jacobian of the transformation J for the triangular element shown in figure. *(VTU-BE, Dec. 2008)*

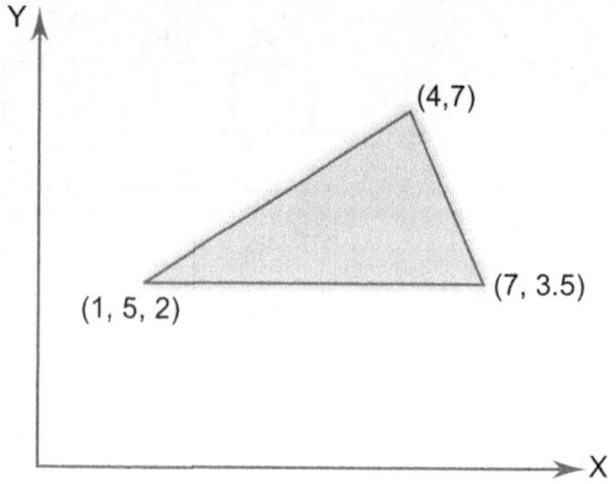

3.3 For the configuration shown in figure determine the deflection at the point of load application using a one element nodel with plane stress condition. Thickness, t = 10 mm, $E = 70 \times 10^3$ MPa, $f = 0.3$. *(VTU-M.Tech., June 2007)*

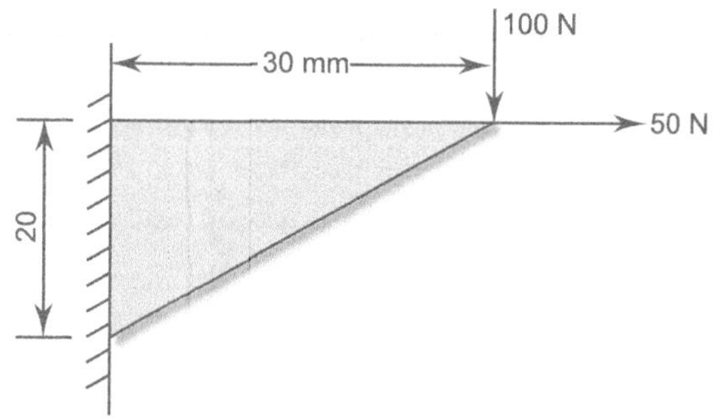

3.4 For a linear quadrilateral element, derive an expression for Jacobian matrix. *(VTU-M.Tech., June 2007)*

3.5 Obtain the shape functions and strain displacement matrix for a CST element. *(VTU-M.Tech., Dec. 2008)*

3.6 Evaluate the shape functions N_1, N_2 and N_3 at the point P in triangular element. *(VTU-M.Tech., Dec. 2008)*

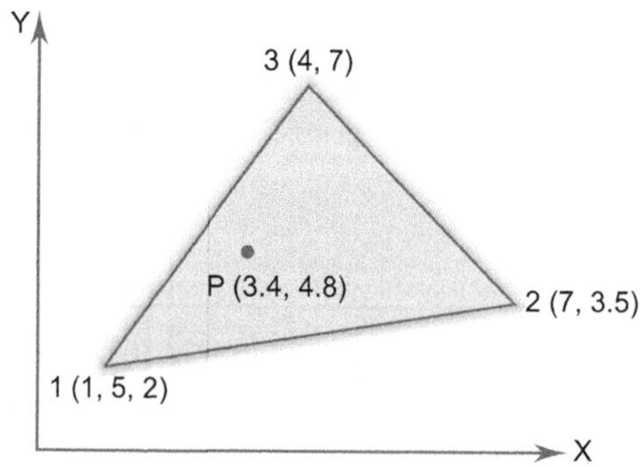

3.7 Derive shape functions for a four noded Quadrilateral element using natural co-ordinates. *(VTU-M.Tech., Dec. 2008)*

3.8 For the four noded quadrilateral shown, the element displacement vector q = [0, 0, 0.2, 0.15, 0.1, 0, 0.05]^T. Find

(i) The x, y-co-ordinates of point P whose location in the master element $\zeta = 0.5$, $\eta = 0.5$.

(ii) U, V displacements of point P (*VTU-M.Tech., Dec. 2008*)

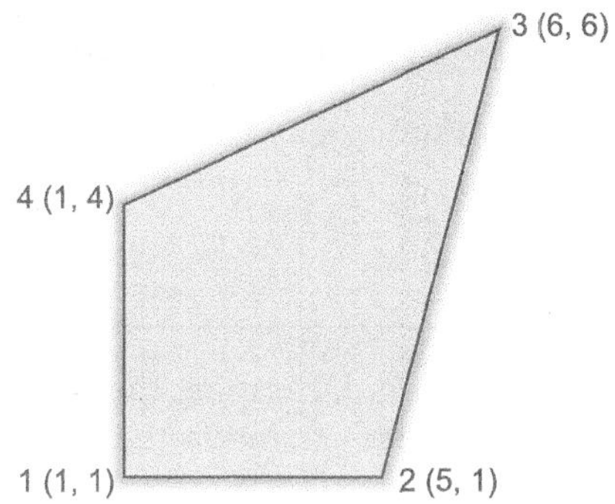

3.9 Derive: (i) Shape Functions, (ii) $[J]$, $[B]$ for 2-D rectangular element. (*VTU-M.Tech., Dec. 2008*)

3.10 Obtain the shape function of a four noded quadrilateral element. Sketch the variation of the shape functions across the element boundaries. (*VTU-M.Tech., Dec. 2008*)

3.11 Derive the shape function for a constant strain triangular element in terms of natural co-ordinate. (*VTU-M.Tech., Dec. 2007*)

3.12 Evaluate the shape function for a triangular element whose nodal co-ordinates are 1(0, 0), 2(50, 0), 3(0, 50). (*VTU-M.Tech., Dec. 2007*)

3.13 A thin plate of 10 mm thickness is loaded as shown in figure. The modulus of elasticity of the plate material is 70 GPa and Poisson's ratio is 0.3. Using the one element model under plane stress condition, evaluate

(i) The stiffness matrix

(ii) Nodal displacement

(iii) Element stresses (*VTU-M.Tech., Dec. 2007*)

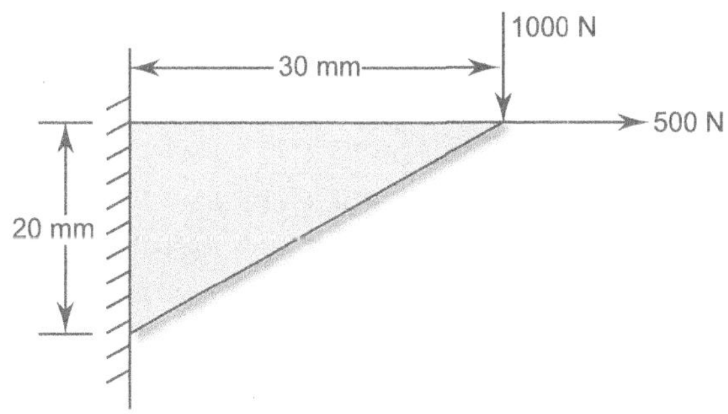

3.14 The nodal co-ordinates of a triangular element are shown in figure. The x co-ordinate of an interior point P is 3.3 and shape function $N_1 = 0.3$. Determine N_2, N_3 and y co-ordinate of point P. (*VTU-BE, Dec. 2007*)

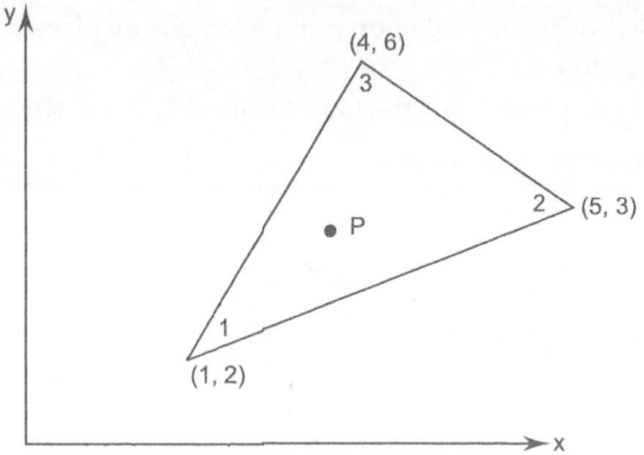

3.15 A CST element is as shown in figure. Determine the equivalent point loads at nodes 2, and 3 for a uniformly distributed load acting on the edge 2-3.

(VTU-M.Tech., Dec. 2008)

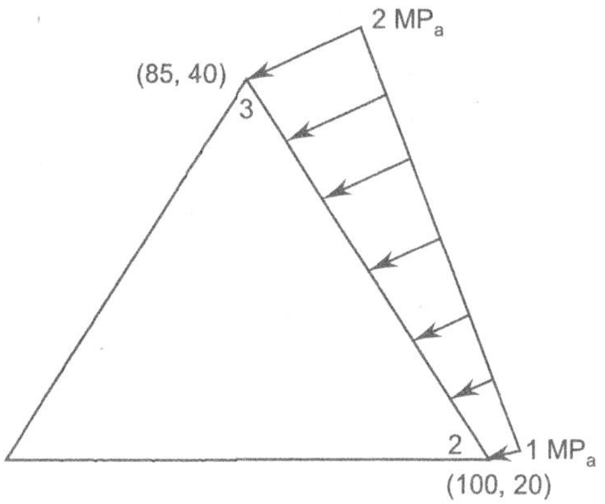

3.16 For a linear quadrilateral element, derive an expression for Jacobian matrix.

(VTU-M.Tech., June 2007)

CHAPTER 4

4.1 Discuss the requirement of a displacement model in order to achieve convergence in finite element method. *(VTU-BE, Dec. 07)*

4.2 Write a note on usefulness of isoparametric element. *(VTU-BE, Dec. 07)*

4.3 Briefly explain isoparametric, sub and super para-metric elements. *(VTU-BE, Dec. 08)*

4.4 What is a higher order element? Obtain the shape functions of nine noded quadrilateral element. *(VTU-M.Tech., June 07)*

4.6 What are the requirements for monotonic convergence solution? *(VTU-M.Tech. Dec. 07)*

4.7 Using the Lagrangian interpolation derive the shape functions for:

(i) Three noded bar element (ii) Four noded rectangular element

(VTU-M.Tech., Dec. 07)

4.8 What is natural co-ordinate system? How does it help in finite element method?

(VTU-M.Tech., Dec. 07)

4.9 What is C_0 and C_1 continuity in FEM? Explain geometric invariance concept. *(VTU-BE, Dec. 07)*

CHAPTER 5

5.1 Using Galerkin's approach, derive an element equilibrium equation for one dimensional steady state heat conduction problem. *(VTU-M.Tech., Dec. 2008)*

5.2 Consider a brick wall shown in Fig. P5.2. The inner surface is maintained at 28°C and the outer surface is exposed to –15. Determine the steady state temperature distribution assuming one dimensional flow. Use a two element model.

$K = 0.7$ w/m°C, $h = 40$ w/m^2 °C, thickness = 300 mm. *(VTU-M.Tech., Dec. 2008)*

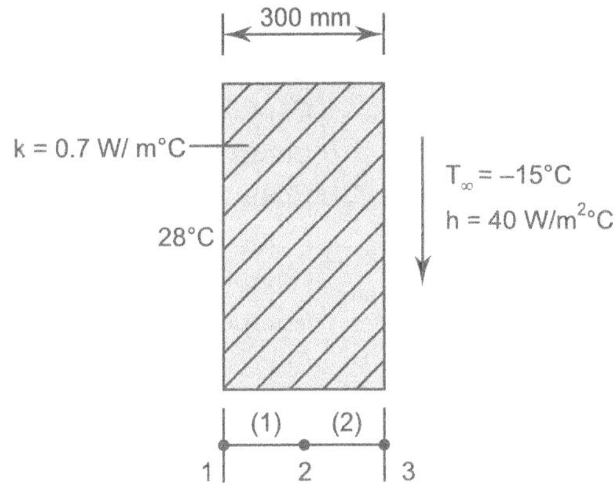

5.3 Solve for the temperature distribution in the composite wall using 1D heat elements. Use penalty approach of handling boundary conditions. Refer Fig. P5.3.

$K_1 = 20$ w/ m°C $\quad K_2 = 30$ w/ m°C $\quad K_3 = 50$ w/ m°C

$h = 25$ w/ m^2 °C $\quad T = 800$°C. *(VTU-M.Tech., May 2008)*

5.4 Consider a brick wall of thickness $L = 0.3$ M. $K = 0.7$ w/m°C. The inner surface is at 28°C. The outer surface is exposed to cold air at –15°C. The heat transfer co-efficient on outside surface, $h = 40$ w/m^2 °C. Determine the steady state temperature distribution within the wall and also the heat flux through the wall. Use two element model.

(VTU-BE, Dec. 2007)

INDEX